An early application of pollen. Relief carving from the Palace of Syrian King Ashuir-nasir-pal II. 883–859 B.C., discovered at Nimrud, the modern Calah, now in the Metropolitan Museum of Art, New York.

The standing figure, a human body with outspread wings, is fructifying the tree, which has the form of a palm. The flowers of the tree are sprinkled with water from the vessel in the left hand. The male palm flowers held in the right hand are used to transfer pollen to the female flowers.

R. G. Stanley   H. F. Linskens

POLLEN

Biology Biochemistry Management

With 64 Figures and 66 Tables

Springer-Verlag New York Heidelberg Berlin 1974

Professor ROBERT G. STANLEY[†]
Formerly: University of Florida
Institute of Food & Agricultural Sciences
Gainesville, FL 32611, USA

Professor H. F. LINSKENS
Botanisch Laboratorium, Faculteit der
Wiskunde en Natuurwetenschappen,
Katholieke Universiteit, Toernooiveld,
Nijmegen, Nederlande

The cover design was kindly supplied by Prof. STANLEY

Library of Congress Cataloging in Publication Data

Linskens, H. F. 1921—
Pollen: biology, biochemistry, and management.

Bibliography: p.
1. Pollen. I. Stanley, Robert G., joint author.
II. Title. (DNLM: 1. Agriculture. 2. Pollen.
QK658 S788p)
QK658. L56   582'.01'6   74–17437

ISBN 0–387–06827–9 Springer-Verlag New York Heidelberg Berlin
ISBN 3–540–06827–9 Springer-Verlag Berlin Heidelberg New York

Preface

Pollen transmits the male genetic material in sexual reproduction of all higher plants. This same pollen is also well suited as a research tool for studying many patterns of plant and animal metabolism. In addition, an increased knowledge of pollen may help plant breeders accelerate efforts to improve the world's food and fiber supply.

This volume focuses upon pollen biology and chemistry; it attempts to integrate these facts with management practices involved in pollen applications.

People have long been involved with pollen. Pollen applications are recorded in the rites of ancient civilizations (see Frontispiece). From the earliest times many benefits have been attributed to the inclusion of pollen in man's diet; also, since the mid-19th century air-borne pollen has been recognized as detrimental to many people's health.

Disciplines concerned with man's cultural history and the earth's changing ecology find pollen a particularly useful and accessible tool. Identifiable parts of pollen have survived over 100 million years. But most books dealing with pollen are generally concerned with the identification of the plant source, an aspect of the science of palynology; other books emphasize the natural vectors transmitting pollen, the pollination mechanisms. Very few works include the biochemistry or biology of pollen. Yet extensive studies by physicians, as well as plant breeders and apiculturists, have contributed a sizeable body of research relating to pollen.

We have endeavored to review these many historical and recent studies and to indicate some areas of pollen biology and biochemistry where critical knowledge is still lacking. These deficiencies in our knowledge, and their relation to improved management practices, present significant research challenges for the future.

The main details of pollen germination and growth to fertilization will be covered in a related volume now in preparation. A second volume will include such topics as incompatibility reactions, stigma responses, population effects, tropism metabolic and cytological changes during growth, and the influence of different chemicals and treatments on *in vitro* and *in vivo* growth. Hopefully, this and the successor volume will encourage increased applications of pollen in research leading to an improved understanding of many basic cell processes, and provide insights to help improve the yields of desired crops.

We would like to express our appreciation for the encouragement and assistance provided by many colleagues during the years this volume was in preparation. Colleagues who have kindly reviewed parts of this volume include: G. BARENDSE, CHARLES A. HOLLIS, III, W. JORDE, EDWARD G. KIRBY, III, MARIANNE KROH, F. LUKOSCHUS, JAMES L. NATION, JOOP K. PETER, FRANK A. ROBINSON, WAL-

TER G. ROSEN and INDRA VASIL. J. BRAD MURPHY patiently reviewed the total manuscript and creatively aided in its correction and improvement.

Great devotion and care in preparing this manuscript was provided by Miss BONITA CARSON assisted by Miss ANN MCLOCKLIN of the School of Forest Resources and Conservation, Institute of Food and Agricultural Sciences at the University of Florida. PATRICIA STANLEY shared or assumed much of the burden of proof-reading drafts of the manuscript.

All these many meaningful contributions to our effort are gratefully appreciated and sincerely acknowledged.

ROBERT G. STANLEY
HANS F. LINSKENS

Postscript

A few weeks after delivering the manuscript to press BOB STANLEY died in a tragic way. This book will therefore be his last work, his ultimate contribution to a field of his special interest, to which he contributed so much.

With melancholy and gratitude I recall the twenty years of our friendship and close scientific cooperation, which were brought to an abrupt end by Bob's untimely death.

R. G. STANLEY †, April 15, 1974   H. F. LINSKENS

Contents

Contents

I. Biology

Chapter 1. Development

Our understanding of the elements and patterns of pollen development represent the outgrowth of extensive light microscopic studies of many 19th century plant anatomists. These insights have been refined, classified and extended by studies with the transmission electron microscope, and most recently by observations with the scanning electron microscope. Some overlapping vocabulary and uncertainty in the meaning of names and labels attached by different workers to various pollen elements has occurred. To help avoid misunderstandings in this volume, the descriptive vocabulary will be clarified in these initial pages. Examples will be given to illustrate a few of the most common patterns of pollen differentiation.

Terminology

A flowering plant, a sporophyte, produces spores. Similar to the reproductive cycle in higher animals, some cells in diploid plants undergo meiotic division, resulting in cell clusters with haploid numbers of chromosomes. In angiosperms the organs which form male spores, the *microspores*, are called *anthers*. Female spores, *megaspores*, are formed in the *ovary* at the base of the *pistil*. In gymnosperms the clusters of male cells are formed on *microsporophylls*; the megaspores are borne on *megasporophylls*. The sporophylls, evolutionarily modified leaves, are usually grouped together in a cone structure, the *strobilus*.

The anther normally consists of two lobes, each with two elongated *microsporangia*, the *pollen sacs*, in which pollen development takes place (Fig. 1-1). The anther lobes, or *thecae*, are fused together by connective tissue which consists of vegetative cells, with a small, central vascular bundle. The *filament*, a stalk containing a single vascular bundle, is the connector which supports and attaches the anthers to the *receptacle* in the flower. The anther plus the filament are called the *stamen*.

Initially, a microspore is not ready to continue the life cycle. After formation of the microspore, one, two, or three mitotic divisions occur, followed by a resting period varying from a few hours to many months. After mitosis, the mature microspore is referred to as a *pollen grain*.

The term microspore should be limited to the uninucleate structures released from tetrads after meiosis. The term pollen grain describes the above structure after mitosis of the microspore nucleus and containing the developed *vegetative* (tube) and *generative cells* or *male cells* (VASIL, 1967).

The pollen grain, ready to germinate and grow, is correctly considered the multicellular *male gametophyte*. *Male gametes*, or sperm cells, are formed in the pollen grain or the *pollen tube*, which forms on germination of the pollen grain. By

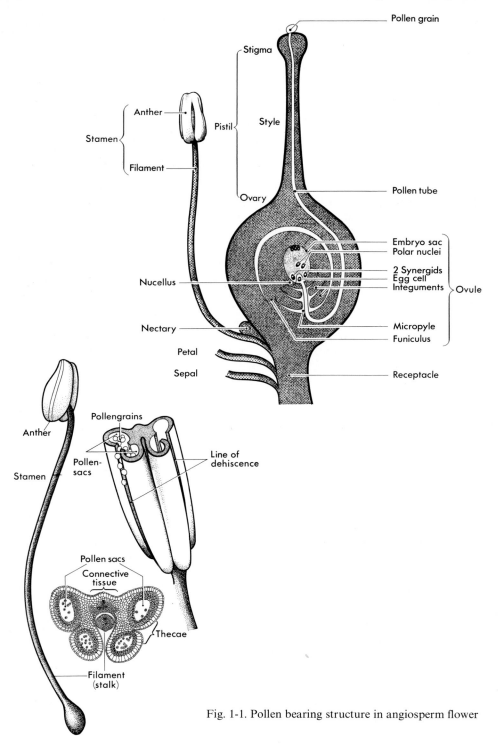

Fig. 1-1. Pollen bearing structure in angiosperm flower

convention, the thecae of the anthers are considered microsporangia and the *pollen mother cells* contained within the microsporangia are termed *microsporocytes* or *meiocytes*.

Young anthers contain *archesporial cells* which differentiate to form the *parietal cell* layer and *sporogenous* tissue. The parietal cells produce the outer wall of the anther, and an innen layer, the *tapetum*. Cells of the sporogenous tissue give rise to many *pollen mother cells* (PMCs) which divide to yield the microspores which mature and are shed as pollen grains.

The multicellular pollen grain transfers the male genome to the female organ by *pollination*. During formation and development, the male gametophyte depends on the parental sporophytic tissue for nutrition. In contrast, the female egg cell in the embryo sac is never independent of the sporophyte. The structure of the pollen grain and growth of the pollen tube are related to the role of conveying the male cells to the egg cell, i.e. the fertilization process, and secondarily, to stimulating development of the fruit or seed which encloses the mature embryo.

In angiosperms, the fused megasporophylls, the *pistil*, has a receptive surface for pollen, the *stigma*. The stigma is connected by the stylar column to the *ovary*, an enlarged basal portion of the pistil. The ovary contains the *ovules*, each with an embryo sac containing several cells. One of these cells is the egg cell with which a male cell must fuse. The male cells, conveyed down the style via the pollen tube, enter the ovary through the *micropyle*.

In pollination of gymnosperms, pollen is transferred directly to the *micropyle* of the ovule. In comparison, angiosperm pollen is transferred to the stigma and must grow through stigma and stylar tissues before entering the micropyle.

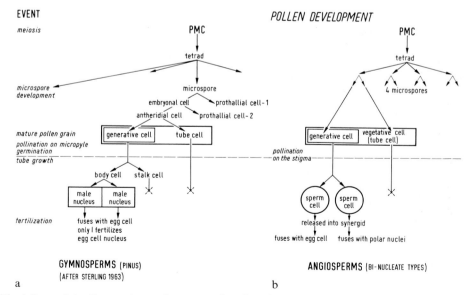

Fig. 1-2a and b. Comparison of pattern of pollen development in (a) gymnosperm and (b) angiosperm

Development Pattern

As already indicated pollen development differs in gymnosperms and angiosperms (Fig. 1-2).

Gymnosperms

Microspores are formed in the microstrobili which generally develop in the axils
of scale leaves near the tips of branches. Each microstrobilus usually bears many
microsporophylls in a spiral arrangement around the central axis, each with two
or more microsporangia on the lower side.

Microspore mother cells undergo meiosis giving rise to the spore tetrad, each
yielding 4 microspores, ultimately the pollen grains. Each pollen grain is sealed in
a double layered wall. The outer pollen wall in many species forms two conspic-
uous wings or sacs which contain air. Such structures reduce the free fall velocity
of these wind-dispersed, *anemophilous*, pollen. Gymnosperm pollens show the
complete gametophytic development pattern; the nuclei divide several times (Fig.
1-2) and resulting mature pollen grains, in *Pinus* for example, contain the following cells: two nonfunctional prothallial cells, a central vegetative or tube cell and
the generative cell, the latter two originating from the *antheridial* initial. With
formation of these cells, gymnosperm pollen grains are shed from the microsporangia.

Angiosperms

Angiosperm pollen development can be separated into three major types
(WULFF, 1939):

The Normal Type. Most commonly in angiosperms, the microspores begin to
enlarge and exine formation is initiated immediately after meiosis. At division of
the nucleus the microspore has already reached a definitive size, vacuoles are
present and account for most of the microspore's volume. Because of the vacuoles,
the cytoplasm is confined to a peripherial layer and the nucleus is in an acentric
position. A chemical or bioelectric potential gradient has also been suggested as
the sources of nuclear displacement (VAZART, 1958). Prior to the first mitosis the
amount of deoxyribonucleic acid (DNA) in the microspore nucleus increases
(BRYAN, 1951). The *first* mitotic division is generally not synchronous among the
developing pollen grains in all the stamen within a flower, or even within the same
anther (KOLLER, 1943). In some species where the walls separating microspores are
very thin, mitotic divisions may be synchronous (MAHESHWARI and NARAYANAS-
WAMI, 1952). The normal type of angiosperm pollen development is found in the
majority of the monocotyledons and dicotyledons, even in species with bi- or
trinucleate pollen.

The Juncus Type. This type of pollen development differs from the normal one
in that division of the primary microspore nucleus takes place before growth of
the pollen; vacuolization starts before formation of the exine is initiated. Also,

after the division of the generative nucleus, further growth of the grain can occur. This type of development is found in the Cyperaceae and the Juncaceae.

The Triglochin Type. The third common type of pollen development is somewhat intermediate. The young microspores, after separation, grow slightly and form a thin exine. But the main growth period, including formation of vacuoles and definitive sculpturing of the exine, starts after formation of the generative cell. The ripe pollen grain is binucleate. This third type of pollen development is found in *Najas, Ceratophyllum, Ruppia, Apomogeton, Triglochin* and others.

The above described types of pollen development are mostly descriptive, but they may be of taxonomic value. Mature pollens, as well as fossil types, can be divided into many classes based on development and wall pattern. This is the subject of comparative palynology and is reviewed in the famous books of ERDTMAN(1952, 1969), WODEHOUSE (1935), KREMP (1965) and others. In this book we are concerned with the biochemical aspects of pollen, and the sources of variation in the maturing pollen grain.

Synthesis and Division

Pollen grains develop through a series of regulated events which occur in a definite time sequence.

Induction of Meiosis

Meiotic division has 3 important features:

1. *Transformation of the chromosomes* by crossing-over processes;
2. *Rearrangement of the genomes* by random distribution of the homologous chromosomes, and
3. *Reduction of the chromosome number* from 2n (diploid) to n (haploid). Up to the present time many textbooks emphasize the third event. If this were the most important aspect of meiotic division, nature could renounce the complex processes of meiosis as a compensation of fertilization and, therefore, sexuality. However (1) and (2) are decisive because these events make recombination possible, which is probably the most important contribution of sexuality to evolution.

DNA Synthesis

The diploid PMC is genetically distinguished at a very early stage of premeiotic division. DNA synthesis takes place during premeiotic interphase (TAYLOR and MCMASTER, 1954). A second period of DNA synthesis takes place during prophase I in late zygonema and early pachynema. A distinct type of DNA is synthesized during this period (HOTTA et al., 1966). This DNA synthesis is a consequence of the break-repair mechanisms and may represent delayed replication of part of the chromosomes. If this interpretation is correct, the major determinant in differentiating meiotic from mitotic cells is the regulation mechanism which delays the reproduction of some critical, essential DNA component. It may be speculated that, functionally, this component is an axial sequence of nucleotides. Crossing over during meiosis may be closely associated with the delayed reproduction of this element. The DNA replicase is inhibited at this division. A special histone (SHERIDAN and STERN, 1967) appears to be involved in meiotic induction. The histone pattern in developing pollen is opposite to that during ribonucleic acid (RNA) and protein synthesis. Low or no nucleohistone staining indicates a high rate of RNA and protein synthesis; high nucleohistone staining corresponds with reduced RNA and protein synthesis.

Histones

Histones apparently play a role in DNA-dependent RNA and protein synthesis in microspores and two-celled pollen grains (SAUTER and MARQUARDT, 1967; SAUTER, 1971). Acidic peptides as well as basic histone proteins may function as derepressor molecules. While the exact mechanism of induction is not known, other chemical changes are recognized. These changes include modified protein and enzyme patterns (LINSKENS, 1966) and periodic changes in sulfhydryl groups (LINSKENS and SCHRAUWEN, 1964).

RNA Synthesis

RNA synthesis is most active just before division (G-2), after the first mitosis and after microspore nuclear division (STEFFENSEN, 1966; LINSKENS and SCHRAUWEN, 1968b). A changing ribosomal pattern accompanies pollen development. Transition from the diploid sporophytic stage to the haploid gametophytic stage is concommitant with formation of new polysomal fractions (LINSKENS and SCHRAUWEN, 1968a). Changes in RNA or histones may also influence or reflect changes accompanying nuclear differentiation in pollen development (GEORGIEV, 1969).

In the transition to meiosis during pollen development it is possible to distinguish (a) transition to the incipient phase of meiosis due to synthesis of a special informational molecule following a signal in the transcription process; and (b) biochemical events which direct the course of meiosis. The latter includes two distinct events: chromosome pairing and crossing over between homologous chromosomes. Both events are dependent on metabolic signals to form enzymes at the right time, in proper amount and sequence, and on the nutrient pool which must deliver energy-rich compounds and chemical building moieties for the synthetic processes.

Induction of Polarity

Generally, no polarity is observed in pollen mother cells. The second meiotic division takes place at an angle of 90° to the first division. Polar differentiation during meiosis and pollen development was investigated in *Tradescantia* (SAX and

EDMONDS, 1933; SCHMITT and JOHNSON, 1938; LACOUR, 1949; BRYAN, 1951). Protein granules in the microspores disintegrate during prophase, generating vacuoles and pressing the nuclei against the ventral wall. Unequal distribution of protein synthesized just prior to mitosis occurs during anaphase. The mass of RNA and protein is shifted to the spindle pole where the vegetative nucleus will be formed. This asymmetric division and cytoplasmic gradient results in polarity. The generative nucleus, in which chromosomes are arranged in the compact, spindle-like, DNA-rich resting stage, is present in a smaller cell with a small portion of its cytoplasm lacking RNA. The vegetative nucleus remains more or less spherical but increases in size and shows a distinct nucleolus as protein synthesis starts.

As will be discussed subsequently, normal pollen development can be strongly influenced by genetic and environmental factors, resulting in abnormally small or large pollen grains (VON WETTSTEIN, 1965). Elevated temperatures, or introduced factors such as X-rays, γ-irradiation or chemicals can disturb polarity and result in male sterility. Pollen grains with nondisjunction of translocated chromosomes can form normal tetrads, but deficiencies in protein metabolism can yield uninucleate grains, with a high degree of sterility (SAX, 1942). Another disorder of polarity reported at asynapsis in *Picea* pollen grains results in formation of one, ring-like, air-filled wing (ANDERSSON, 1947).

In some cases of sterile pollen it is difficult to determine the source of abnormality. Among the 39 different species studied by ZIELINSKI and THOMPSON (1966) the five species of *Pyrus* producing sterile pollen are a good example of the difficulty in recognizing the origins of pollen degeneration. All five species had two normally appearing meiotic divisions. The breakdown in development occurred somewhere during the maturation process after liberation of the microspores from the tetrads and before anthesis. This post-meiotic phase is one in which tapetal activity is particularly critical. External factors may influence the tapetal cells and if tapetal cell metabolism is drastically upset, microspores do not develop normally.

Differentiation of Pollen Grain Nuclei

After formation of the microspores, a resting period generally occurs before the first division of the nucleus in the microspore. The length of this dormant stage varies from a few hours to several months (DAHLGREN, 1915; FINN, 1937). During this period, pollen is particularly sensitive to temperature (KOLLER, 1943). The mitosis which follows results in two distinct cells: a vegetative cell (tube cell) with the vegetative nucleus, and a generative cell with the generative nucleus. Differentiation seems to depend on unequal division of basic protein of the pre-mitotic nuclei (MARTIN, 1960). The problem of heteropolarity associated with the pollen mitosis is still under discussion (STEFFEN, 1963).

The products of the division of the microspore nucleus, the vegetative and the generative cells, are qualitatively different. The generative cell is smaller, hyaline and contains less RNA; its associated cytoplasm is separated from the vegetative cell cytoplasm by a plasma membrane. This has been demonstrated by light

(HOFMEISTER, 1956) and electron microscopy (BOPP-HASSENKAMP, 1959; BAL and DE, 1961; SASSEN, 1964; LARSON, 1963, 1965). Thus, cytoplasmic dimorphism exists within the cells of the pollen grain. Differentiation also can be shown by X-ray absorption (DAHL et al., 1957). In the vegetative nucleus, nucleoli are usually larger than those in the generative nucleus; decreased staining ability of the vegetative cell is believed to be due to the lower DNA content of the nucleus (LACOUR, 1949). The protein content of the vegetative nucleus is about twice that of the generative nucleus (STEFFEN, 1963), and the proteins seem to be more acid (BRYAN, 1951). The histone staining pattern also differs in the vegetative nucleus and generative cell nucleus. In the vegetative nucleus of *Lilium candidum* pollen, RNA was detected bound to the histone, a condition not observed in the generative cell nucleus (JALOUZOT, 1969).

Two hypotheses have been proposed to explain the *mechanism for the morphological and physiological differences* between the genetically alike cells of the pollen grain. Theory I assumes that difference in the DNA content are the cause of differentiation. The generative nucleus DNA increases over that of the vegetative nucleus (TAYLOR and MCMASTER, 1954). That would mean that all other differences in assembly of subcellular organelles, e.g. plastids, mitochondria, and spherosomes, result from nuclear differentiation (STEFFEN and LANDMANN, 1958; RICHTER-LANDMANN, 1959). Theory II accounts for differentiation by the fact that after division the two nuclei are in different cytoplasmic environments. One explanation for these differences in cytoplasm is based on the quantity (GEITLER, 1935), the other on the quality (LACOUR, 1949). Qualitative differences would also result from a quantitative difference in RNA content of the surrounding cytoplasm or from difference in hydration of the nucleus and viscosity of the cytoplasm (PINTO-LOPES, 1948). One reason for concentrating research on the pattern of the spindle in pollen development is because asymmetry of the spindle mechanism may be a key factor in establishing nuclear differences. The mechanism and time of differentiation of the two nuclei is a complex and intriguing problem.

Origin of the Sperm Cells

In angiosperm pollen development, the nucleus in the generative cell subsequently divides once more, forming two sperm nuclei. This division may take place (a) in the anthers, so that the ripe pollen contains three functional cells (COOPER, 1935), (b) in the mature pollen grain after pollination but before germination (CAPOOR, 1937), (c) just after germination on the surface of the stigma, which is the most common case in angiosperms, or (d) at the time the pollen tube reaches the embryo sac (D'AMATO, 1947).

BREWBAKER (1957) advanced a theory relating the nature of the pollen grain nuclei at maturity to the mechanism of incompatibility. He suggested that two-celled pollen grains occur in homomorphic plants which express pollen incompatibility reactions in the style and are generally linked to gametophytic incompatibility. In pollen which is three-celled at maturity, the stigmatic incompatibility reaction is most commonly observed, representing the sporophytic incompatibility. When pollen is shed in the three-celled stage, the incompatibility reaction is

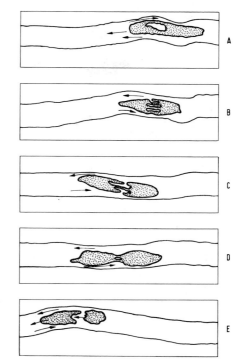

Fig. 1-3 A-E. Generative cell division in the pollen tube. Two male cells separate (E) but are retained in generative cell cytoplasm. Arrows indicate direction of cytoplasmic streaming and nuclear movement

stronger and occurs earlier. Three-celled pollen is viable for much shorter periods after dehiscence than pollen shed in the two-celled stage.

In the development of gymnosperm pollen, the antheridial initial cell gives rise to the tube cell and generative cell as it does in the angiosperms. However, in gymnosperms the generative cell divides to produce a *spermatogenous cell* (or *body cell*) and a *sterile cell* (Fig. 1-2). The latter, called the "stalk cell" by STRASBURGER (1892) and the pollen "wall cell" by GOEBEL (1905), is a terminal cell which disintegrates in the mature pollen grain. The two male cells in gymnosperm pollen arise from the spermatogenous cell. A pollen grain in most extant species of gymnosperms and angiosperms produces just two male gametes. However, in cycad pollen, particularly *Microcycas*, as many as 24 male cells may be produced (FAVRE-DUCHARTRE, 1963).

Factors contributing to formation of sperm cells in the pollen grains while still in the anthers are not known. Division of the generative cell can be experimentally induced by changing the water availability (GEITLER, 1942; PODDUBNAYA-ARNOLDI, 1936). BREWBAKER (1957) speculates that during the second mitosis, the generative cells deplete the sugars and other energy-supplying metabolites in giving rise to sperm cells. Yet corn and other trinucleate grass pollens are among pollens with the highest known percent endogenous carbohydrates at maturity (Table 9-1).

The two sperm cells produced during germination generally assume a spindle-like or lappet shape (Fig. 1-3). Each sperm nucleus is surrounded by cytoplasm

containing cell organelles, including mitochondria, ribosomes, and small but autonomous plastids, all contained within the plasmalemma (RENNER, 1934; KOS-
TRIUKOWA, 1939; KAIENBURG, 1950; SASSEN, 1964).

When formation of the sperm cells takes place in the pollen tube, division of
the generative nucleus presents a special spacial problem. While in many species
the spindle fibers are visible in a phase contrast microscope, in some cases a
normal spindle cannot be observed. In such cases chromosomes often appear
arranged in a row or in a tandem arrangement. Cytoplasmic microtubules,
usually found in plant cells, appear as spindle fibers attached to the chromatids
during anaphase. Although they are more easily observed in developing microspores (ROWLEY, 1967; ROSEN, 1971) they are also present in pollen tube cytoplasm
(FRANKE et al., 1972). Fixation technique can easily destroy or obscure microtubules in prepared sections.

The nuclei of the sperm cells complete their division in the pollen tube relatively slowly; they remain in delayed telophase with strong chromatization and
general absence of nucleoli. Finally, the two male sperm cells separate within the
generative cell membrane (Fig. 1-3 E) and ultimately become separate sperm cells.
Occasionally, abnormal nuclear divisions may occur increasing the number of
sperm cells (MAHESHWARI, 1949).

A complete male gametophyte, ready to fulfill its function in fertilization,
consists of three fully organized cells, each with cytoplasm and organelles which
can undergo independent mutation (RENNER, 1922). The sperm cells participate in
double fertilization in angiosperms, one forming the 2n zygote, the other usually
giving rise to the 3n endosperm. In gymnosperms one sperm cell generally disintegrates, the other produces a 2n zygote. Figure 1-2 presents a diagramatic comparison of the development of the male gametophyte in the angiosperms and
gymnosperms.

Chapter 2. Wall Formation

Pollen wall composition can be elucidated by various techniques. These include chemical analysis, histochemical stains, optical methods such as double birefringence, and electron microscopic observations. All these methods have contributed to our detailed knowledge of the chemical composition, organization and ontogeny of the pollen wall. The wall of the mature pollen grain differs chemically from that of the developing microspore.

Pollen Mother Cell Wall

Wall material in higher plants generally consists of cellulose, pectin, and some hemicellulosic components. In the PMC wall the substance, callose, is present (MANGIN, 1889). The pioneering observations and work on callose are reviewed by ESCHRICH (1956). Callose, a β-1,3-polyglucan, can be detected around the PMC during initiation of meiosis. It forms a layer between the cytoplasm and pollen mother cell wall (BEER, 1906). Callose is formed initially by the parent microsporocyte cytoplasm. The genesis of the callose walls surrounding the PMCs is given in Fig. 2-1. Additional callose is formed after the second meiotic division and isolates the young microspores (HESLOP-HARRISON, 1968a), effecting a macromolecular block between all future pollen cells. As soon as the formation of the outer pollen wall begins, the callose starts disappearing (ESCHRICH, 1966). The emphasis in microsporogenesis has generally been placed on chromosome behavior in meiosis. Yet the transformation from mitosis to meiosis is also accompanied by a transition in carbohydrate and wall metabolism. Although much less clearly understood, it is almost as profound a metabolic change as the nuclear events (STERN and HOTTA, 1968). The sculptural pattern of the final pollen wall, the exine, is established while the microspores are enclosed within the callose wall. Nuclear information controlling the exine pattern moves through the cytoplasm to the callose layer.

Development of pollen grains in angiosperms proceeds through the following general stages (Fig. 2-2): 1. *transformation of the microspore* including disintegration of the tetrad, formation of the pollen membrane, increase of the volume with increase of cytoplasm, and formation of the central vacuole; 2. *first mitotic division* and origin of the vegetative and generative cells; 3. *differentiation* of the vegetative cell and division of the generative cell to form two male cells (CAILLON, 1958). Pollen development in gymnosperms, e.g. *Pinus* and *Ginko*, is similar to the angiosperm pattern except that the generative cell divides to yield a sterile cell and a spermatogenous (body) cell; the latter divides to yield two male cells (STERLING, 1963) (Fig. 1-2a).

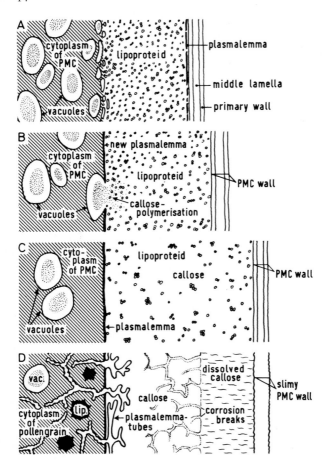

Fig. 2-1 A-D. Formation of partition wall in developing microsporocytes.(After ESCH-RICH, 1964)

When the anthers are only differentiated as tapetal and sporogenous tissues, the two groups of cells are structurally distinct without direct cytoplasmic connections between the two groups. However, elaborate connections may exist between cells within each group. The PMC nuclei are very large just before meiosis and formation of the tetrad of microspores. Just prior to meiosis of the enlarged PMCs a layer of callose forms an amorphorous wall around each PMC. The callose layer acts as a barrier probably excluding the penetration of informational macromolecules from the dividing PMC. Each of the new microspores in the tetrad is also isolated by a callose layer.

Pollen development is accompanied by certain parallel changes in the synthetic and degradative activity of the surrounding tissue, the tapetum, which develops in mutual relationships with the pollen (VASIL, 1967). The tapetum differentiates as one of three types, based on cell nuclei: 1. *cellular monocaryotic,* 2. *cellular polycaryotic,* or 3. the *periplasmodial* type with increased number of nuclei. In all three types there is a strong increase in chromatic material due to multiplication of the nuclei and formation of restitution nuclei during pro-, meta-,

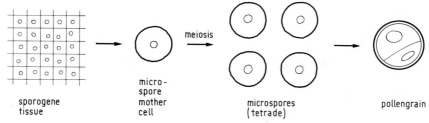

Fig. 2-2. Sequence of pollen grain development from pollen mother cell and microspore

ana-, or telophase (CARNIEL, 1963). The tapetum may also be characterized as either secretory or amoeboid, depending upon the pattern by which they impart protoplasm to the developing pollen grains (ECHLIN and GODWIN, 1968).

While cytological and anatomical changes during pollen development are fairly well known, the biochemical events and their time sequence are still subjects requiring further investigation. During pollen formation changes occurring in the tapetum probably regulate the sequence of biochemical and developmental events occurring in the pollen. The tapetum supplies substrates for incorporation as nutrient reserves and cell wall materials during pollen grain formation. Also, the deoxyribosides necessary for DNA synthesis in the developing microspores are hydrolyzed and released by the somatic tapetal tissues (STERN, 1961; STERN and HOTTA, 1968).

Partition Walls in Microsporocytes

Callose, the unbranched β-1,3-linked glucan (ASPINALL and KESSLER, 1957; KESSLER, 1958) is deposited in an amorphous mass without microfibrillar organization (FREY-WYSSLING et al., 1957). It can be built up and broken down very quickly (ESSER, 1963) and may act as substrate for the developing microspores and can influence membrane transport and levels of materials secreted by the cells (ESCHRICH et al., 1965). BARSKAYA and BALINA (1971) suggest that callose around developing microspores protects them from dehydration and provides water when plant absorption is insufficient.

Partition walls in the PMCs are formed primarily by the cell plate method (REEVES, 1928; MÜHLDORF, 1939). The callose covering of the microspores is dissolved by an externally supplied enzyme, callase, β-1,3-glucanase (ESCHRICH, 1966).

Plasma Connections between Microsporocytes

Massive plasma strands, similar to plasmodesmata but much larger, interconnect the meiocytes in the anthers during prophase of meiosis I (HESLOP-HARRISON, 1964, 1966a, b). These plasma strands called cytomictic channels (ECHLIN and GODWIN, 1969) bridge the callose walls of the PMCs (ESCHRICH, 1962, 1963). The channels are absent between PMCs of some gymnosperms such as *Podocarpus*

(VASIL and ALDRICH, 1971). They are initiated in the pre-leptotene period and persist throughout meiotic prophase. But the plasma channels between the PMCs are interrupted at the end of prophase.

Movement of nuclear chromatin material has been suggested and seen by several authors (GATES, 1911; KAMRA, 1960; TAKATS, 1959). Extrusion of nuclear material from one cell to the other was observed in electron micrographs. This secreted nuclear material is sometimes surrounded by a double membrane and passes through the cytomictic channels to the neighboring cell (BOPP-HASSEN-KAMP, 1959). The plasma connections also serve as pathways for intercellular movement of solutes, which in normal tissue diffuse through permeable cellulose walls. The intermeiocyte connections permit synchronization of meiotic events which would not be possible with the cellular isolation imposed by impermeable callose walls (HESLOP-HARRISON, 1966 b). This is shown by the fact that as soon as the plasma strands are interrupted, subsequent cytological events become less synchronous. The interdependence of the pollen mother cells can also explain the fact that attempts to culture isolated meiocytes *in vitro* at an early stage, before onset of prophase, have not been successful (HESLOP-HARRISON, 1966 a).

Wall Development

The mature pollen grain is enclosed in a double wall structure; the inner one called the intine, the outer, the exine. Pollen wall, or sporoderm, terminology is an area of active discussion (HESLOP-HARRISON, 1963). Some people want to distinguish a third distinct layer, the medine (SAAD, 1963). It is suggested that the medine, being hygroscopic and lamellate-globulate, helps to protect the apertural region, to accommodate changes in pollen diameter, and to initiate tube elongation. Various names and classifications have been applied to different regions of the mature pollen grain (Fig. 2-3). The different terminologies often reflect the

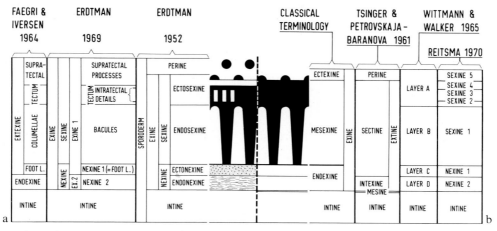

Fig. 2-3. Names and classifications applied to the sporoderm pollen wall. (For literature see: WITTMANN and WALKER, 1965; BRONCKERS, 1968; REITSMA, 1970)

various purposes for which the pollen is being examined, i.e. taxonomic, morpho-genetic or developmental.

·Pollen wall development may be considered to take place in two phases, each of which can be divided into several stages.

Phase One: The Tetrad

With the pollen tetrad still surrounded by the callose wall, primexine formation takes place. The principal features can be described in the following steps (HES-LOP-HARRISON, 1963, 1968c):

Wall-less Stage. In this stage the spores are isolated from each other and from the organelles of the tapetum by callose. The PMC wall-like lipoprotein is in direct contact with the plasmalemma, which consists of a simple membrane (Fig. 2-1). Elements of the endoplasmic reticulum (ER) of the developing pollen grain approach the plasmalemma without fusing with it. The ER complex in developing *Beta vulgaris* pollen was observed as connected to the nuclear membrane, lending further support to the concept that the ER functions in synthesis or movement of materials through the pollen cytoplasm (HOEFERT, 1969).

Early Wall Growth Stage. The first evidence of individual microspore wall formation is the appearance of a narrow halo, presumably cellulose, around each cell. Wall growth is not uniform over the whole surface; the primexine is not laid down where there is an underlying plate of ER. Primexine consists of cellulose microfibrils penetrated by radial rods of electron-dense material (HESLOP-HARRI-SON, 1968d).

Definition of the Pore Region. The disjunction in the cellulose primexine matrix sheath foretells the aperture pattern, pores and interpores of the final exine. As the primexine continues to thicken, eminences are formed in the plasmalemma opposite the plates of ER. Each aperture is associated with an underlying plate of ER.

Development of a Pattern in the Interporal Region. In the interporal region, primexine grows uniformly in thickness but is not homogenous. At intervals, probably limited to areas opposite tubules of ER-like material (SKVARLA and LARSON, 1966) it is traversed by columns of granular material, the probacula. This amorphous material ultimately extends over the outer face of the primexine forming a foot layer at the base of the probaculum. The tops of the probaculum may also be consolidated to form roofed chambers; the final roof with sporopollenin is generally called the tectum. But at this stage there is no deposition of sporopollenin although all major features characteristic of the mature spore wall have already been established. The deposition of sporopollenin takes place during phase two.

Phase Two: Post-tetrad, Free Spore

In general, the PMC callose wall is dissolved and the spores are released into the tapetal fluid which fills the space of the locule. The microspores are not independent, individual cells; their further growth and differentiation utilize primarily the substances in the tapetal fluid. Deposition of sporopollenin, the chemical material characteristic of the exine, begins immediately upon separation of the tetrad.

Exine wall development shows the following steps:

Early Exine Deposition. Deposition of sporopollenin begins on the outer surface of the primexine, as a discontinuous layer in the interporal region, where it extends over the probacula. The early-formed sporopollenin, deposited soon after release of the spores from the tetrad, has a chemical reactivity somewhat similar to lignin (Chapter 9). However, this specific lignin reaction is lost as further deposition occurs. Microspores in lilies and other plants rapidly expand and incorporate nutrients after release from the callose. No visible thinning of the primexine, which is already present, occurs during this rapid expansion; this suggests that new wall materials are rapidly added during such increases in spore volume.

Accumulation of the Exine Material. This continues first on the outer and then on the inner surface of the primexine until a complete coating is established. In the terminology of palynologists the exine wall is composed of two layers: the sexine and nexine. These sporopollenin layers are linked by bacula or columellae (Fig. 2-3). The inner nexine probably accumulates from the inner spore cytoplasm. The association of the cytoplasmic membranes with the wall pattern is now lost.

During exine formation, changes in the aperture regions take place. A thin layer over the plasmalemma marks the region of each pore; it is not continuous with the primexine and can be considered as the first manifestation of the intine. The material is primarily cellulosic in nature. When probacula are absent, there is no linear extension of rods of sporopollenin, but instead there is a deposition of irregular masses.

Intine Growth. During exine development the formation of intine proceeds in waves, a fact suggested by the concentric laminations that can be seen in it. In the pore regions a local accumulation of amorphorous cellulosic material appears at the plasmalemma inside the spore, which is then incorporated into layers of the intine. The intine is laid down after the sporopollenin is deposited.

Pollenkitt. In the final period of wall development, additional lipoidal and pigmented substances may accumulated on and within the outer exine. This material, called Pollenkitt (PANKOW, 1957), may impart color or odor to the pollen and may cause the pollen grains to adhere together during dehiscence. Lipid globuli, the pre-Pollenkitt, form in an inner layer of the tapetum; the globuli are infused with carotinoids and other pigments just prior to release of the spores from the tetrads (HESLOP-HARRISON and DICKINSON, 1969).

Deposition of Sporopollenin

The chemical nature of sporopollenin is not definitively known (Chapter 9). It belongs to the lipid group, like cutin and suberin, and is characterized by a high resistance to acids and enzymatic degradation. Sporopollenin is isotropic (SITTE, 1959) and shows no intrinsic birefringence (AFZELIUS, 1955). In some cases sporopollenin can be accompanied by other molecules, so that the exine may become positive in birefringence (FREYTAG, 1964). The empirical chemical composition, worked out by ZETZSCHE and VICARI (1931), suggested sporopollenin might be a polyterpene. Chemical analysis of pollen walls led SHAW and YEARDON (1966) to conclude that while the intine is largely cellulose, the exine sporopollenins are not terpenoid in nature but rather are oxidative polymers of carotenoids or carotenoid esters (SHAW, 1971). The chemical nature of sporopollenin may be resolved by a more vigorous analysis of sporopollenin, after separation from the pollen grain, or by a study of the metabolic precursors incorporated during exine formation. Sporopollenin thickness varies in some species in response to temperatures prevalent during microsporogenesis (KAWECKA, 1926). This temperature effect may provide a tool to help study chemical pathways in exine formation.

Sporopollenin is transported in a highly polymerized form-as monomers from the tapetum to the exine. Electron micrographs (ROWLEY, 1963, 1964; ROWLEY and SOUTHWORTH, 1967) indicate that isolated agglomerations of sporopollenin, surrounded by a unit membrane, Ubish bodies, are found at a considerable distance from the microspore. Pro-Ubish bodies, more correctly called proorbicular bodies, have been recognized in the tapetum of developing *Allium cepa* microspores (RISUENO et al., 1969; HESLOP-HARRISON and DICKINSON, 1969). The endoplasmic reticulum widens to form pockets in which electron-dense materials accumulate along the plasmalemma. These are the nuclei around which spherical laminae of sporopollenin granules, orbicules, will be deposited. Once the proorbicular bodies pass through the plasmalemma they are rapidly coated with sporopollenin (ECHLIN, 1969). The orbicules are transported across the tapetum cytoplasm into the locule of the pollen sac before being positioned on the microspore outer wall as exine.

Sporopollenin seems to be arranged in bundles of anastomosing strands with an initial diameter of 80 to 150 Å. Approximately four superimposed bundles of sporopollenin in the early wall separate to produce an exine composed of two shells, each having about two superimposed bundles in cross sections (ROWLEY, 1963). Later, between meiosis and pollen mitosis, sporopollenin has a homogenous appearance. In some cases, when pollen grains remain together in groups (tetrads, polyads) exine stratification leads to fusion of the surfaces, resulting in the cohesion of pollen grains (SKVARLA and LARSON, 1966).

The origin of the lamellae membrane upon which the sporopollenin is deposited and exine formed has not been clearly established. ROWLEY and DUNBAR (1967) suggest there are four possible ways for synthesis of the exine supporting membrane. Data on developing *Populus* pollen suggest that in this species the lamellae may arise from pre-existing vesicle membranes in both the tapetum and microspore cytoplasm. In *Anthurium* the lamellae membrane may arise by *de novo* formation in the inner exine, the nexine 2 region (ROWLEY and DUNBAR, 1967).

Two interesting questions remain: what factor(s) determine the pollen exine pattern which is so genetically stable for the different species that it can be used for species identification in recent and in fossil pollens (SITTE, 1953; MÜHLTHALER, 1953; 1955; HESLOP-HARRISON, 1968d) and what is the function or significance of such distinct patterns?

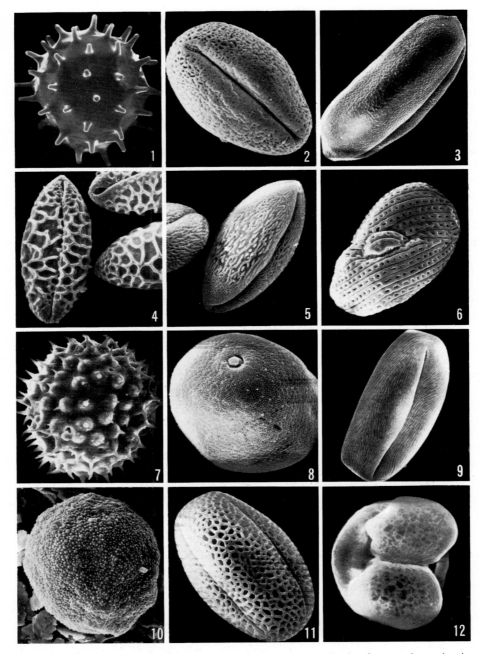

Fig. 2-4. Pollen grains illustrating exine wall patterns, differences in size, form, and germination pore. (1) *Hibiscus* 420 × ; (2) *Vitis* 2000 × ; (3) *Tradescantia* 1500 × ; (4) *Lilium* 400 × ; (5) *Petunia* 1350 × ; (6) *Sanchesia* 450 × ; (7) *Gossypium* 500 × ; (8) *Paspalum* 2200 × ; (9) *Pyrus* 1000 × ; (10) *Populus* 1700 × ; (11) *Jasminum* 2000 × ; (12) *Pinus* 940 × . (Scanning electron micrographs by courtesy of M. M. A. SASSEN and A. W. DICKE)

It seems clear that the primary sporoderm pattern is related to the deposition and location of ER. This suggests the genotype of the microspore directs the pattern of formation. On the other hand, the material for synthesis of the pollen wall must be delivered by the tapetal cells .from outside the pollen cells. This suggests that a sporophytic control mechanism can be interposed. The substrates or precursors for the exine wall are derived from the locule tapetum (ROWLEY et al., 1959). The observations of HESLOP-HARRISON (1963, 1966a) suggest the directions for the laying down of tapetum-supplied polymers is controlled by localized RNA or ER material from the microspore nucleus. The question becomes more complicated by the detection in the exine of channels, which can serve as a route for the transfer through the pollen grain wall. In some developing microspores, e.g. *Ipomoea purpurea*, the exine patterning is first detected in the callose wall enclosing the developing microspores (WATERKEYN and BIENFAIT, 1971). The enclosing callose acts as a negative template. Definitive experiments must still be done to resolve the mechanism of how exine formation is genetically controlled.

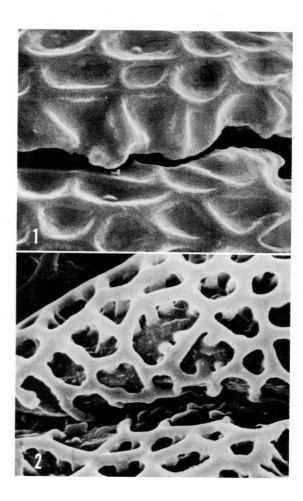

Fig. 2-5. Scanning electron micrograph of *Brassica oleracea*. (1) Intact; (2) After brief washing with chloroform — germination not affected (ROGGEN, 1974). (Photographs courtesy of H.P. ROGGEN)

The role of sculptured exine can only be speculated upon. Protection of the sensitive nuclei, and a method of facilitating tube germination are obvious benefits (HESLOP-HARRISON, 1971). There is a constant, genetic control in the size range (Chapter 3) and wall pattern (Fig. 2-4). Yet, as CRANG and MILLAY (1971) observed in *Lychnis alba*, some characteristics such as exine spines and pits vary in size and number. Indented, plated surface patterns may facilitate the bonding and positioning of viscin threads, protein or lipoidal materials which coat the exine and are beneficial in germination. More obvious are the alternate thickenings and thinnings in sporopollenin which, by affording places of less pressure and allowing the intine to expand and contract easily, define points of tube release from the aperture.

Surface-localized materials usually disperse quite readily when pollen is placed in contact with a solution. Thus, such components can act as wicks to imbibe and conduct water throughout the pollen wall, or release materials to influence or modify the surface where the tube must grow. As noted in the scanning electron micrographs of *Brassica* pollen (Fig. 2-5) the exine may be covered with a lipoidal, oleaginous coating, Pollenkitt in PANKOW's (1957) terminology, tryphine to others (DICKINSON and LEWIS, 1973) or pollen coat in the descriptive term of ROGGEN (1974). This coat is readily removed by solvents, or may be partially dislodged under vacuum and high temperature fixation for the SEM (PARTHASARATHY, 1970), revealing the tectum and supporting columellae, colonnades and exine surface.

Pollen grains with large areas of apertures can more readily interact with the external environment by releasing and absorbing materials more rapidly than those with fewer apertures. Some pollen need to extend their tube rapidly due to short-lived female receptor cells, or the long distance they must grow to reach the micropyle and egg cell. Such species have generally evolved, or been selected, with thinner exine walls and a greater number of apertures, in comparison to species that can tolerate or require longer periods between pollination and fertilization. This hypothesis might help explain, for example, wall differences observed between grass and pine pollen, to compare the extremes.

Pollen Wall as a Living Structure

The solidity and resistance of the pollen wall to erosion and chemicals has led to the common opinion that the living content of the microgametophyte, enclosed in a mature pollen grain, has no contact with the environment. Contradicting this opinion is the observation that when dry pollen is put into an artificial medium within one to five seconds it releases protein with enzymatic capacity (STANLEY and SEARCH, 1971). Within a few minutes many isoenzymes are released by the pollen (STANLEY and LINSKENS, 1964, 1965; LEWIS et al., 1967). Under such conditions pollen grains swell very quickly and increase in diameter. Furthermore, there is evidence that pollen can secrete amylase, pectinase and cutinase (LINSKENS and HEINEN, 1962) to hydrolyze external stigmatic substrates. In some cases, such as grass pollen, there is direct evidence that when pollen reaches the stigma it secretes a liquid drop within a few seconds (WATANABE, 1955).

Direct histochemical evidence for the presence of protein in the pollen wall, including the intexine and mesine, is given by TSINGER and PETROVSKAYA-BARA-NOVA (1961) and more recently by KNOX and HESLOP-HARRISON (1970) and SOUTHWORTH (1973). These proteins, which are primarily concentrated at the intine pore region (Chapters 9 and 11), include some enzymes (Chapter 14), and effectively permeate the whole of the pollen grain including the outer surface of the sporoderm.

In spite of the high chemical resistance and hydrophobic properties of the pollen wall, it can in many cases be penetrated by proteins and plasmatic strands. These proteins may be readily activated by elements in the environment, in particular water and metabolic substrates, and upon activation play an important role in the interchange between the pollen grain and the female tissues.

Chapter 3. Dehiscence, Size and Distribution

Dehiscence

The mechanism of dehiscence, opening of anthers or microstrobili and release of pollen, varies in the different plant families. In gymnosperms it is released following a simple parting of the microstrobili sporophylls. Dehydration causes a retraction of the bract scales, freeing the pollen to be dislodged by wind or shaking. With the exception of the cycads, all gymnosperm pollen is distributed by wind (anemophilous).

Temperature primarily controls the rate of maturation of strobili in temperate pine zones and strongly influences when dehiscence occurs. The simple, older degree-hrs concept which correlated time of dehiscence with accumulation of a given number of hrs of temperature above 0° or 5°C (DANCKELMANN, 1898; SCAMONI, 1955) is seldom used today. The fact that a group of plants change their dehiscence time from year to year and in relation to another species in the same environment indicates that the heat sums approach is difficult to apply to the complicated sequence of changes between induction and dehiscence.

In *Pinus*, low temperatures will both delay and extend the period of dehiscence (BOYER, 1970). However, for any one species at a given location, a regression coefficient can be derived to predict the day of maximum pollen shed from the degree-hour heat sums if sufficient data is available. BOYER (1973) found that time of maximum pollen shed in *Pinus palustris* in south Alabama ranged from February 23 to April 3 over a 10-year period. The heat sums above 10°C occurring after January 1, provide a good index of pollen shedding time. A definite cold period must precede the cumulative degree-hrs to promote strobili development in this pine, as is probably the case in most temperature zone species.

In angiosperms, pollen is commonly shed following one of several opening patterns of the anther-sac walls. In longitudinal dehiscence, the most common form, a slit-like opening occurs; in porose dehiscence discharge is through pores, a cap generally opening at the distal ends of the sac (Fig. 3-1). Some dehiscent forms are intermediate between the pore and longitudinal slit (VENKATESH, 1956).

Change in atmospheric humidity is the most frequent causative agent leading to hygroscopic shrinkage and anther wall rupture. A layer of structurally weak fibrous cells, the endothecial layer, is the common site of dehiscence in anther sacs. Enzymes may also be involved in the weakening or breakdown of these cells (RICHTER, 1929; BECQUEREL, 1932). Pollen may be released at once, or may gradually escape depending upon the species and environmental conditions. Thus, pollen from each of 4 anther sacs on one filament may be released too far apart to mix; or a common slit opening between two sporangiate may facilitate mixing at dehiscence. In some species, e.g. *Cassia*, the outer whorl of anthers dehisce by

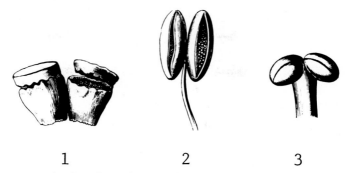

Fig. 3-1. Common dehiscence mechanisms in angiosperm anthers (KERNER, 1904). (1) Pore, circumscessile slit in *Garcina* sp.; (2) Longitudinal slit in *Calandrinia compressa*; (3) Distal slit in *Calla palustris*

slits; the inner whorl by pores. In addition to wind, bees, beetles, wasps, moths, birds or bats may be involved in pollen transfer (Chapter 7) from dehisced anthers (entomophilous).

Some anthers are nondehiscent. In many such cases co-evolution has occurred with insects acting as vectors to transfer pollen or entire anthers to the stigma surface (MEEUSE, 1961; FAEGRI and VAN DER PIJL, 1966). In a few tropical flowers, pores are actually forced open by insects squeezing on the anthers; or insects may cut or manipulate the filament to expose the pollen. The interdependence of the yucca flower and yucca moth, *Pronuba yuccasella*, is a classic illustration of co-evolution. The moth larve feed on *Yucca* seed. The female moth facilitates seed formation by collecting and depositing a ball of pollen on the stigma as it oviposits.

Water plants have evolved several mechanisms for pollen release (hydrophilous). The whole male flower may float and on contact with the female flower catapult pollen onto the stigma e.g. *Zostera* and *Vallisneria* (Fig. 3-2) or, dehiscence and pollination can occur under water by a balloon-like attachment on the pollen, e.g. *Ceratophyllum*. A modified flotation mechanism occurs in certain *Narthecium* and *Ranunculus* where water accumulating in the flowers floats the pollen up to the stigma (HAGERUP, 1950). In the sea-shore tidal plant *Plantago*, pollen floats on the water and is carried by the rising tide to the stigmas on the elevated spike pistil (HENDERSON, 1926). It will be interesting to learn more of the salt tolerance and surface properties of such pollen.

Some angiosperm plants produce cleistogamous flowers which do not dehisce, or which dehisce and remain closed to all pollinating agents. The adaptive capacity of some flowers to close in overcast or rainy weather is beneficial in limiting anther exposure to times when pollination is most lilely to succeed. The number of cleistogamic florets increased in wheat under adverse conditions of rain and low or extremely high temperatures (DE VRIES, 1971). On a single plant or tree, dehiscence will frequently occur over a longer time period than female flowers are receptive (ILLY and SOPENA, 1963). Seasonal and genetic variations in time of dehiscence also occur. Red pepper, *Capsicum*, dehisced in June between 8 A.M.

Fig. 3-2. Filiformous pollen grains of *Zostera marina* L packed in the anther. (Scanning electron microscopical photo, courtesy of W. A. SCHENCK, magn. about 80 ×)

and 12 A.M.; in August, most flowers dehisced before 8 A.M. However, among the six varieties tested, one persisted, even in August, in a 10–12 A.M. pattern of shedding (HIROSE, 1957). The times when individual species of flowers open and dehisce and are available for collection (Fig. 7-5), have a minimum and maximum during the day (Table 4-2). The diurnal pattern of pollen flight in pine may correspond to relative humidity and the drying-retraction of the microsporophylls. However, if continuous rain occurs after the pine microstrobili mature, they open sufficiently for release into the rain (BUSSE, 1926). Pollen in such rain drops has a reduced capacity to grow. Most plants dehisce in the early morning, some at two peaks during the day, and a few dehisce at night, fortunately for certain moths and bats (Chapter 7).

The distance pollen moves after dehiscence and time of flight is dependent on atmospheric conditions, nature of the vector (transmitting agent) as well as pollen size and surface characteristics.

Size Range

Grain size, when measured, is affected both by chemical treatment (CHRISTEN-SEN, 1946) and mounting media (ANDERSEN, 1946). Reproducible size measurements require that determinations be made under identical conditions. Measurements are most reliable when made in humid conditions, where the pollen is not

Table 3-1. Variations in pollen size

Species	Dimensions in microns (μ)			Volume in 10^{-9} cm^3	Weight 10^{-9} g
	Length	Width	Height		
Abies alba	97.8	102.9	62.7	499.4	251.6
Abies cephalonica	97.1	98.6	86.2	422.6	212.2
Picea abies	85.8	80.5	66.3	278.2	110.8
Pinus sylvestris	41.5	45.9	36.0	35.5	37.0
Larix decidua	76.0	72.0	50.0	180.2	176.3
Pseudotsuga taxifolia	84.8	81.1	54.8	219.2	188.8
Acer saccharum	32.5	23.6	24.6	16.5	6.6
Aesculus hippocastanum	31.0	16.4	18.2	4.8	0.9
Alnus glutinosa	26.4	22.8	13.7	4.4	1.4
Betula verrucosa	10.1	10.1	16.8	2.9	0.8
Fagus silvatica	55.1	40.5	41.1	50.3	26.0
Quercus robur	40.8	26.1	21.5	13.3	5.7
Tilia platyphyllos	40.5	40.1	20.6	15.0	6.5
Ulmus laevis	33.4	32.7	17.7	12.8	6.8
Zea mays	116.3	107.3	107.3	702.4	247.0
Cucurbita pepo	213.8	213.8	213.8	5,117.0	1,068.0

swollen, but is no longer dry. The size is not only dependent on external moisture conditions but also on swelling induced by other means (JAESCHKE, 1935). A complete list of mounting media (BJÖRK, 1967) suggests that two of the best are: silicone oil (ANDERSEN, 1946) and glycerine jelly. The latter is prepared from 5 g gelatine, 21 ml water, 30 ml glycerine, 5 g crystal phenol (REITSMA, 1969). Pollen which is not clumped and lacks viscin threads or a sticky surface can be sized in a Coulter Counter (BROTHERTON, 1969).

Size varies over a broad spectrum, ranging from a mean diameter of 5 μ in *Myosotis* to 177 μ in *Mirabilis jalapa* (AMELUNG, 1893) and greater than 200 μ in *Cucurbita pepo*. In the majority of anemophilous plants, pollen grain size is within the limits of 17–58 μ (WODEHOUSE, 1935). Species with smaller or larger grains are generally zoophilous or entomophilous. If pollen is embedded, then the imbedding technique plays an important role in size and shape determinations (SCHOCH-BODMER, 1936). Dimensions of single grains of a few species together with their volume and weight are given in Table 3-1, from POHL (1937) and EISENHUT (1961), supplemented by measurements made in the laboratory at Nijmegen. Size distribution of a collection of European grains is given in Fig. 3-3. Complete tables of size and month shed in the United States are given by WODE-HOUSE (1945) and for southern Europe by CEFULU and SMIRAGLIA (1964).

Sources of Size Variation

Pollen size and volume data must be viewed critically because of methods used during measurement and the actual status, i.e. water content, turgidity, and age of the grains at the time of measuring (FAEGRI and IVERSEN, 1964). Inherent variation also occurs. Thus, many factors contribute to size variability.

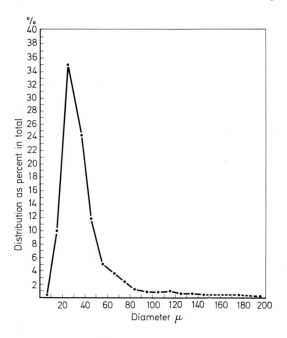

Fig. 3-3. Size distribution of 586 European pollen species

Internal Factors

Chromosome Numbers. Pollen size is related to chromosome numbers and generally is fairly constant (RENNER, 1919). Significant geographic variation was found in some species of *Betula* (CLAUSEN, 1962), but some closely related taxa of grasses showed little variation while others varied widely (BRAGG, 1969). A conventional method used in experiments for polyploidization is the measurement of the diameter or volume (SCHWANITZ, 1952). Pollen size is more than a parameter to indicate ploidy level; it is also considered to be a reliable taxonomic characteristic (MÜNTZING, 1928; SPASOJEVIĆ, 1942). Highly significant positive regression coefficients were obtained within various species (NISSEN, 1950; LÖVE, 1952; GOULD, 1956; BELL, 1959; KAPADIA and GOULD, 1964). In *Petunia* and in many other plants, size as an indicator of ploidy, or as a taxonomic characteristic, must be used with caution (BELL, 1959). Not only is size influenced by the chromosome number, but the number of germination pores is as well. In auto- and allopolyploid plants the pore numbers increase with increasing ploidy level. An exception seems to be the grasses where even tetraploids have a constant single pore (FUNKE, 1956).

Flower Character. Size in some cases varies from flower to flower (HARRIS, 1956), or even from anther to anther, on individual plants (KUPRIJANOV, 1940). Generally pollen is largest in terminal flowers or in flowers formed just below the apex (SCHOCH-BODMER, 1940; KURTZ and LIVERMAN, 1958). Variation in mean size was also found from spiklet-to-spiklet and from floret-to-floret in wheat (OVCINNIKOV, 1951, 1953). In certain families, e.g. the Fagaceae, intraspecific variation in size is greater than interspecific variation (HARRIS, 1956a). In *Petunia*, there was greater size variation from flower to flower on the same plants than among flowers from different plants of the same clone grown on different nutrient solutions. Cultivated species of rice seem to have larger pollen than wild species (SAMPATH and RAMANATHAN, 1951), suggesting that size differences are a combination of genetic and nutritional conditions. *Betula* did not significantly vary in size with age of the tree (SZAWBOWICZ, 1971), or from the upper to lower part of the tree, or from apex to base of individual catkins (CLAUSEN, 1960). Species with grains larger than *Betula* would probably show greater size variations.

Changes in pollen sizes over the flowering period have been reported (PIECH, 1922; KAWECKA, 1926; KRUMBHOLZ, 1926; SCHWANITZ, 1952). In *Oenothera* size decreases towards autumn (RENNER, 1919).

Heteroanthery. Within a species, various size classes of pollen can occur in various height classes of anthers. The special cases of heteroanthery are usually linked to heterostyly. Yet the two phenomena, grain size and position of the anthers, are genetically independent (BODMER, 1927; SCHOCH-BODMER, 1937; 1938). In the dioecious genus *Cannabis*, pollen size is correlated with sexual expression (HERICH, 1961).

External Factors

While the characteristics of sexual organs are, in general, more genetically stable than vegetative characteristics, pollen characters may vary. External characters such as diameter, volume, number of pores and furrows, and sculpturing pattern of the exine can be influenced by climatic conditions.

Temperature. High temperature has often been stated to favor the formation of large grains (KAWECKA, 1926; STOW, 1930); repeated cooling of flower buds during microsporogenesis causes formation of abnormal pollen with a variable number of germination pores (MICHAELIS, 1928). MIKKELSEN (1949) concluded from comparative experiments that the relation between size and temperature cannot be assumed as proven, since every temperature change will, at the same time, affect the overall general growth pattern of a plant. Experiments with tomato and *Xanthium*, grown in the Phytotron under controlled conditions, show that pollen diameter is influenced by night temperature in a pattern quite similar to the general growth of the plants under these conditions. There was no variation of the exine character, but the diameter decreased at high (above 20° C) and low (below 10° C) night temperatures, and at high day (30° C) temperature (KURTZ and LIVERMAN, 1958).

Mineral Nutrition. Although several workers have suggested that nutrition could influence pollen size (JONES and NEWELL, 1948), few experimental data have been reported. WAGENITZ (1955) and BELL (1959) modified size under different mineral nutrient levels. However, no correlation was detected between the pattern of variation in pollen size and the type of mineral deficiency. Moreover, different clones react differently to nitrogen or potassium deficiency. In *Sinapis*, nitrogen fertilizer increased pollen size (SCHWANITZ, 1950).

Water Conditions. Although water stress increases the amount of sterile pollen (ANIKIEV and GOROSCENKO, 1950), soil water conditions influence size only within narrow limits. Size is affected much less than the dry weight production of vegetative plant organs (WAGENITZ, 1955). Water conditions can be important at the early stages of pollen development, especially during the onset of meiosis. Once the production of gametophytes is initiated, the plant may very well reduce water and nutrients going to other organs since the metabolic demands of the reproductive organs probably exert a dominant influence. Fewer mature grains may be produced, suggesting a more general depression of synthesis and growth rather than a specific effect of water deficiency on pollen formation (FEDOROVA, 1955).

Quantity Produced

Distribution patterns primarily reflect size, density, quantity of pollen available and conditions during dehiscence. The yield varies with the species. The quantity produced is usually measured in cubic centimeters (cc). Examples of pollen yields, compiled by SNYDER and CLAUSEN (1973), are given in Table 3-2.

Table 3-2. Pollen yields (SNYDER and CLAUSEN, 1973)

Genus	Number and type of flower	Approx. cc
Gymnospermae		
Larix	100 strobili	0.3
Pinus	100 strobili	150
Pseudotsuga	100 strobili	2
Angiospermae		
Alnus	100 catkins	4
Betula	100 catkins	12
Fagus	100 inflorescences	1.3
Liquidambar	100 flowers	25
Populus	100 catkins	75
Ulmus	100 flowers	0.3

Presence or absence of wings and pollen size and moisture content influence packing volume. Quantities produced by pine strobili (Table 3-2) depend indirectly on nutrition and directly on light exposure, i.e. side of the tree, position on the tree and age of the tree. Highest yields occur from strobili located at the top part of the crown in *Pinus elliottii* (CHIRA, 1966). Strobili growing at the top and outside of the tree are generally larger than those on lower or inside areas. In plants flowering over a large part of the growing season, the numbers of anthers dehiscing often decreases toward the end of the growing season. This may reflect the decreasing availability of nutrients. Estimated cumulative 50 year yields from anemophilous forest tree species (Table 3-3) suggest that many light, small packing volume pollen is among the most abundantly produced.

Comparing pollen volume yields per strobilus or per flower can distort the significance of productivity, if such assays are separated from the functional role of yield, survival and plant reproduction (KUGLER, 1970). Anemophilous species almost always outproduce entomophilous species; the hydrophilous species usually produce very few grains per plant. Quantities of 10^7 grains per plant are

Table 3-3. Pollen quantities shed by forest trees (BROOKS, 1971)

Species	kgms of pollen/tree produced in 50 yrs
Picea abies	20.0
Fagus sylvatica	7.6
Pinus sylvestris	6.0
Corylus avellana	2.8
Alnus sp.	2.5
Betula verrucosa	1.7

typical in *Zea mays*, while *Vallisneria* produces 72 to 144 grains per plant (POHL, 1937). But a more meaningful index, from the ecological viewpoint, is to measure numbers of grains per viable ovule. Considered on the basis of pollen produced per ovule the ratios for wind and insect pollinated species are quite similar. To complete the ecological loop, the contribution of pollen to ground biomass, dry matter, must also be recognized. FIRBAS and SAGROMSKY (1947) computed that stands of alder, hazel or rye plants deposit about 10 kgm per hectare, a contribution of about 1 kgm of protein and an equal amount of fat to the soil microbial community.

Sources of variation occur in the amount of pollen produced on an individual plant; differences can be quite great in the amount produced on the same plant in successive years. Thus, quantities measured in one year are not always a good index of a given plant's producing capacity in successive years. But the capacity to produce pollen is primarily under genetic and physiological control. In comparing 22 varieties of wheat, BERI and ANAND (1971) found grains per anther varied from 581 to 2,153. The taller varieties produced larger florets, and longer filaments, with bigger anthers. Plant height and anther size correlated with the amount shed. In varieties of *Pyrus communis*, counts show Winesap consistently low, at about 400 grains per anther, while Delicious usually produces about 7,000 grains per anther. In anthers of var. Delicious 85% of the microspores matured, while 85% of the microspores aborted in var. Winesap (OBERLE and GOERTZEN, 1952). The potential to produce higher quantities probably exists during initial

Table 3-4. Annual variation in pollen productivity (HYDE, 1951)

Genus	Average of catch 1943—1948 (= 100%)	Catch by year as % of 1943—1948 averages					
		1943	1944	1945	1946	1947	1948
Pinus	386	92	121	72	127	40	146
Alnus	385	45	226	120	55	34	118
Fraxinus	675	330	25	24	159	54	8
Fagus	273	23	140	11	143	8	275
Betula	620	57	160	42	106	22	214
Ulmus	4,579	84	87	180	50	48	150
Quercus	2,776	58	90	42	280	10	117

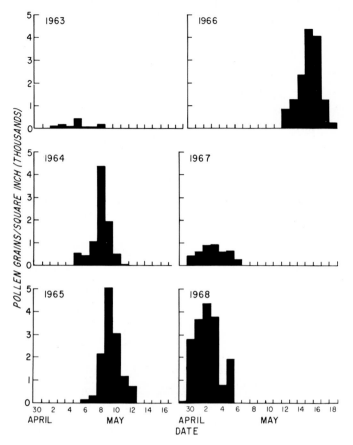

Fig. 3-4. Pollen dispersal pattern in one stand of *Pinus echinata*. Trees mean age of 26 years (BRAMLETT, 1973)

phases of microsporogenesis in Winesap and other low yielding varieties, but some block due to a genetic control or metabolic mechanism apparently inhibits viable pollen production in such varieties.

Climate, in addition to influencing time of deshiscence, also influences pollen quantities produced. Variations in quantities in 7 genera over a 6-year period in Great Britain (Table 3-4) assayed by impact slide counts (Chapter 4) reveal dissimilar yields in any two successive years (HYDE, 1951). Some trees, e.g. *Fraxinus* and *Ulmus* have a high yield about every third year, others, e.g. *Fagus silvatica*, give high yields every other year. Not only will annual yield vary, but time of maximum dispersal will also vary from year to year. This is illustrated by the bar graphs of Fig. 3-4 which represent pollen release over a 6-year period in one 5-acre stand of *Pinus echinata* (BRAMLETT, 1973). Obviously, when a breeding goal is to select consistently high pollen producers, final selection should depend on analysis of several years data.

Distribution

Insects, particularly bees, carry pollen in fairly well recognized patterns and limited distances (Chapter 7). How anemophilous pollen responds to air movements has also been widely studied, particularly in species that are focal points of plant breeding programs, or which induce allergenic reactions in man (Chapter 12). Distance and patterns of shedding and movement are the common indices of dispersal. The principal characteristic studied is sedimentation rate.

Sedimentation

The free-fall velocity of grains is determined by their specific gravities and is dependent on the air resistance, which is related to their size and form. Measurements have rarely been made under standard environmental conditions. A few examples from EISENHUT (1961) are given in Table 3-5.

Table 3-5. Free fall velocity of pollen (cm/sec)

Species	BODMER (1927)	KNOLL (1932)	DYAKOWSKA (1937)	EISENHUT (1961)
Abies alba			38.7	12.0
Larix decidua	12.5—22.0	9.9	12.3	12.6
Picea abies		8.7	6.8	5.6
Pinus sylvestris	2.9— 4.4	2.5	3.7	3.7
Taxus baccata	1.1— 1.3		2.3	1.6
Abies incana	1.7— 2.2			2.1
Betula verrucosa	1.3— 1.7	2.4	2.9	2.6
Carpinus betulus		4.5	6.8	4.2
Quercus robur		2.9	4.0	3.5

Pollens with air sacs are adapted to a slower free fall than those without such appendages. The air sacs retard falling velocity and facilitate dispersion. Generally, the sedimentation velocity of types with the same construction decreases with decreasing pollen size (Fig. 3-5). A few species in the families Ericaceae and Pyrolaceae release grains in tetrads; in *Leschenaultia*, octads are often released (ERDTMAN, 1969). Such pollen conglomerates would not be characterized by sedimentation values in Fig. 3-5.

Clumping may occur as often as 85% of the time in some anemophilous species (Table 3-6). This clumping leads to a modified sedimentation velocity and distribution pattern (ANDERSEN, 1970). The number of grains per clump usually varies between 2 and 9, depending on species and weather condition at dehiscence. Air moisture can modify such factors as electrical charges which may influence pollen separation or clumping. Obviously, individual size alone is a poor index of pollen sedimentation and dissemination potential (STANLEY and KIRBY, 1973).

Pollen can spread over wide distances. Under the influence of atmospheric convection currents, pollen can attain a relatively high altitude. The highest density in the atmosphere during day-time occurs at heigths between 350–650m, with

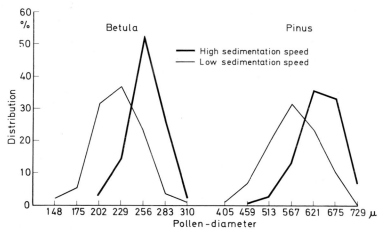

Fig. 3-5. Relation of pollen size to sedimentation velocity (FIRBAS and REMPE, 1936)

no selection according to the grain size. At night, when convection currents are reduced, a separation of the grains by their sedimentation velocity takes place. That means that most species, present up to a height of 700 m in the atmosphere and having a free fall velocity of less than 2 cm/sec, fall to the surface during the night. Many observations of long distance transport have been made (SCHMIDT, 1918; HESSELMANN, 1919; REMPE, 1937; DENGLER and SCAMONI, 1944; SCAMONI, 1949, 1955). They all show that large numbers of grains can be transported for long distances (Table 3-7). Atmospheric conditions, including wind and thermal air currents, are important in influencing the distance wind-born pollen is transported.

The probable range of flight is reasonably predictable by applying formulas KOSKI (1967) devised. His calculations treat pollen as a spherical atmospheric particle of a given size and density. They are limited to a given set of meteorological conditions. But not formula can cope with the changes in wind velocity during a rain storm that propelled pine pollen a distance of 1750 km (BUSSE, 1926). Sedimentation values, calculated as velocity of deposition, are derived by measurements of pollen concentration at a given height and on the ground at a given distance (Fig. 3-6). This provides a reasonably good comparison of distribution.

Table 3-6. Pollen clumping (ANDERSEN, 1970)

Species	% of pollen clumped	Average number grains/clump	Sedimentation velocity cm/sec
Alnus glutinosa	34	3.2	1.7
Betula verrucosa	34	2.9	2.4
Corylus avellana	37	3.9	2.5
Fagus sylvatica	24	2.4	5.5
Quercus spp.	50	2.7	2.9
Tilia cordata	85	9.0	3.2
Ulmus montana	58	3.1	3.2

Table 3-7. Relation between pollen size, falling speed and the average dispersion distance at a given wind velocity (Data from KNOLL, 1932; DYAKOWSKA and ZURZYCKI, 1959)

Species	Average weight of one pollen grain in 10^{-9} g	Average diameter (longer axis) in μ	Sedimentation in motionless air in cm/sec	Average dispersion at windspeed of 10 m/sec in km
Picea exelsa	93.2	162.0	6.84	22.2
Pinus sylvestris	30.08	59.0	3.69	267.8
Quercus robur	18.1	24.8	3.96	199.0
Alnus spec.	9.37	24.6	2.77	546.7
Corylus avellana	9.45	24.2	2.9	267.8
Dayctylis glomerata	21.85	33.3	3.1	174.2

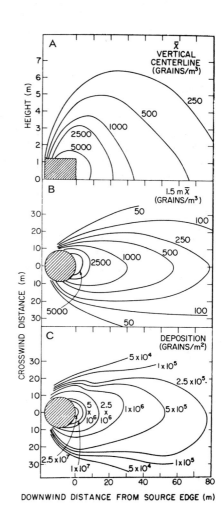

Fig. 3-6 A-C. Patterns of distribution of timothy *(Phleum pratense)* pollen. Concentration in (B) is measured at 1.5 m height (RAYNOR et al., 1972)

Pattern of Distribution

Typical patterns of distribution show that anemophilous pollen moves along concentration gradients with a rapid decrease in deposition occurring with distance from the source. The greatest bulk, in the case of grasses, lands within 3 m of the source (Fig. 3-6). Less than 1% of all air-borne grass pollen reaches 1 km from the source (RAYNOR et al., 1972). Trees near the edge of forest stands often disseminate pollen great distances because of wind updrafts that skew normal distribution. An interesting example is that of pollen from coastal stands of *Betula* in Finland which accumulated at a higher concentration on an island 20 m away, than it did at the forest edge (SARVAS, 1955). This extreme case was induced by the prevailing wind pattern. Most *Pinus sylvestris* pollen in Sweden is normally deposited within 700 m of the source tree (PERRSON, 1955). Within stands, trees can filter out the pollen drastically and modify the deposition pattern depending upon the species, height and density (DENGLER, 1955). Another skewing factor, in addition to stand and meteorological condition is reflected when clumping (Table 3-6) increases deposition near the source. Pollen from trees on an upslope will tend to concentrate in a valley below. SILEN and COPES (1972) counted twice the concentration in a narrow valley as from similar orchard tree species, yielding like amounts, but growing in a flat area.

Most speculations on the ability of pollen to float in air are based on the dimensions (KNOLL, 1932) and the sedimentation velocity (FIRBAS and REMPE, 1936). Although both factors are important in long distance transport, absolute and relative weight of the grains undoubtedly also influence the time it will float in air (DYAKOWSKA and ZURZYCKI, 1959), and the area of its dissemination. Furthermore, the functional role of rain drops as a dispersal agent must be taken into consideration (BRODIE, 1957). Bioelectrical forces may also be involved in pollen sedimentation (McWILLIAM, 1959). A high percentage of pollen carry a negative electrostatic charge (ARABADZHI, 1973). While there is no evidence to support the view that electropotential gradients between pollen and receptive organs are involved in pollination, the possibility does exist that long distance transport is influenced by electrical relations.

II. Management

Chapter 4. Collection and Uses

Improvements in techniques for collecting and preserving pollen have facilitated research and commercial uses of pollen. Tomb engravings from Egypt (Frontispiece) indicate that by 1000 B.C. whole branches of date palm, *Phoenix dactylifera*, were transported by hand to pollinate female flowers (WODEHOUSE, 1935). In the 20th century many different methods of collecting and handling pollen have been developed. These vary with the pollen species and purposes for which the pollen is to be used.

Purpose and Quantities Collected

Hybrid Production

When specific progeny are desired, known parental sources of pollen are used. In general, only limited quantities of pollen are needed for plant breeding programs. A few milligrams to several grams of pollen are generally sufficient for small breeding studies with flowers of most species. In collecting pollen for such purposes precautions must be taken that the pollen is free from contamination and that genetic purity is assured. Standard microscopic evaluation is used to determine the purity of such pollen samples (HELLSTRÖM, 1956).

Artificial breeding of forest trees, i.e. spruce and pine, was first initiated by SYLVÉN in 1909. By mid-1950, programs in reforestation and agronomic crop breeding utilized controlled pollinations to produce millions of seeds of improved or selected strains (WRIGHT, 1962). Kilogram pollen lots from known parent trees or from a herbaceous genotype must often be collected and preserved for subsequent use. In some agronomic crops, i.e. corn or sorghum, where successive crops for hybrid breeding are desired as rapidly as possible, the pollen is usually used shortly after collection. In such programs, plants are grown to maturity under controlled greenhouse conditions, or in two different plant development zones. In the latter process, breeders in the United States often grow one or two crops in south Texas of Florida during winter and spring, and another crop at a location in Nebraska or elsewhere in the north during the summer. Such intensive breeding programs generally store the germ plasm as seeds rather than as pollen. An additional reason for this procedure in the case of Gramineae is that a good method for extended storage of this pollen has not yet been developed.

Increased Fruit Yields

The Potential. In many fruit orchards pollen is artificially applied to assure an abundant crop (JOHANSEN, 1956). Mechanical application of the pollen to trees can supplement natural pollination. In the fruit growing regions of the United

Table 4-1. A partial list of commercial sources of pollen

Australia	
A. S. Jackel	Olive St. Wangaratta 3677, Victoria. (Bee supplements)
N. Redpath	Waverley Rd., Mt. Waverley, Victoria. (Bee supplements, human foods)
Canada	
Doyan & Doyon	2720 Rue Dischesne, St. Laurent, Montreal, Quebec. (Ag. pollen)
G. Vant Haaff	710 Violet Avenue, Victoria, B. C. (Bee supplements)
England	
Pollen Products	299 Ballards Lane, London N 12. (Human foods)
France	
Ph. Co. Heudebert	79 Rue Henri-Barbasse, 92 Nanterre, Seine. (Dry allergens and extracts; Human supplements)
Valentin	22 Bd. Jules-Guesde, Saint-Dnsis, Seine. (Ag. pollen; Bee and human supplements)
Sweden	
A. B. Allergon	Välinge 55, S-26200 Engelholm. (Dry allergens and extracts)
A. B. Cernelle	Vegeholm 6250. (Human foods and health supplements)
USA	
Abbott Labs.	Noth Chicago, Illinois. (Dry allergens and extracts)
Antles Pollen Supplies	P. O. Box 1243, Wenatchee, Washington. (Ag. pollen)
Berkeley Biologics	2nd and Hearst Streets, Berkeley, California. (Dry allergens)
C. G. Blatt Co.	10930 E, 25th Street, Independence, Missouri. (Dry allergens)
Center Lab.	5 Channel Drive, Port Washington, New York. (Dry allergens and extracts)
Dome Labs.	400 Morgan Lane, West Haven, Connecticut. (Dry allergens and extracts)
Firman Pollen Co.	126 Highland Avenue, Chelan, Washington. (Ag. pollen)
Greer Drug Co.	P. O. Box 800, Lenoir, North Carolina. (Dry allergens and extracts)
Hollister-Steir	500 Industrial Park Drive, Lansdowne, Pennsylvania. (Dry allergens and extracts)
Karel Rehka Products	880 Northwood Drive, N. E., Salem, Oregon. (Bee and human supplements)
Miles Lab.	1127 Myrtle Street, Elkhart, Indiana. (Dry extracts)
Pollen Commerce	P. O. Box 1001, Placerville, California. (Ag. pollen)
Pollen Products	1439-H Street, Bakersfield, California. (Human food; Ag. pollen)
Prana Pollen	Box 747, Los Altos, California. (Human food)
Sharp & Sharp	P. O. Box 8, Everett, Washington. (Dry allergens)
Smith Pollen	Route 3, Box 3315, Wenatchee, Washington. (Ag. pollen)
Stemen Lab., Inc.	1205 N. E. 18 Street, Oklahoma City, Oklahoma. (Dry allergens)
Wonder Pollen	11711 Redwood Highway, Wonder, Oregon. (Ag. pollen)
West-Germany	
Allerogopharma	Volkers Park 10, 2057 Reinbek b. Hamburg. (Dry allergens and extracts)
J. Ganzer	

States and France several commercial pollen collecting companies now supply large quantities of pear, apple and almond pollen for this purpose (Table 4-1). Alternatively, the quantity of pollen transported from the anthers to styles can be increased by introducing bee hives into orchards at flowering time (Fig. 4-1).

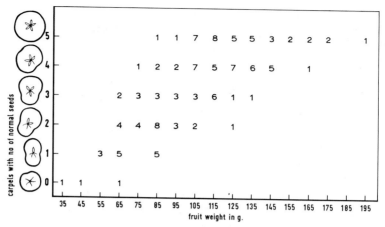

Fig. 4-1. The size of the apple fruit and the fruit weight depends directly on the intensity of pollination, which again depends on the number of bees/hives in the orchard. (According to F. LUKOSCHUS, pers. comm.)

Bee flights and pollen dispersal may be restricted by adverse weather conditions. Pollen distribution by bees is thus not as certain as good mechanical application. Low temperatures, heavy rainfall, poor bee activity, and unequal wind distribution patterns may contribute to poor pollen dispersal at the critical period of stigma receptivity.The best way to assure a high crop yield or a known genetic progeny is to collect and supply the pollen artificially to the plant. This conclusion is now accepted by an increasing number of plant breeders and orchardists who find that supplemental pollination is often necessary to assure good yields (STEUCKARDT, 1965).

The most widespread application of pollen is to increase seed or fruit formation. Also, some biochemical reactions induced by pollen can be studied only after controlled application of pollen to the female tissue. These uses all require direct application of pollen to the plant. We will, therefore, briefly review some of the methods used in artificial pollination.

The Methods. The simplest way to apply milligram quantities of pollen is to use a needle or a fine camel hair brush to transfer pollen to the angiosperm stigma or open megasporophyll in gymnosperms. Under experimental conditions, anemophilous species can be pollinated by moving plants and flowers smoothly to and fro. In greenhouses, insect pollinated flowers can be pollinated by placing flies, *Musca domestica*, or bumble bees, *Apies*, in the isolation enclosure. These insects promote seed set by tripping the stamen (KOBABE, 1965) or transferring pollen on their bodies to stigma areas. Where pollen is applied by hand or insects, for controlled breeding or studies of pollen germination in the style, care must be taken to protect the style from contaminating pollen.

When intermediate levels of pollen, i.e. 2 to 100 gms, are to be distributed, then a milliliter hypodermic syringe can be adapted by replacing the plunger with a rubber air atomizer bulb (CUMMINGS and RIGHTER, 1948; VAN VREDENBURCH and VAN LAAR, 1967). This permits directing the dry pollen to a specific area. For large volume applications in orchards or fields pollen can be sprayed with an air pressure hose, a pole duster (ANTLES, 1965), or from a back pump in water (ALLAN and SZIKLAI, 1962). In mass pollination programs, application can be achieved by helicopter or airplane (BULLOCK and OVERLEY, 1949). In such large distribution procedures the pollen is generally diluted with an inert carrier, i.e. talc, dead pollen (CALLAHAM and DUFFIELD, 1961) or *Lycopodium* powder (AN-TLES, 1965).

In outdoor applications of pollen by bees, hives are rented to fruit growers and farmers by apiarists. Insert containers of pollen can also be attached at the hive entrance so that the bees are covered with the pollen as they leave the hive and their pollination capacity is increased (BURRELL and KING, 1931; TOWNSEND et al., 1958; DICKSON and SMITH, 1958; CAILLAS, 1959).

Non-seed Forming Uses

Pollen is frequently collected for purposes other than plant breeding, hybrid production, or increased fruit yield in commercial orchards. Alternative uses will be listed to indicate the spectrum of potential roles for pollen in and out of the laboratory.

Chemical

Chapters 8 to 16 will provide details on pollen composition. Occasionally, pollen may provide a good source for specific compounds. Such chemical extractions generally employ quantities of 100 mg to several kgs.

Physiological Applications

a) Cell Metabolism

In pollen growing under proper conditions, the metabolic processes common to most living cells occur. Pollen can also be preserved for extended periods, often up to several years (See: Storage). Many readily obtained pollen species are a storable source of potential metabolically active tissue. It can be easily extracted, manipulated and used for the study of such problems as cell wall formation, protein and RNA biosynthesis, and carbohydrate metabolism. Free "naked" protoplasts can also be prepared from pollen during its development (BHOJWANI and COCKING, 1972). In such studies, quantities of a few mg to 100 mg are generally employed in each experiment.

b) Toxic Inhibitors

Growing pollen tubes are often rapidly sensitive to toxic compounds (SADASI-VAN, 1962). The problem in such assays is to find the pollen which is most sensitive to a particular compound. Pollen of one species may require a particular element such as boron to germinate, while other pollen is able to germinate normally without added boron, and added boron may, in fact, prove toxic. Replicate pollen samples of 30 μg germinated in spot tests of 100 μl of solution can provide a test sufficiently accurate to determine the effectiveness of inhibitor compounds.

c) Cytoplasmic Organelles

The cytoplasm of pollen, although lacking chloroplasts, contains the other subcellular units common to plant and other living cells, i.e. plastids, mitochondria, golgi bodies, endoplasmic reticulum, vesicles and ribosomes (LARSON and LEWIS, 1962). In addition, a terminal tube nucleus and a generative cell nucleus with the capacity to divide (or 2 male nuclei) are also present (Fig. 1-2). Starch granules and amyloplastids are actively formed during tube growth (HELMERS and MACHLIS, 1956; BAL and DE, 1961; LARSON, 1965; KAIENBURG, 1950). Spherosomes are also abundant in many pollen tubes (TSINGER and PETROVSKAYA-BARANOVA, 1965). Some subcellular constituents of pollen can be isolated and studied *in vitro*, e.g. mitochondria (STANLEY and YEE, 1966), ribosomes (LINSKENS, 1967; LINSKENS et al., 1970), vesicles (VAN DER WOUDE et al., 1971; ENGELS, 1973) and nuclei (LA FOUNTAIN and MASCARENHAS, 1972; SHERIDAN, 1972). Because of the very rapid extension of the tube and the high metabolic levels in germinating pollen (STANLEY, 1962) the system readily lends itself to cytological studies at both the light and electron microscopic level. Pollen in μg levels is sufficient for the microscopic *in situ* studies. However, for isolation of subcellular organelles, 1 to 5 g quantities are required.

d) Fertilization

Fertilization and factors which affect this process are frequently studied by artificial pollination experiments. Microgram to mg amounts of pollen are generally placed on the stigma; the response of the pollen to the female tissue, and vice versa, is then observed. Pollen application may be coupled with treatments which overcome incompatibility reactions or chemical-physical factors associated with blocking germination or penetration of the pistil tissues. Methods utilized can involve chemicals, mixtures of different pollen or physical treatment of the style or stigma with high temperatures or ultraviolet light. ROGGEN et al. (1972) used the heat produced when electrically conducting hand forceps covered with pollen were applied to the stigma to overcome the wax barrier on the *Brassica* papillas, a procedure now known as electric-aided pollination (EAP). Such work has led to significant progress in our understanding of the mechanism of incompatibility in plants and the metabolic conditions which precede fertilization (LINSKENS, 1955; LINSKENS and KROH, 1967).

Medical Applications

The collection of select species of pollen, particularly grasses and the ragweed, *Ambrosia*, is necessary for the study of hayfever and other pollen allergies (KELLY, 1928). For diagnostic tests and allergy treatments, commercial pharmaceutical firms (Table 4-1) prepare a suspension of pollen or an extract which is stable for months of storage. A common preparation involves extracting pollen in a 1:20 saline solution and precipitation of protein from the pollen extract with tannic acid and $ZnCl_2$ (NATERMAN, 1957) (see: Chapter 12, Pollinosis). The aqueous extract is often made from kg quantities of pollen. For experimental testing individual extracts can be prepared from gm levels of pollen. Pollen mascerates are often injected into test animals to determine sensitivity. Micro-allergenic reactions in humans are generally determined by a "scratch test", involving direct application of mg samples of pollen in solution to the broken skin surface, or injections of small amounts of extracts under the forearm epidermis (WERNER and RUPPERT, 1968; LINSKENS and VAN BRONSWIJK, 1974).

In most cases of pollen sensitivity, the allergic response occurs when pollen contacts the mucous membrane. In skin testing, the reaction is cutaneous. To desensitize a person to pollen, extracts of a particular pollen such as *Ambrosia* are pepared as a buffered alum-precipitated pyridine-complex which can be ingested as a pill or given by injection. Commercial pharmaceutical companies extract 100 gm to kg quantities; individual laboratory test extracts are generally prepared from 100 mg to 1–2 gm samples. Details of problems with medical applications of pollen are discusses in Chapter 12.

Bee Diet Supplements

To insure a long life for hives which lack natural pollen sources, artificial supplements containing pollen can be supplied (MAURIZIO, 1954). These preparations generally involve only gm quantities of pollen. As discussed in Chapter 7 (Nutritive Role) about 145 mg of pollen are required to raise one bee larva to maturity. There is limited use of pollen for this purpose.

Human Diet

Pollen is included by bees in honey removed by man from the hive. The amount of pollen in honey varies with the particular pollen species and source of nectar collection by the bees (Table 7-2). Limited amounts of pollen, principally date palm, *Phoenix dactylifera* L., and cattail, *Typha elephantia*, are collected for direct human consumption. Pollen candy, a mixture of pollen with honey or molasses and chocolate, is occasionally sold in health food stores in the United States. Many health benefits are attributed to such candy and to capsules containing pollen removed from the hind legs of bees as they enter their hives (Fig. 7-6). Sweden and England have many advocates of this application of pollen (BINDING, 1971).

While pollen, or its equivalent, may be irreplaceable in the bee diet, we fail to see a correlation with suggested benefits to man. Certain human steroid hormones are present in date palm pollen (Chapter 10), but many other pollens and plant tissues also contain trace amounts of these compounds compared to the levels naturally synthesized by the mammalian body. Our opinion of pollen candy and other pollen additives in the human diet is that they are interesting, dry in taste, and something we and others consuming a balanced diet can perfectly well do without.

Pollen Analysis

Geological, ecological and ethnological studies often characterize pollen from a given source. Vertical core samples from peat or sediment deposits are analyzed from the surface downward. Pollen in different layers is identified by comparison with known samples. This can provide information about the vegetation, climate and foods of earlier periods in the source area. Ethnological interpretations of pollen residues in Indian settlements in the southwest United States have provided an index to the diet of these early Americans (MARTIN and SHARROCK, 1964). Pollen trapped in bark of aged trees such as 3200 year old *Pinus aristata* (ADAM et al., 1967) and sediment core samples have helped characterize the migrations and successions of plants over wide areas.

Pollen analysis compares grains on the basis of form and surface characteristics of the exine. Groups in Scandinavia led by LANGERHEIM, VON POST and ERDTMAN pioneered these techniques (PRINTZ-ERDTMAN, 1963; ERDTMAN, 1969) based on the suggestions of VON MOHL in 1835 and FRITZSCHE in 1837 (WODEHOUSE, 1935). Early studies recognized that exine markings were constant enough to be used for species characterization. However, pollen exine markings from field samples must be compared with known contemporary and fossil pollen. This requires that an index of pollen samples be prepared for microscopic observations and descriptive classification keys (see Fig. 2–4).

Procedures for palynological preparations are numerous and frequently modified. Progress in this field of characterizing pollen is chronicled primarily in four journals, *Pollen et Spores*, *Grana Palynologia*, *Review of Palaebotany and Palynology*, and the *Japanese Journal of Palynology*. A relatively complete bibliographical list of pollen publications is maintained by the Musée National d'Histoire Naturelle in Paris and published in *Pollen et Spores*. Although the emphasis of their indexing is on physical characteristics of pollen and fossil reports, they do index some physiological and chemical reports relating to pollen.

Palynological laboratories maintain a large collection of prepared slides showing samples of different pollen grains. Preparations of these research tools, while laborious, only require a few grains of each pollen. The chemical composition of some pollen exines precludes the survival of those species as fossils. Knowledge of pollen wall chemistry (Chapter 9) and the effect of climate on the pollen wall during aging are of considerable significance to palynologists.

Conclusions based on pollen analysis alone may lead to many errors (DAVIS, 1963). The indexers must not only identify the pollen and its frequency in a given deposit, but must also know something about the pollen distribution mechanism and pattern. Thus, the distance a pollen is carried (Chapter 3) and the ability of a pollen to survive can determine the possibility of a single or infrequent sample being typical of the site. Such factors must be carefully evaluated by the palynologist.

Species Effect on Pollen Collection

In collecting pollen from the field, the quantity available, season or period it can be obtained, and method of handling and storing the pollen depend on the plant species.

Abundance of Available Pollen

From Anemophilous Species

These plants have evolved a wind dependent pollen dispersal mechanism and produce pollen in great abundance (Table 3-2). They include the catkin-bearing trees, *Pinus*, *Alnus* and *Betula*. Plants with this type of pollen-producing structure are generally quite prolific, and the pollen can be collected in abundance and with ease.

From Entomophilous (Insect Pollinated) Flowers

These plants produce open or closed flowers. The open types, i.e. *Rosa* and *Pyrus*, have a group of stamens massed about the style. The insects, generally bees, which visit such flowers may do so primarily to collect an attractant secreted from nectar glands at the base of the style, or the bees may actually collect pollen from the dehisced anthers. Pollen in such flowers is limited to single stamens. The anther in each stamen may contain two or four pollen chambers (Fig. 1-1) but relative to the anemophilously disbursed pollen, extraction is more laborious and considerably smaller amounts of pollen are obtained. However, these flowers present no problems of mechanical handling, and the pollen is readily removed in simple laboratory procedures.

The closed flower type is more difficult to manipulate and to extract its pollen mechanically. Examples of such species include *Pisum*, *Phaseolus* and other Leguminosae. Extracting pollen from these flowers requires opening the petals or uncovering the anthers or pollinia before separating them from the flower.

Factors Controlling Pollen Availability

Pollen quantities produced by individual plants are endogenously established and environmentally modified.

Genetic Controls. The most important factors determining the time of pollen development and dehiscence are inherited. Genotypic variation and blocks to natural hybridization are often established when pollen dispersal does not coincide with stigma receptivity of plants within pollinating distance (STANLEY and KIRBY, 1973). The time of pollen shedding in pine species and many other plants has been carefully followed and related to geographic distribution and elevation (DUFFIELD, 1953; SCAMONI, 1955). Pollen shedding in *Pinus radiata* during a single season in Australian plantations at one elevation occurred over a period of six weeks; the bulk of the pollen was dispersed during the second to fourth week (FIELDING, 1957). *P. pinaster* in Europe sheds pollen over a two- or three-week period (ILLY and SOPENA, 1963).

Table 4-2. Periodicity of pollen availability

Species	Time when pollen is available pollen volume at:	
	Maximum	Minimum (approx.)
Papaver rhoeas	5:30—10 a.m.	
Papaver somniferum	6 — 9:30	11 a.m.
Verbascum thapsiformae	6 —10	5 p.m.
Verbascum phlomoides	6 — 9:30	2 p.m.
Rosa arvensis	6:30—10:30	11 a.m.
Rosa multiflora	6:30—10:30	1:30 p.m.
Verbena officinalis	7 —11:30	2 p.m.
Convolvulus tricolor	8 — 2 p.m.	4 p.m.

Dehiscence time depends in part on meteorological conditions which change from day to day. Under identical conditions every species has its specific time of anther dehiscence (OGDEN et al., 1969). Collecting insects are well adapted and quickly aware of this time. As an example, presentation time when pollen was available on a clear, sunny day is given in Table 4-2 (KLEBER, 1935). The relation of time of flowering to pollen collection by bees is also illustrated in Fig. 7-7.

In most plants pollen quality, as measured by percent germination in vitro, remains about the same throughout the period of natural dehiscence. Pollen isolated at the beginning and the end of the three weeks dehiscence period of a single cultivar of Arachis hypogaea showed no difference in viability (DE BEER, 1963). But pollen from potato plants showed optimum germination and viability at mid-flowering period (ERVANDYAN, 1964). In this latter study, the in vitro tests and the same sucrose concentration throughout the growing season. Variation frequently occurs in the external sugar concentration required to obtain optimum germination of pollen from the same plan at different times of the year. The results with potato plants can be interpreted to indicate, not that the mid-point of the pollen dispersal period gives pollen of maximum viability as tested in vitro, but that pollen growth requirements and endogenous nutrient status change during the plant's development period. Determinations of relative pollen viability most often require testing over a range of conditions.

Moving a pine tree from its normal geographical range to another latitude, at the same elevation, thus modifying the photoperiod, has relatively little effect on the time of pollen development and shedding (WRIGHT, 1962). This suggests pines are relatively day-neutral with strong genetic control of flowering time (MIROV, 1967). The same relative day-neutral response is not observed in all plants; many do not produce pollen when moved to shorter or longer photoperiods than that in which they evolved. While in most plants dehiscence occurs in the day, in Lagenaria vulgaris and some other species, anthers dehisce in the evening or at night (PERCIVAL, 1950; VASIL, 1958).

External Factors. a) Temperature and moisture are the primary elements affecting pollen development on mature plants. If one is collecting pollen from a given plant species and misses pollen dehiscence at one elevation or planting, it may be possible to find a similar genotype at a higher elevation or colder location where

the pollen has not yet been shed. MILLETT (1944) compared two genetically similar plantings of *Pinus radiata* separated by 1,800 feet of elevation in Australia.
Pollen at the higher elevations started to shed after dehiscence at low elevations
was completed. MILLETT (1944) also discussed the effect of an extended period of
low temperature or high moisture in delaying dehiscence. Variations in temperature or moisture can shorten the interval of pollen dispersion of these trees from
seven weeks to two weeks. The maximum dissemination of pollen in an oak
plantation occurred on days when meteorological conditions were such that average day temperatures were not lower than 15° C, relative humidity was between
30 and 80% and wind velocity was 2 m/sec above the canopy in sun (ROMASHOV,
1957).

Low or high temperatures during the development period can adversely affect
the quantity and germination response of mature pollen. CHIRA (1963) attributed
low pollen production and fertility in *Pinus sylvestris* to low temperatures during
development. Low temperature resulted in pollen abortion and inhibition of
stamen development in *Pennisetum clandestrium*, although stigma development
continued normally at the same low temperature (YOUNGER, 1961). This suggests
that in this case the cool weather increased plant sterility by blocking pollen
development, and that different flower organs, and in particular the stamens and
pollen, are more sensitive to temperature during development than other parts of
the flower. DE BEER (1963) observed that in peanuts, *Arachis hypogaea*, pollen on
plants growing at a constant 33° C was only 10% viable, compared to 40%
viability for pollen developed in greenhouses under a normal day-night temperature cycle. Rice plants grown with a root temperature of 28° C yielded pollen with
higher viability than plants grown with roots at 23° C or 33° C (YAMADA and
HASEGAWA, 1959).

Moisture and nutritional status of the plant can also affect pollen viability and
abundance. DOMANSKI (1959) found that barley plants, subjected to water stress,
or grown in sub-optimal phosphorus levels, developed less pollen with decreased
viability. Boron deficiency in the media or soil can also reduce viable pollen
production.

Low light intensities or non-normal photoperiods during microsporogenesis
can reduce sugar pools in leaves and anthers as indicated by chemical analyses,
and a visible shift in the quantity of flavonoid pigments in anthers. In *Ornithogalum caudatum*, shading of leaves during pollen development resulted in a decreased quantity of viable pollen (Goss, 1971). Temperature also influences substrate metabolism in maturing anthers and the amounts of pollen maturing (KI-
YOSAWA, 1962).

b) Collecting time. When pollen was isolated from peanuts at different periods
throughout the day it was observed that in the morning 35% of the pollen
germinated, but 10 hrs later only 3% was viable. Similar observations of percent
germination differences, although less dramatic, were made on *Acer*, *Crataegus*
and *Tilia* trees near Leningrad. Pollen isolated in the early morning germinated
better than that collected at other times of the day (KAUROV, 1957). While the
reasons for such deviations have not been ascertained, they probably are related
to metabolic transitions and moisture stresses. It may be meaningful, in this
regard, to follow translocation of sugars in developing pollen.

c) Radiation can also exert a negative influence on pollen development. Exposure of pollen to U.V. light, such as occurs when pollen is shed in sunlight, decreased viability of some pollen as much as 80% while only slightly injuring others. Wind-dispersed pollen was the most resistant to the effects of sunlight (WERFFT, 1951; ASBECK, 1954). Pollen is best dried, or induced to shed, in darkness or shade. Drying or storing pollen in sunlight will decrease viability (SEDOV, 1955). Diffuse light, while not as good as darkness for storing pollen, is preferable to direct light.

Chronic irradiation from a gamma source in the environment of developing *Quercus alba* flowers caused a delay in the time of dehiscence and decrased pollen viability (STAIRS, 1964). *Zea mays* (FROLIK and MORRIS, 1950), *Pyrus* (TARANOVA, 1965) and *Pinus densiflora* (WATANABE and OBA, 1964) were also shown to be similarly sensitive to gamma rays during pollen development.

d) Chemicals applied to plants for insect control or leaf defoliation generally adversely affect pollen development. Bordeaux spray, $CuSO_4$, applied to grapes or apples at blossom time can decrease pollen viability and seed set (MACDANIELS and HILDEBRAND, 1939). Fungicides applied to control cherry blight also markedly inhibit pollen germination (EATON, 1961). Complete anther loss may occur with more extreme chemical treatments such as defoliants applied to cotton (SCOTT, 1960). The methyl fumigant compounds, used to control fungal infection, inhibited flower opening and lowered pollen viability in peaches (JOLEY and HESSE, 1950). In general, plants exposed to herbicides or pesticides applied as vapor or liquid sprays during flower development, will yield damaged, less viable pollen (KRAL'OVIC and KAULOVA, 1971; DUBEY and MALL, 1972).

e) Industrial waste gasses such as SO_2 inhibit pollen germination, while at the same time doing little apparent harm to female flower parts (DÖPP, 1931). Pollen development and activity are among the more sensitive indicators of adverse factors in the botanical environment. This led to the proposal that the inhibited development and growth potential of pollen, in response to various toxic atmospheric contaminants, e.g. ozone, is a good index of air quality (FEDER, 1968; MUMFORD et al., 1972).

To collect viable pollen from field-grown plants one must be careful to select plants exposed to a minimum of adverse chemical and environmental factors. An awareness of these potential sources of pollen growth inhibition can also help the biochemist or plant breeder to recognize the source of any unusual deviations in the pollen behavior.

Potential Problems in Pollen Collecting. Species of pollen differ in the length of time they remain after dehiscence or collection. Storage procedures necessary to preserve pollen will be discussed in Chapter 5. Here we will review several factors which can affect pollen collecting and modify the biochemical potential of collected pollen.

Surface characteristics and dehiscence mechanisms can influence the method and ease of pollen collection. Most pollen grains from insect pollinated plants stick together by an oil-like substance which fills the hollows of the exine (FREY-TAG, 1958) (Chapter 10). In *Oenothera*, pollen is bound in a sticky viscid thread-like mass at the time of dehiscence. In other cases, i.e. in *Typha, Asclepias* and *Acacia,* pollen tetrads and polyads remain together because of cohesive cement on their surfaces. As will be discussed under Collecting Methods, these pollens present special problems in collecting and handling.

Pollen grains may be shed in either the 2- or the 3-celled stage. The observations of BREWBAKER and EMERY (1962) relative to radiosensitivity of different pollen, and the normal brief survival quality of certain pollen, suggest that pollen shed in the 3-celled stage is more radiosensitive and remains viable for a shorter period than pollen shed in the 2-celled stage. Pollen containing 3 cells at maturity is found in the Graminales, Caryophyllales, and over a dozen other plant families (BREWBAKER, 1959).

The presence of an additional functional nucleus, as a result of the mitotic division of the generative nucleus to yield two male cells, increases the DNA-sensitive volume in the 3-celled pollen. Pollen containing 3 cells at dehiscence is more sensitive to lowering of the water content (Chapter 6) and is probably more metabolically active than pollen shed with two cells. Also, the 3-celled pollen may have advanced evolutionarily to where it no longer develops an exine wall with the multiple layers or pigment components which characterize less radiation-sensitive pollen. The important point is: different species present different survival potentials and problems in collecting and handling.

Distance and traveling time to pollination may modify the viability and sensitivity of pollen to handling and natural environmental conditions. Winged pollen of the conifers is adapted for traveling many miles, and each species has its own degree of bouyancy in water and air (Chapter 3).

In a water flotation test, only 8.8% of *Pinus resinosa* pollen sank after 3 hrs, while most deciduous, angiosperm pollen sinks within minutes (HOPKINS, 1950). The weight and density factors of pollen exert a marked influence on plant distribution (EISENHUT, 1961). Yet, because of density and weight differences, some deciduous tree pollen is carried greater distances by wind than are the winged gymnosperm pollens, i.e. pollen of *Ulmus americana* will drift 1,000 feet versus an average of 130 feet for *Picea abies* pollen (WRIGHT, 1952). However, more than wind is involved in successful pollination. Presence of receptive female flowers, rain, humidity, and as already indicated, the susceptibility of pollen to sunlight and radiation can also modify viability.

Collecting Methods

Pollen is collected to characterize plant species in an area, to determine the volume and kinds of pollen in the air, and to facilitate other uses and applications already discussed. Because there are many environmental factors that have a destructive influence on production of viable pollen, special techniques have been developed for collecting pollen.

1. Air sampling ascertains the nature of medically important pollen which may be prevalent in an area at a particular time of the year. Many reviews on pollen frequency in the air over certain areas are published for the information of allergists who must treat patients suffering pollinosis (WYMAN, 1872; WODEHOUSE, 1935; HARSH, 1946; HYDE, 1950; PARAG et al., 1957; CEFULU and SMIRAGLIA, 1964). Air pollen samples are collected by placing microscope slides coated with petroleum jelly (EDGEWORTH, 1879), glycerin or polyester tape at various locations (GRANO, 1958). At given intervals the slides are collected; the pollen and spore types are counted and classified by comparison with standard slides of individual species. An automatic rotating drum with 24 replaceable slides has been developed for field use (VOISEY and BASSETT, 1961). Special spore traps have been devised (HIRST, 1952; SCAMONI, 1955; GRANO, 1958; KOSKI, 1967) to permit more extensive monitoring and directional determination of the pollen source. Many of these newer methods determine the density of pollen per unit volume of air rather than grains deposited per unit surface area.

Vaseline or grease slides, used for deposition assays, may present several problems in pollen sampling. Heavy dew or rain can wash off the grease layer, or pollen samples may sink into the coated layer and be difficult to identify; also, slides are fragile and easily broken in transport. A pollen trap useable in wet weather has been devised for sampling in areas of high precipitation (KENADY, 1968). The unit requires a small circulating pump which, in remote locations, might be battery operated. By applying a very thin layer of silicon grease on a plastic globe surface, which could be lowered into a protective cover, KOSKI (1967) was able to sample air pollen patterns effectively in forests. Pollen is easily counted by projecting light at the top of the detachable globe. However, what is probably the simplest, least elaborate pollen sampler, uses a plastic plant label tag, or aluminum foil, with holes cut through, behind which cellulose tape is placed with the sticky side facing out of the holes. These simple collectors facilitate exposure, collection and pollen counting; rain is not as destructive to these devices as to the grease coated glass slides (HOEKSTRA, 1965).

Plant breeders are frequently interested in knowing the prevalence of a given pollen in an area because stray, genetically unknown pollen of similar species must be excluded from control plants. Before plant breeding plots are established in an area, local pollen samples are generally used as part of the site selection criteria. The usual sampling procedures must be corrected when relating survey data to stands of forest trees if skewing of the pollen distribution occurs. Pollen can accumulate at a distance from its stand and not necessarily along a decreasing concentration gradient from the stand (WRIGHT, 1952; LANNER, 1966).

The range and magnitude of pollen sampling has been extended to both polar regions and over the oceans. Airplanes are frequently used for such wide-spread sampling (POLUNIN, 1951). Such studies have indicated that pollen is often carried several miles up and hundreds of miles away from the dispersal point. These distributions are not very significant in terms of plant populations mixing because of the many other barriers to fertilization. But the settlement point of the high-flying pollen may give a false representation in plant analyses by pollen recovered from soil deposits.

2. *Quantity collections from the field* are the most important and easily obtained sources of pollen. Different techniques have been devised for the various species. In general, the procedure involves isolating the pollen source, i.e. flower or branch, to avoid contamination by other species of plants; inducing or waiting until dehiscence and gathering the desired quantities after the pollen is shed.

In the case of amentiferous strobili—or catkins—the general procedure is to remove a group of mature catkins from the tree a few days before opening. The closed catkins are allowed to open in a controlled environment. The optimum time to remove the catkins is recognized by certain morphological and color changes on the catkins. The period of dehiscence varies over a whole tree or even within an individual compound florescence. Pollen shedding in trees generally starts at the top or outside branches which receive high sun exposure and proceeds basipetally towards the low branches. Minimum catkin flower opening will generally occur at night and early morning when temperature is lowest and relative humidity highest. Closed catkins can be easily removed from the branches in kilogram lots and brought inside to open.

The rapidly respiring catkins are spread on well aerated trays with fine mesh wire on the bottom, then set over paper or cellophane sheets for drying; or they can be placed on screen bottomed trays in a drying cabinet with air gently circulated through the trays (LEITCH, 1971). After dehiscence, shaking the tray deposits pollen onto the paper below from where it is transferred to storage containers. In forest tree breeding programs utilizing large quantities of pollen, a more elaborate funnel type drying apparatus has been devised (CUMMINGS and RIGHTER, 1948). This 4 to 6 foot unit moves a stream of drying air through the catkins in the funnel. When the catkin mass is dry, the pollen is collected in glass containers at the bottom vortex of the funnel. Pollen is usually sold commercially by gm quantities; however, plant breeders often measure collected pollen in cm^3 (Table 3-2). One liter of *Quercus* catkins will yield about 4 cm^3 of pollen while a liter of pine catkins (strobili) yields a pollen volume of about 150 cm^3 (SNYDER and CLAUSEN, 1973).

In *Rosa*-like species the general procedure is to gather flowers and dry them on a wire screen tray. Gently fragmenting the flowers releases the anthers from the stamens. These are then collected below the tray. The pollen can further be shaken free of the anthers and collected through a 200 mesh screen. The anthers can also be shaken free of pollen by placing them in a dry polyethylene container. The electrostatic charge between the pollen and the plastic container wall holds the pollen on the container wall while the anther residue collects at the bottom. Clean pollen can then be scraped off the plastic container wall with a spatula.

Pollen with wing adaptations, such as Pinaceae, can usually be fractionated after collection. An air compressor is easily adapted to separate viable from non-viable shrunken or damaged grains (WORSLEY, 1959; LESKOVCEVA and KORCKOV, 1965).

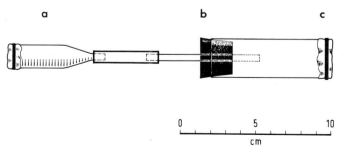

Fig. 4-2. Apparatus for vacuum line pollen collector. (a) Eye dropper glass tube; (b) removable stopper; (c) nylon cloth or gauze trap

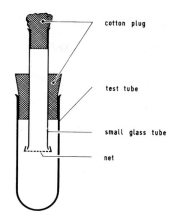

cotton plug

test tube

small glass tube

net

Fig. 4-3. The anthers are placed in the inner tube on the net after surface disinfection. The dehiscens takes place in the small sterilized test tube, the pollen grains fall through the openings of the net into outer test tube. (PETRÚ et al., 1964)

Some flowers, such as cotton, are collected prior to anthesis and opened in the lab. The pollen is easily collected by shaking the inverted open flower over a container (MURAVALLE, 1964). Where limited amounts of pollen are desired, as from *Petunia*, the anthers can be removed with forceps or a fine comb, placed in a petri plate and allowed to dry and dehisce. The pollen is shaken loose of the anthers and collected from the plate (VISSER, 1951). Another way of collecting large quantities from individual anthers is to use a vacuum hose; a simple home-type vacuum cleaner with tube attachments reduced down to eyedropper glass tubing size (LEWIS, 1944; KING, 1955) as seen in Fig. 4-2 can be used. A piece of nylon cloth or gauze can be inserted in the collecting hose extension to trap the pollen or anthers pulled into the hose by the glassdropper tip.

Several problems encountered in collecting procedures can render pollen use-less for some experiments. The principal problem is contamination by microorganisms. This can be avoided by collecting the pollen from anthers or catkins which have been surface washed with hypochlorite ("Chlorox") or a similar disinfecting solution, then rinsed in sterilized distilled water before drying under aseptic conditions (TULECKE, 1954) (Fig. 4-3) or, by extracting the sterilized anthers in a small vacuum sterile filter (PETRÚ et al., 1964).

Often a large amount of already extracted pollen is all that is available and the desired uses require that it be sterile. We have found that placing non-sterile pollen on a sintered funnel under mild vacuum, and washing the thin layers of pollen with sterilized water frees the pollen surface from most contaminating bacteria. Such pollen has proved useful in long-range germination and respiration experiments where antibiotics are not feasibly added to the pollen media to control bacterial and fungal contaminations. Pollen grains can, however, transmit bacterial and viral diseases (GOLD et al., 1954). Thus, in experiments where contaminant-free pollen is desired, flower maturation and extracting conditions must be carefully controlled throughout development. Low concentrations of antibiotics and fungistats can also be used in the germination medium to control the growth of contaminating microorganisms.

Laboratory maturation and extraction of pollen allows control over time of shedding and can provide non-contaminated pollen. This technique is particu-

larly useful when pollen collection must be accelerated. In such cases, meiosis and
dehiscence is forced to occur earlier than it would on the plant.

Individual branches with pollen-bearing strobili or flowers can be removed
and placed in water in a warm greenhouse (JENSEN, 1943; JOHNSON, 1945). The
viability of pollen produced from such branches is generally as high as occurs on
the tree (MASKIN, 1960). Placing *Tsuga canadensis* branches in wet sand under a
plastic cover and exposing them to extended daylight, accelerated pollen development in comparison to branches excised and exposed to normal daylight (SANTA-
MOUR and NIENSTAEDT, 1956). Special greenhouses have been designed to hasten
the development of pollen on excised branches (BARNER and CHRISTIANSEN,
1958). Using such a controlled environment, BARNER and CHRISTIANSEN dehiced
Alnus catkins a month before the pollen was shed outside. However, not all parts
of the flower develop equally when it has been excised. Development of the stigma
in *Quercus* and other flowers will not continue to maturity in such an *in vitro*
arrangement (MASKIN, 1960).

A few genera in which pollen dehiscence has been accelerated by exposure of
detached branches to controlled conditions are listed in Table 4-3 (SNYDER and
CLAUSEN, 1973). Catkins on detached branches of *Picea excelsa* developed best
when exposed to alternating temperatures of $+7°$ C and $+20°$ C rather than to a
constant temperature. The constant temperature led to cessation of further pollen

Table 4-3. Acceleration of pollen dehiscence under controlled conditions

Genus	Temperature °C	Light			Weeks detached before natural shedding	Reference
		Power W	Distance Meters	Time Hrs		
Abies	27	600	1	20	3	WORSLEY, 1960
Acer	22	500	1	15	12	LARSON, 1958
Alnus	27	600	1	22	—	WORSLEY, 1960
Betula	22	500	1	15	12–15	LARSON, 1958
Betula	27	600	1	22	—	WORSLEY, 1960
Cedrus	27	600	1	20	4	WORSLEY, 1960
Chamaecyparis	27	600	1	20	3	WORSLEY, 1960
Corylus	10	—	—	—	3–4	COX, 1943
Fagus	27	600	1	22	—	WORSLEY, 1960
Fraxinus	—	—	—	—	3	WORSLEY, 1960
Fraxinus	16–22	500	1	15	12–15	JOHNSON, 1945
Larix	20	200	1.5	3	3	WORSLEY, 1960
Picea	27	600	1	20	8	WORSLEY, 1960
Pinus	27	600	1	20	4	WORSLEY, 1960
Populus	20	500	1	15	12–15	JOHNSON, 1945
Prunus	16–22	—	—	—	12–15	LARSON, 1958
Pseudotsuga	27	600	1	20	4	WORSLEY, 1960
Quercus	20–22	500	1	15	1–3	JOHNSON, 1945
Salix	16–22	500	1	15	12–15	JOHNSON, 1945
Sequoia	15	60	1.5	24	—	WORSLEY, 1960
Tsuga	25	400	1	20	8	WORSLEY, 1960
Ulmus	16–22	500	1	15	12–15	JOHNSON, 1945
Vaccinium	20	—	—	—	1–2	MERRILL and JOHNSON, 1939

development on the excised branches (CHIRA, 1963). In gymnosperms, KONAR (1962) reported that viable pollen will occasionally develop *in vitro* even though the male strobili do not reach normal size. These results suggest that a supply of nutrients and maintenance of photosynthesis and translocation at certain critical levels is apparently essential for completing pollen development on detached branches. TAYLOR (1950) was able to determine the essential nutrient requirements in *Tradescantia* flowers placed on different *in vitro* nutrient solutions at specific stages of microsporogenesis, thus correlating observed growth patterns with biochemical pathways essential for development.

Maturation depends on nutrients which are available from the parent plant or in synthetic culture media. The research group at the University of Delhi, India, has pioneered plant organ culture techniques for developing anthers and female flowers (MAHESHWARI, 1958; VASIL, 1967). Development will seldom take place in anthers excised earlier than pachytene (VASIL, 1958). Anthers removed from the parent plant at earlier stages apparently lack essential information or substrate(s) required by the pollen mother cells to undergo meiosis. Addition of kinetin and gibberellic acid increased the percentage of microspores in excised *Allium cepa* anthers developing to pollen tetrads (VASIL, 1967).

Culture of anthers in vitro to the stage of pollen dehiscence permits the controlled collection of pollen and also provides a means of incorporating chemical compounds into the developing microspores (TAYLOR, 1950; KOSAN, 1959). These proceede to other parts of the plant. Although uptake of radioactive solutions into excised branches, or into a tissue flap below the developing microspores is easily done, tissue culture procedures are more direct and readily controlled. Time tables for incorporating radionucleotides, ^{32}P and ^{35}S, at specific stages of pollen development have been reported for *Solanum nigrum* (SAMSONOVA and BOTTCHER, (1966). Radioactive pollen produced by these procedures can be used in studying metabolism during pollination or distribution patterns on dehiscence (Chapter 3).

With planning, each pollen-collecting procedure can supply the quantity of pollen needed for immediate use after dehiscence. But the application of pollen in future experiments, or distant places, requires the proper methods of storage.

Chapter 5. Storage

Reports on the storage and transportation of date palm pollen were among the earliest concerned with pollen viability. The male inflorescences of *Phoenix dactylifera* were prominently mentioned in trade contracts of the Hammurabi period about 2000 B.C. When storage of male flowers in a dark, dry place was first recognized as prolonging fertilizing capacity is lost in antiquity. Systematic research on pollen storage started at the end of the 19th century.

Observations on the pollen viability of about 80 species stored at low humidity were reported in the late 19th—early 20th centuries, (GÄRTNER, 1844; MANGIN, 1890; RITTINGHAUS, 1886; HOFFMANN, 1871; MOLISCH, 1893; GOFF, 1901; SANDSTEN, 1909; PFUNDT, 1910). ROEMER (1915) found that pollen storage at low temperature preserved germination capacity better than high temperature storage. Literature on pollen storage has been reviewed by VISSER (1955); BREWBAKER (1959); PRUZSINSZKY (1960); JOHRI and VASIL (1961), and LINSKENS (1964).

Factors Affecting Viability in Storage

The maintenance of pollen germination capacity depends on conditions of storage. Critical external factors include relative humidity, temperature and the atmosphere surrounding the pollen.

Relative Humidity (R.H.)

The humidity of the air during storage decisively affects pollen longevity (PFUNDT, 1910). Most species maintain their pollen viability best at low relative humidity. It is difficult to generalize regarding the exact optimum R.H. because a great number of the results reported for different species are not comparable (MANGIN, 1886; RITTINGHAUS, 1886; MOLISCH, 1893; HOLMAN and BRUBAKER, 1926).

Experiments with homogeneous pollen samples stored at different, controlled humidities have been reported (KESSELER, 1930; KAIENBURG, 1950; VISSER, 1955; PRUZSINSZKY, 1960; LANNER, 1962). These results indicate that pollen longevity, in general, is negatively correlated with the R.H. during storage. Longevity usually increases with the reduction of R.H. to about 6% during storage. Many pollen species lose viability at either very high or very low humidities. Longevity is best at a R.H. between 6 and 60%, and for most species the optimum condition is below 60% (GOLLMICK, 1942). However, low moisture levels must be avoided in the storage of certain pollen. In species of *Tulipa, Plantago medja, Clivia, Aescu-*

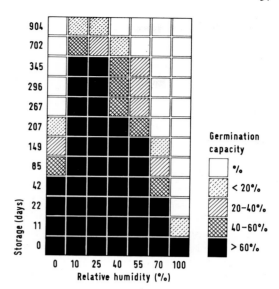

Fig. 5-1. Effect of moisture content during storage at 3°C on pollen germination *in vitro*, *Pyrus communis* var. Clapp's favorite (VISSER, 1955a)

lus and *Parnassia* the water content of the pollen cannot be brought below a critical R.H. level of 40%. The effect of R.H. on storage of pear pollen is shown in Fig. 5-1.

An exception to any form of drying is Gramineae pollen. This pollen loses viability within one day under dry conditions. This loss is slightly slower at low temperatures (ANDRONESCU, 1915; ANTHONY and HARLAN, 1920). DANIEL (1955) stored *Zea mays* pollen for 10 days at 60% R.H. at 7° C; viability rapidly decreased after in the 6th day. Studies by PFAHLER and LINSKENS (1973) of two *Zea mays* varieties showed, by *in vitro* tests, an improved ability to germinate after 24 hrs at 2° C or after 12 hrs at 20° C (Table 5-1). The increased capacity to germinate exhibited by one species after 3 hrs at 35° C was lost after 6 hrs at that temperature. Also, tube lengths *in vitro* significantly increased after storage for one day in comparison with fresh harvested pollen. The more recently developed technique of freeze drying (see below) has facilitated storage of corn pollen for up

Table 5-1. *Zea mays* pollen percent germination after storage at different temperatures (PFAHLER and LINSKENS, 1973)

Time hrs.	Storage temperature					
	2° C		20° C		35° C	
	———————— variety ————————					
	W (%)	K (%)	W (%)	K (%)	W (%)	K (%)
0	33.0	38.9				
3	62.2	71.1	54.5	54.5	27.0	60.4
6	75.8	82.4	62.1	65.1	0	slight
12	81.4	87.3	56.7	56.4		
24	84.0	90.5	48.2	44.7		
96	0	48.4				

to 14 days. But even utilizing all known techniques, the Gramineae are still short-lived when compared to many gymnosperm and angiosperm pollen which can survive years in storage.

The short survival period of Gramineae has been attributed to the thinner, less resistant morphological structure of the grass pollen. The exine gives no protection against desiccation during storage. Their brief viability after dehiscence has been correlated with the advanced development of the nuclei at the time the pollen is shed. BREWBAKER (1959) suggested that since mature Gramineae are trinucleate and contain two male nuclei, a decreased metabolic pool probably exists in those pollen grains; they are adapted to an abbreviated time interval between pollination and fertilization.

Alternatively, many pollen species store optimally at extremely low R.H. PFUNDT (1910) and HOLMAN and BRUBAKER (1926) found that of about 22 different pollen species tested, about half of them retain germination capacity better at 0.005% R.H. over concentrated sulfuric acid, than at 30% R.H. Species of Gramineae and Ranunculaceae, in particular, were unable to survive storage without at least 30% R.H. Some species like *Delphinium* survived fairly well at both low and high R.H. Frequent fluctuation of relative humidity during storage causes a quick loss of viability. Apparently, pollen cannot withstand extreme moisture variations which induce phases of high metabolic activity to alternate with phases of low activity (BULLOCK and SNYDER, 1946).

The degree of humidity is the most important single factor affecting viability during storage. The following hypothesis has been offered to explain the resistance or tolerance of pollen to damage by desiccation. Metabolic processes are severely retarded and respiration is reduced as a consequence of the low water content in mature pollen (O'KONUKI, 1933). The pattern of inhibition of biochemical reactions involved in pollen respiration can also be explained on the basis of low water content (BUNNING and HERDTLE, 1946). Dehydration of proteins would reduce enzyme activity (HAECKEL, 1951). Likewise, ordinary dehydration of trinucleate pollen would probably damage the male cell nuclear components and thus reduce viability.

Temperature

Another important environmental factor which affects stored pollen is temperature. The results for 36 species listed in Table 5-2 indicate that viability can be substantially extended at a temperature of about $0°$ C (McGUIRE, 1952; MORGANDO, 1949). However, water content also plays an important role. OLMO (1942) observed that germination capacity of pollen preserved at $-12°$ C was higher at 28% R.H. than at 56% R.H. For many pollens, pre-drying is essential before storage at temperatures below $0°$ C (ALDRICH and CRAWFORD, 1941). X-ray analysis of ice formation in pollen stored at low temperature confirmed the importance of lowering pollen moisture before or as the temperature is lowered (CHING and CHING, 1966). Viability can be extended beyond 1 year and up to 3 years or longer for some pollens by storage at $0°$ C or $-15°$ C, with relative humidities between 10 and 50% (Table 5-2; NEBEL and RUTTLE, 1937; NEBEL, 1939; PFEIF-

FER, 1936; NEWCOMBER, 1939; GOLLMICK, 1942; STONE et al., 1943; HOWARD, 1958; HESSELTINE and SNYDER, 1958; DUFFIELD and CALLAHAM, 1959).

Gramineae pollen is again an exception to the above storage pattern. A low moisture content and temperature below $-17°$ C are unfavorable for storing *Zea mays* pollen (KNOWLTON, 1922; LIEFSTINGH, 1953; PONCOVA, 1959; WALDEN, 1967).

At the extremely low temperatures, from -180 to $-271°$ C, obtained by liquid gases, it may be assumed that cytoplasmic activity is reduced to nearly zero (BEQUEREL, 1929). Theoretically, germination capacity and the ability to survive for a relatively unlimited time would result. Observations of BREDEMANN et al. (1948) seem to confirm this hypothesis. These authors calculated that *Lupinus* pollen preserved at $-180°$ C should remain unchanged and live for about 1 million years. It seems that pollen such as *Lupinus* is also characterized by a high resistance to desiccation, i.e. maximum viability occured at low humidity. It would be interesting as a further test of this hypothesis to search in the Pleistocene ice deposits of Northern Europe for pollen grains possibly trapped in pre-historic periods. Time stands relatively unchanged for pollen when all biochemical activity is stopped.

Gas Atmosphere

The influence of the composition of the atmosphere on pollen storage has been investigated. An increase in percent carbon dioxide in the atmosphere, such as occurs when the pollen is stored over dry ice, prolongs viability (KNOWLTON, 1922; ANTLES, 1920); storage in pure oxygen shortens longevity (GRIGGS et al., 1950). Pollen of *Pinus*, sealed in nitrogen gas at room temperature after freeze drying, remains viable for at least an initial 100 days (KING, 1959), and probably remains viable considerably longer with this method than with other storage procedures (see below: freeze drying). The initial germination percent of *Citrus* pollen increased during storage at high concentrations of carbon dioxide (RESNIK, 1958). Storing pollen in flower buds with the pedicles dipped in a 50% sugar solution (VENGRENOVSKII and DZHELALI, 1962) doubles the viability of some pollen. This is attributed to an accumulation of CO_2 around the pollen as a consequence of increased carbohydrate metabolism, but it could also be due to increased nutrient substrates accumulating in the pollen.

Oxygen Pressure

Reduction of the partial pressure of oxygen can prolong viability of some pollen. When pre-dried under reduced pressure, the pollen of *Citrus* (KELLERMAN, 1915), *Lilium* (PFEIFFER, 1936; 1938), *Pyrus* and *Malus* (VISSER, 1955) retains a high percentage of germination during shipment under vacuum. By comparison, some pollen species, including *Hordeum* (ANTHONY and HARLAN, 1920), *Antirrhinum* (KNOWLTON, 1922), *Saccharum* (SARTORIS, 1942) and *Cinchona* (PFEIFFER, 1944), die when placed under reduced pressure. *Pinus nigra* and *Betula verrucosa* showed increased longevity when placed in glass ampoules and sealed to reduce pressure

Table 5-2. The effect of storage conditions on pollen germination (*in vitro*)

Species	Storage conditions Temperature (°C)	% Relative humidity	Air*	Longevity (days)	% Germination in vitro before storage	% Germination in vitro after storage	Reference
Azalea mollis	++2	10	—	169	80	2	VISSER, 1955
Betula lutea	?+18	25	—	30	40	3	JOHNSON, 1943
B. verrucosa	++5	0	v	920	60	20	JENSEN, 1964
Cinchona ledgeriana	++10	30–50	—	365	45–65	5–10	PFEIFFER, 1944
Cucurbita moschata	−17	0	v	30	98	98	GRIGGS et al., 1950
Cydonia sp.	++2−+7	25	—	550	50–60	46	KING and HESSE, 1938
Ginkgo biloba	++5	0	—	700	90	35–45	TULECKE, 1954
Haemanthus katherinae	++20	100	—	1	80	60	PRUZSINSZKY, 1960
H. katherinae	++20	100	—	3	80	8	PRUZSINSZKY, 1960
H. katherinae	++20	15	—	1	80	15	PRUZSINSZKY, 1960
H. katherinae	++20	48	—	3	80	60	PRUZSINSZKY, 1960
Hevea sp.	++6	70–80	—	19	90	37	DUKMAN, 1938
Hordeum vulgare	++10	30, 60–90	·	1	40	0	ANTHONY and HARLAN, 1920
Lilium henryi	++0, 5	35	—	194	69	53	PFEIFFER, 1955
Lupinus polyphyllus	−190	30–70	—	93	78	78	BREDEMANN et al., 1948
Lycopersicum esculentum	+2−+4	10	—	252	47	10	VISSER, 1955
Medicago sativa	−17	0	v	34	88	73	GRIGGS et al., 1953
Olea sp.	−17	0	v	374	33	32	GRIGGS et al., 1953
Pinus banksiana	?+2	25–75	—	365	90	90	JOHNSON, 1943
P. nigra	++5	0	v	920	80	20	JENSEN, 1964
P. nigra	++5	60–80	—	378	80	10	JENSEN, 1964
P. strobus	?+18	25	—	413	93	20	DUFFIELD and SNOW, 1941
P. strobus	?+18	15–35	—	365	90	70	JOHNSON, 1943
P. resinosa	?+4	25	—	413	93	70	DUFFIELD and SNOW, 1941
P. resinosa	?+0	50	—	413	93	91	DUFFIELD and SNOW, 1941
P. resinosa	?+2	25–75	—	365	90	90	JOHNSON, 1943
P. sylvestris	?+2	25–75	—	365	90	90	JOHNSON, 1943
Prunus armeniaca	++2−+8	50	—	912	60–80	20–30	NEBEL and RUTTLE, 1937
P. avium (cerasus)	−17	0	v	410	53	37	GRIGGS et al., 1953
P. avium (cerasus)	+0	25	—	550	57	55	KING and HESSE, 1938
P. avium (cerasus)	+2−+8	50	—	1460	60	20	NEBEL and RUTTLE, 1937
P. communis	−17	0	v	346	91	82	GRIGGS et al., 1953

Species							Reference
P. communis	−17	0		1130	91	24	Griggs et al., 1953
P. communis	+2	0	v	550	70	53	King and Hesse, 1938
P. domestica (insititia)	−17	0		435	69	60	Griggs et al, 1953
P. domestica (insititia)	+10−+30	0	v	400	76	5	Manaresi, 1924
P. persica	0	50		550	85	42	King and Hesse, 1938
Pyrus salicina	−17	0	v	439	40	35	Griggs et al., 1953
P. communis	−17	0	v	419	77	65	Griggs et al., 1953
P. communis	+10−−30	0		400	55	3	Manaresi, 1924
P. communis	+2−−4	10		662	66	42	Visser, 1955
P. communis	+2−−4	0	v	1032	64	15	Visser, 1955
P. communis	−20	0	v	1032	64	50	Visser, 1955
P. communis	−190	0	v	662	64	50	Visser, 1955
P. communis	−17, −37	5		3287	—	1	Ushirozawa and Shibukawa, 1952
P. malus	−17	0	v	385	92	64	Griggs et al., 1953
P. malus	+10−+30	0		400	93	7	Manaresi, 1924
P. malus	+2−+8	50		1461	70−80	20	Nebel, 1939
P. malus	+2−+8	10	v	673	76	70	Visser, 1955
P. malus	−20	30−70	v	673	76	63	Visser, 1955
Rhododendron catawbiense	+2−+4	30−70		252	30	2	Visser, 1955
R. catawbiense	−20	30−70		662	30	25	Visser, 1955
R. catawbiense	−190	30−70		662	30	25	Visser, 1955
Rosa damascena	+18	0		50	85	5	Zolotovich et al., 1963
Saccharum spontaneum	+4	90−100		8	90	70−90	Sartoris, 1942
Tulipa hybrida	+20	99		4	60	10	Pruzsinszky, 1960
T. hybrida	+20	95		4	60	35	Pruzsinszky, 1960
T. hybrida	+20	82		4	60	15	Pruzsinszky, 1960
T. hybrida	+20	60−		4	60	5	Pruzsinszky, 1960
T. hybrida	+20	90		10	40	15	Pruzsinszky, 1960
T. hybrida	+20	48		10	40	2	Pruzsinszky, 1960
Trollius europaeus	+20	71		1	80	75	Pruzsinszky, 1960
T. europaeus	+20	95		3	90	2	Pruzsinszky, 1960
T. europaeus	+20	6		3	98	70	Pruzsinszky, 1960
Vitis vinifera	+10	25		365	43	10	Olmo, 1942
V. vinifera	+2	25		730	43	7	Olmo, 1942
V. vinifera	−12	28		1461	43	12	Olmo, 1942
Zea mays	+4	90−100		8	90	60−70	Sartoris, 1942

* —— = normal air

v = vacuum

to -0.15 mm (JENSEN, 1964). Temperature was highly correlated to survival in these experiments. Pollen of pine and birch germinated after $2^{1}/_{2}$ years storage under vacuum at $5°$ C, while pine pollen stored at $5°$ C without vacuum treatment survived only 378 days.

Pollen of related species may respond differently in vacuum dried storage. Vacuum dried *Citrus* pollen did not survive storage at room temperature, but all survived well at $0°$ C, except *Citrus sinensis* var. Parson Brown, which rapidly deteriorated in storage regardless of the pretreatment or storage temperature (LEAL, 1964). Alfalfa *(Medicago sativa)* was viable after 11 years storage under vacuum at $-21°$ C (HANSON and CAMPBELL, 1972).

Freeze-drying of Pollen

Lyophilization or freeze-drying can preserve the life of many vegetative and generative tissues and has been especially useful in preserving fungal and bacterial spores. Freeze-drying of pollen, followed by storage at various temperatures, is as effective for preserving many pollen as it is with lower organisms. Generally, freeze drying is a workable technique for the treatment of pollen; the viability of lyophilized pollen after prolonged room temperature storage, while below that of the freshly harvested pollen, is in excess to the percent required for good fertilization levels (WHITEHEAD, 1962; WOOD and BARKER, 1964).

Lyophilization consists of freezing at $-60°$ C to $-80°$ C, then evacuation from 50–250 mm Hg to remove the water by sublimation. Freeze-dried pollen can then be stored at room temperature under nitrogen or in vacuum. Pollen not frozen before drying, but dried under vacuum and frozen during the process by virtue of rapid evaporation, is correctly referred to as vacuum dried pollen, not lyophilized, freeze-dried pollen. In lyophilization it is essential that the initial rate of freezing be slow, so that crystallization may be wholly or predominantly extracellular before rapid rates of freezing are reached and crystal nucleation becomes general (GRIGGS et al., 1953; PFIRSCH, 1953; PFEIFFER, 1955; HESSELTINE and SNYDER, 1958; KING, 1959; 1961; WHITEHEAD, 1962; SNOPE and ELLISON, 1963; LAYNE and HAGEDORN, 1963; LAYNE, 1963; CHING and CHING, 1966). Fresh pollen of many species stored at $-183°$ C or $-78°$ C still retain nearly normal germination capacity after 14 months (LICHTE, 1957). In some cases, a preliminary drying treatment is necessary before freezing, but even this is not always satisfactory for some pollen, i.e. Gramineae. X-ray diffraction analysis of ice crystal formation in pollen of *Pseudotsuga menziesii* and *Pinus monticola* indicate that membrane permeability and organization can be destroyed, and thus lower viability occurs under improper drying and freezing conditions (CHING and SLABAUGH, 1966).

Poor viability results when lyophilized pollen is germinated immediately after the evacuated storage tubes are opened. This is due to the low moisture content of the pollen grains. A critical precaution in using lyophilized material is that the pollen be first rehydrated. Rehydration was previously found to be effective in restoring germination capacity of some pollen dried at room temperature (GOTOH, 1931; PFEIFFER, 1939; LICHTE, 1957). The physiology of microspore recovery

Table 5-3. Organic solvent storage of pollen, *Camelia japonica*, tested after 3 days at 5° C (IWANAMI, 1972)

Solvent	% Germination 20 hrs, 28° C	Tube length (mm)
Control (fresh pollen)	98.9	8.2
Acetone	98.6	10.4
Acetonitril	58.4	8.6
n-Amyl acetate	98.5	9.4
Ethyl acetate	98.3	10.4
Methyl acetate	78.6	7.0
n-Amyl alcohol	98.6	10.2
Benzyl alcohol	0	—
n-Butyl alcohol	98.8	9.7
iso-Butyl alcohol	98.9	10.2
Ethyl alcohol	0	—
Methyl alcohol	0	—
iso-Propyl alcohol	82.3	8.0
Paraldehyde	90.0	8.2
Benzene	99.0	10.8
Petroleum benzene	98.3	10.4
Carbon tetrachloride	89.9	10.2
Chloroform	23.9	4.8
Dioxane	52.4	5.8
sym-Dichloroethane	48.7	7.0
1,1,1-Trichloroethane	96.8	9.8
iso-Amyl ether	98.3	8.9
Diethyl ether	97.9	10.7
Petroleum ether	98.6	10.9
n-Heptane	98.3	8.5
Hexane	58.8	5.9
Dichloromethane	9.7	0.8
n-Pentane	92.2	10.5
iso-Pentane	98.9	10.8
Pyridine	42.8	7.0
Toluene	98.6	9.8
Xylene	90.1	9.9

from the freeze-dried state is still unknown. Freeze-dried pollen has less viability than fresh pollen, as well as a reduced biological effectiveness for bees for reasons not yet understood (MAURIZIO, 1958a).

Storage of pollen in organic solvents avoids the problem of maintaining a specific relative humidity and may be a useful technique for transporting pollen without dry ice or refrigeration (IWANAMI, 1962, 1971). The technique has not yet come into general use. But in some cases it holds promise of affording an easier method of storing pollen, particularly where hand pollination is planned and the pollen has a sticky surface or connecting viscin threads. The storage method does not work for all organic solvents (Table 5-3), nor for all pollen. *Brassica* pollen germinated very well after solvent treatment, but the treated pollen did not set seed (ROGGEN, 1973). Since acetone, benzene and petroleum ether (Table 5-3) are readily available, inexpensive chemicals, the method is attractive where applicable. However, response of individual species to storage will have to be tested for each organic solvent.

Causes of Decreased Viability in Storage

The reduction in germination capacity as measured by *in vitro* tests of pollen during storage should be distinguished from the influence of storage on the ability of pollen to set seed (CRAWFORD, 1937; HAGIAYA, 1949). These are often differentiated as tests for "germinability" as compared to "fertility". However, germination as measured by *in vitro* viability tests can be positive while, in fact, the pollen is unable to form normal tubes *in vivo*, to penetrate the female tissues, or to complete normal zygote formation (STANLEY, 1962). *In vitro* germination tests are generally sufficient to determine the effectiveness of pollen storage procedures when the pollen is commercially applied in field pollinations (COOK and STANLEY, 1960).

Little is known about the influence of changes during storage on the genetic capacity of pollen (AIZENSTAT, 1954). It has been suggested that old pollen gives rise to a higher number of mutations. The primary reason for decreased pollen viability in storage is probably related to enzyme activities which decrease respiratory substrates. The mechanism by which pollen retains its viability during storage is related to the intracellular rates of respiration, e.g. the conversion of sugars to organic acids (STANLEY and POOSTCHI, 1962). This is confirmed by the observation that stored pollen requires higher concentrations of sugar for normal germination than fresh pollen (VASIL, 1962). The increased sucrose concentration required to obtain optimal germination has also been attributed to a decrease in pollen permeability (KÜHLWEIN, 1937). The respiration rate also decreases with age, while boron sensitivity of the stored pollen increases (VISSER, 1955).

Most pollen is poorly protected against desiccation. This is suggested by the observed shrivelling of grains stored at very low humidity and by the rapid uptake of moisture on exposure to humid environments. This poor protection may also explain the improved viability which results from storage of certain species at higher humidity. Acitvity of enzymes and changes in endogenous growth hormones (Chapter 16) contribute to the decrease in viability (NEBEL and RUTTLE, 1937; NIELSEN, 1956; ZOLOTOVICH et al., 1963). The observation that certain pollen, after a short exposure to below freezing temperatures, shows an increase in germinating capacity compared to the highest percentage obtained by initial germination at room temperature (Table 5-1), can also be interpreted as an increase in, or liberation of, endogenously bound enzymes. Reduction of germination capacity under certain storage conditions can, therefore, be interpreted as an inactivation of enzymes and metabolic substrates essential for germination. Furthermore, an accumulation of secondary metabolic products, such as organic acids (STANLEY and POOSTCHI, 1962) during storage, may inhibit subsequent pollen growth.

Other inherent factors may influence pollen survival in storage. A decrease in germination capacity was related to changes in oil deposits in the pollen exine (PFEIFFER, 1955). As discussed under solvent storage, ROGGEN (1973) observed that exine localized lipophilic substances in *Brassica* can be removed by gently washing the surface with organic solvents without decreasing germinability, but the fertility decreases. Mineral nutrition of the plant during pollen development also influences longevity (GREBINSKIJ and ROLIK, 1949). The influence of bacterial or viral contamination during storage should not be neglected. Viability of pollen from virus infected flowers decreases more rapidly than that of pollen from healthy flowers. This influence can be direct or indirect via colloidal chemical changes or inhibition of such enzymes as amylase and invertase (TULECKE, 1954; SCHADE, 1957). The destructive effect of ultraviolet light on pollen viability (PFAH-LER, 1973) already discussed in relation to pollen collecting, should be kept in mind, particularly when U.V. is used to prevent fungal infection (WERFFT, 1951).

Pollen placed in storage may be derived from secondary collecting sources. However, pollen collected in pellets by bees, and removed by means of traps at the hives, generally does not germinate (FERNHOLZ and HINES, 1942). Therefore, in selecting pollen for storage the fact that chemical additions by the bees or an influence of bee's bodies (Chapter 7) can cause complete inhibition of germination should be kept in mind. GRIGGS et al. (1951) observed that quick-frozen bee-collected apple pollen, stored in a dry ice container, showed some germination ability.

Percent of germination determined after storage is often related to the type of viability assay used. This requires that, in handling pollen, consideration be given not only to the optimum storage requirements for each particular species and the source, but also that the viability assay method be selected with cognizance of the variables and planned use for the pollen.

Testing Methods

The storage and handling procedures applied to collected pollen can modify pollen viability. Establishing which storage conditions are best depends partly upon the assay method used to determine pollen quality before and after storage. The different methods and problems in determining the quality of stored pollen is reviewed in Chapter 6. In general, chemical tests using dyes which react with specific enzymes, and germination of pollen *in vitro* or *in vivo* are used as viabilitiy assays.

The effect of storage conditions on the normal function of pollen grains can be tested by comparing germination capacity *in vitro* to the ability of pollen to induce normal fruit and seed formation. However, these criteria may not be identical (PASSEGER, 1930). Pollen samples with a high germination capacity *in vitro* may not produce tube lengths sufficient to reach the embryo sac on pollination. Results also indicate that stored pollen which shows a reduction in germination ability by *in vitro* tests may not always be non-viable. Some pollen of *Gossypium* or *Pennisetum* is non-viable by *in vitro* tests, however, when used for pollination may give a small but satisfactory seed set (VASIL, 1962). Fertilizing ability measured by seed set in *Zea mays* was retained after 6 days in storage at 2° C even though *in vitro* tests indicated the pollen was non-viable after 4 days. The retention of capacity to set seed was related to specific pollen genotypes

(PFAHLER and LINSKENS, 1972). Obviously, *in vitro* germination tests are not always conclusive indices of pollen growth potential after storage.

Often stored pollen fails to germinate or test as viable in one standard test, but gives a high percentage of viability in the same or in another test a few days later. Irregularities and great variations are often observed when pollen is collected from seemingly equally mature flowers on different days or at different localities (HOLMAN and BRUBAKER, 1926).

In some cases germination percent *in vitro* is higher after a few days of storage than in fresh, ripe pollen tested at dehiscence. These variations may be due to an after-ripening metabolic process, to a lack of uniformity in the samples, or to changing conditions used for the test procedure. It is often useful when pollen is removed from storage over a desiccant to place the pollen in a water-saturated atmosphere for a few hours before testing for germination (GOTOH, 1931; LICHTE, 1957). Ultimately, the criteria of how effective the storage has been is the ability of the pollen to perform its normal role in embryo-seed formation.

Chapter 6. Viability Tests

After pollen is collected or removed from storage, an assessment of the capacity to germinate and grow normally is often desired. This assay is usually done before the pollen is used in the field for seed and fruit production or in the laboratory in biochemical experiments.

Since a considerable investment in time and money is involved in preparing and pollinating a field planting, or in setting up and running a biochemical experiment, precautions must be taken so that the pollen used will grow reasonably well. Appearance of the grains alone, even at collection time, is not always a good index of viability. Low temperature may prevail during pollen development and render a great percentage of the grains unable to germinate or form viable seed although the external appearance is not drastically altered (STEINBERG, 1957). Moreover, as reviewed in Chapter 5 (Storage), the percent of viable grains in some plant families such as Graminae rapidly declines under most storage conditions, but the physical appearance of pollen grains may not change greatly. In some tree species, i.e. *Picea omorica, Salix alba* and *Quercus robur*, non-viable pollen can be easily detected under the microscope (JOVANCEVIĆ, 1962). Non-viable grains differ from viable grains by abnormal shape, size and exine color.

In an *in vivo* assay, a long time period may elapse between pollination and development of viable seed. But such an *in vivo* assay may not be valid; for example, incompatibility reactions may inhibit pollen growth in the style even though the pollen may germinate normally. Viability tests which rapidly determine the growth potential have been developed. Standard tests of viability or fertility involve pollen germination *in vitro*, occasionally *in vivo*, and the direct assay of viability of nongerminated grains.

Germination Tests

In vitro Assays

Most pollen viability tests germinate a small sample of the pollen and observe, under a microscope, the percent of grains producing tubes after a given time. This percent is considered an index of viability of the pollen sampled. Such tests assume that the optimum conditions have been established for the *in vitro* test so that germination approximates that on the plant. However, most pollen tubes cultured *in vitro* stop growing before they reach the size normally attained in the style and the rate of tube growth is seldom as rapid as *in vivo*. This suggests that optimum growth conditions are not often established in *in vitro* media.

Table 6-1. Germination conditions for pollen viability assay

Species	Temp. °C	Nutrients	Media	Reference
Gymnosperms				
Pinus canariensis	27	None	H_2O drop	DUFFIELD, 1954
P. ponderosa	27	None	H_2O vapor	DUFFIELD, 1954
P. elliottii	25	EDTA-Fe	H_2O	ECHOLS and MERGEN, 1956
P. edulis	30	2% suc., B[a]	1% agar	CHIRA, 1967
Larix sibirica	27	B, Ca, Mg, KNO_3	H_2O	Ho and ROUSE, 1970
Angiosperms				
Monocots				
Zea mays	27	15% suc., $CaNO_3$, B	0.7% agar	SEN and VERMA, 1958; PFAHLER, 1966
Secale cereale	27	25% suc., B	0.6% agar	PFAHLER, 1965
Areca catechu	28	0.75% suc., B	H_2O	RAGHAVEN and BARUAH, 1956
Caryota spp.	29	12% lactose, .02% colchicine, B	0.5% gelatin	READ, 1964; FORD, 1968
Lilium candidum	25	12.5% suc., B, Ca, NAA	H_2O	SEN and VERMA, 1958
Lilium spp.	20	10% suc., B	H_2O	TSUKAMATO and MATSUBARA, 1968
L. regale	25	White's nutrients + 7 ppm GA_3	H_2O	SEN and SAINI, 1969
Dicots				
Beta vulgaris	20	30% suc., B	6% gelatin, pH 5.6	GLENK et al., 1969
Chrysanthemum barbankii	20	30% suc., B	H_2O	TSUKAMOTO and MATSUBARA, 1968
C. cinerariifolium	30	40% suc., B	$CaPO_4$ buffer, pH 6.8	HOEKSTRA, 1972
Corylus avellana	30	18% suc.	H_2O	HOEKSTRA, 1972
Eucalyptus	30	20% suc., B	1.5% agar	BODEN, 1958
Fraxinus excelsior	21	15% suc.	2% agar	NIKOLAEVA, 1962
F. ornus	21	5% suc.	2% agar	NIKOLAEVA, 1962
Juglans nigra	27	20% suc., B light	H_2O	HALL and FARMER, 1971
Nicotiana alata	30	10% suc., B	H_2O	HOEKSTRA, 1972

[a] Boron is usually added as the boric acid at a concentration of 0.01% B. suc. = sucrose.

Another assumption made in *in vitro* tests is that the micro-sample of pollen tested is typical of the total source. This assumption is permissible provided care is taken to mix the pollen batch before drawing the sample. No significant variation was recognized in pollen shed in alfalfa heads throughout the period of dehiscence on a single plant; but highly significant variation was noted between pollen from different plants (SEXSMITH and FRYER, 1943). Plant-to-plant variation must be guarded against in drawing a sample from bulk pollen for viability testing.

Germination Procedures. Factors influencing *in vitro* growth include the species of pollen, time of collection, season of the year, method of collection, and storage history. Table 6-1 lists the assay conditions suggested as optimum for the germination of many pollen species.

References prior to 1932 concerning viability assays of most pollen by germination are of questionable value. In 1932, SCHMUCKER reported that boron stimulates pollen germination *in vitro*. He made this highly significant observation by discerning that stigmatic fluid which had long been known to stimulate growth of pollen (MANARESI, 1921) contained high levels of boron. SCHMUCKER (1932) substituted boron in place of the *Nymphaea* spp. stigmatic fluid and obtained the same marked stimulation as with the natural stigma solution. Thus, unless a survey for optimal growth conditions for pollen germination *in vitro* includes boron, the conclusions may not be valid. Hence, in compiling Table 6-1 we have omitted most work done prior to 1932.

Many chemical and physical factors are now known to influence optimal pollen germination *in vitro*. Some chemicals stimulating germination, e.g. boron, calcium and magnesium, were first noted as similar or identical to factors found in the style tissue or stigmatic fluid in which the pollen naturally germinates. Older tests on pollen germination *in vitro* often added stigma solution (MANARESI, 1921). This test procedure is still frequently employed to detect chemicals stimulating pollen growth. Slices or extracts of the style or stigma are added to the pollen germination media (MIKI-HIROSIGE, 1961; SCHILDKNECHT and BENONI, 1963). Use of such tissue slices may induce a tropic response in the pollen and give misleading results. Also, extraction procedures may concentrate in stigmatic solutions many chemicals which pollen growing *in vivo* encounters only at very low concentrations or not at all. In this discussion we will confine ourselves to *in vitro* growth procedures used to assay viability and not focus upon techniques used in discerning chemotactic growth responses.

The Assay Method

The specific germination method used will vary with the pollen species and the accuracy or purpose for which the determination is made. Five simple methods regularly used are the hanging drop, spot test, well, agar and membrane assays. The five test arrangements are shown in Fig. 6-1. Other specialized tests have been developed and are occasionally used (EIGSTI, 1940).

The hanging drop technique, first introduced by VAN TIEGHEM (1869), involves a drop of media containing pollen inserted in a circular chamber or closed area suspended over water. This arrangement (Fig. 6-1A) restricts evaporation of the media and permits direct microscopic observation. This simple method, or a modification of it, is the most commonly used procedure for germinating micro-amounts of pollen to determine viability.

The hanging drop method is generally limited to one drop per chamber; to test the effect of many variables on germination a great number of chambers are required. Other disadvantages relate to the inaccessibility of the pollen to manipulation, and to decreased germination because of concentration of the pollen on the apex of the drop due to surface tension (SATO and MUTO, 1955); or conversely, if the pollen breaks through the surface tension it may float to the top of the drop, against the glass, and grow in a relatively anaerobic area which may inhibit germination. Inability to mix the pollen evenly may also make counting of the concentrated mass of grains and pollen tubes germinated in this method quite difficult. The emulsifying agent polyoxyethylene (Tween 80), at about 0.3–0.5%, has occasionally been used in the germinating medium as a pollen dispersing agent.

IN—VITRO GERMINATION TESTS

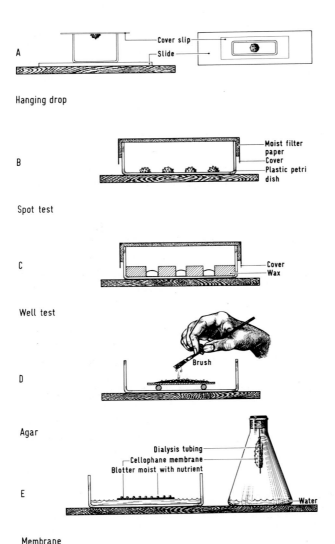

A

Hanging drop

B

Spot test

C

Well test

D

Agar

E

Membrane

Fig. 6-1 A-E. Methods of germinating pollen *in vitro*

A slight modification of this procedure involves arranging the germination solution on microscope slides with the drops inside wax or petroleum jelly circles. These slides are then inverted and rested on glass or wooden rods over water in a petri dish. This arrangement allows the advantages of several drops per slide and of adding chemical fixatives to the pollen to preserve them in a permanent mount for future reference. However, this method of germinating pollen involves considerable handling of the slide mounts and samples are often lost.

Spot tests avoid the risk of losing a sample in one of the preparation steps; they also allow a rapid comparison of many samples of pollen. This germination assay was initially performed by placing a series of petroleum jelly rings on the bottom of a glass petri plate, Fig. 6-1B (DUFFIELD, 1954). The end of a 5 mm glass tube is a good applicator for making circles of petroleum jelly. The germinating solutions are placed in the circles, and pollen is sprinkled on the drops with a camel's hair brush, or tapped on from a spatula. Attempts to pipette pollen are generally not successful because of the tendency of pollen to float or quickly sink, and to attach to the walls of the pipette. Moistened filter paper is usually placed in the top of the petri plate to raise the vapor pressure over the drops and reduce evaporation of the solutions. If the pollen grains have a tendency to sink, the plates can be inverted. Cover slips can be set over the drops for observing and counting the germinated pollen. The petroleum jelly tends to spread under pressure of a cover slip, often into areas of the solution drop where it interferes with the assay.

A simpler procedure for spot tests is to use plastic petri plates. Drops of 100 or 200 µl can easily be placed on the bottom of a plastic petri dish. Surface tension properties of the plastic-solution interface maintain the solution drop integrity. The number of drops that can be placed on a plastic dish vary with the size of the plate and drop. Inverting a plastic dish with free hanging drops is not as easy as inverting a dish with germination drops contained in petroleum jelly circles. But plastic dishes with drops can be inverted by a rapid movement which does not permit the free-standing drops to slide over the plastic surface. A moistened filter paper insert in the cover of the dish is again recommended to prevent drying of the germination solution.

If germination tests are to be run for many days at a temperature above 28° C, then the filter paper in the top of the dish must be frequently remoistened. The simplest procedure is to seal the plate with a rubber band or tape to eliminate evaporation. The oxygen in the closed plate is sufficient for pollen germination. A cover slip can be readily placed on the drop when it must be assayed; or the drop may be viewed without a cover slip, although distribution of pollen on the apex or curvature of the drop makes observation of the non-flattened drops difficult.

Well tests supply a repository for the pollen germination drop. The simplest procedure is to use a multiple depression microscope slide. Such slides cannot be inverted and this creates a problem in assaying pollen which sinks. The slides have the advantage of ease of handling and a microscope cover slide can be placed directly over the well depression when it is time to count the pollen.

Pollen which does not require a nutrient solution to germinate, e.g. pine pollen (BROWN, 1958), can often be sprinkled directly onto a dry microscope slide and then placed in a moist chamber. This procedure was actually the basis for one of the first *in vitro* observations made on germinating pollen tubes. VON MOHL (1835) observed that pollen formed a tube when placed on a glass slide in a saturated atmosphere.

A more elaborate procedure involves pouring wax into a petri plate and boring holes with a cork borer, Fig. 6-1C (RIGHTER, 1939). This rigid wax prevents diffusion of liquids and pollen between test samples. Pollen placed in a hole spreads over the solution surface in that hole and the plate can be readily inverted. This method was initially designed for pine pollen which generally is tested in distilled water without nutrients, but is grown for 48 hrs at 25–29° C.

Contaminating fungi interfere with germination assays requiring long time periods. The fungal hyphae grow and occlude the pollen tubes. Chemical or antibiotic fungistats are occasionally used with germinating pollen to reduce contaminating growth; the antibiotics may also restrict the pollen respiration (HAVIVI and LEIBOWITZ, 1958).

Where there are many wells on each plate, replicate samples can often be set up without antibiotics. One or two of three replicates will usually be free of contamination and allow easy counting of germinated pollen (RIGHTER, 1939). The wells with pollen can be viewed directly from the bottom or top of the petri plates.

Agar or gelatin is frequently used for germinating pollen. The agar supplies moisture at a constant relative humidity and various carbohydrates or other pollen growth stimulants can be readily incorporated in it (KUBO, 1955, 1960). The thickness of the agar, as well as the agar and sugar concentrations affect moisture availability and germination (KUBO, 1955). Use of a sugar, such as the animal sugar lactose which pollen can generally metabolize very readily, can provide the required osmotic environment and minimize growth of fungi (BISHOP, 1949). An additional advantage of agar slides (Fig. 6-1D) is the ease with which they can be handled and the possibility of preparing permanent mounts (BECK and JOLEY, 1941; EIGSTI, 1966). Aerobic conditions are very good on the surface of agar slides or plates.

Unlike agar and gelatin which dry out unless a relatively high humidity is maintained, silica gel can be used and will retain the moisture essential for germination. An aseptic silica gel medium of 16% sucrose has been adapted for the long-term germination of pollen of Abietacae (DURRIEU-VABRE, 1960). Pollen is also easily fixed for permanent microscopic slides in this preparation.

Membrane supports have been used for germinating pollen. A collodion membrane floating on water, or a dialysing tube hanging in a moist chamber, offer the advantage of being transferrable to many different solutions during pollen growth; also, they can be fixed for permanent observation, or for viewing at some later time (SAVAGE, 1957; NARASIMHAN, 1963). Our experiments with this germination procedure indicate that not all pollen germinates well on membranes; however, some pollen gave higher germination percentages on membranes than in any other in vitro tests. Pollen tubes generally grow straighter on a dializing membrane than in solution or on agar (Fig. 6-1 E).

Very often one method of testing may give poor results with a particular pollen which on retesting by another in vitro procedure shows a higher percentage of viable pollen (KLAEHN and NEU, 1960). Other pollen, such as cotton, which are difficult to germinate in vitro, germinate best at a liquid-filter paper interface (MIRAVALLE, 1965) or in an oil medium (KLYUKVINA, 1963). Other sources of error may occur in testing stored pollen. Higher optimum sugar concentration for germination in vitro is frequently required as pollen ages in storage (MOFFET, 1934). It is possible to fractionate some pollen by air. This removes damaged or aborted grains which differentially sink in an air column and ordinarily would bias and lower the observed percent germination (WORSLEY, 1959).

Recording Data

Sources of Error. A false germination count can result under certain conditions. The quantity of pollen in a spot can affect the capacity of pollen to grow. A "mutual growth stimulation" occurs at a certain concentration of pollen (BRANSCHEIDT, 1939; SAVELLI, 1940; LINSKENS and KROH, 1967). In germination tests, a certain minimum concentration of pollen grains must be placed in a given volume of solution if maximum germination is to be attained. The mutual stimulation effect is not as strong in pollen spread evenly over agar. Conversely, in both the spot and agar tests high pollen concentrations can inhibit germination. Viability tests are most valid when pollen is grown in an optimal concentration range. The addition of Ca^{++} at approximately 2×10^{-3} M to the germination media tends to overcome the mutual stimulating effect (KWACK, 1965).

Care must be exercised in viewing and counting the pollen samples. Centers of high and low pollen concentration can exist in a single drop. Generally, the concentration of grains is high towards the center and low towards the outside. Because of the mutual stimulation effect, counting pollen in non-typical areas can lead to distorted viability values. Pollen that has sunk to the bottom of a solution drop during growth may also yield distorted values, since sunken grains may represent damaged pollen, or pollen germinating under anaerobic conditions.

When a cover slip is placed on a solution drop containing germinating pollen, ungerminated grains which are not entangled in germinating tubes frequently move to the outside edge as the cover slip is lowered into place. Counts made in these regions of the slide can give spurious values. The easiest way to avoid this source of error is to stir the pollen gently throughout the drop with a fine glass needle before applying the cover slip.

Another way to minimize the problem of localized variations in the pollen germination medium is to increase the volume of liquid media and quantity of pollen grown. These larger volumes can be germinated on a shaker in a controlled temperature environment to assure even mixing. Alternatively, a volume of germination media with pollen can be gently mixed by bubbling air through the solution (SCHRAUWEN and LINSKENS, 1967). One such apparatus for facilitating the growth or testing or larger quantities of pollen is shown in Fig.6-2. These units can be built to accommodate any desired volume of solution and still allow evenly diffuse samples to be periodically withdrawn by pipette for germination assay. Mass culture methods usually increase the percent germination and tube length attained *in vitro*.

Growth inhibiting or stimulating chemicals (Chapter 16) must also be recognized as potential sources of error in assaying pollen germination *in vitro*. Inhibitors can diffuse out of pollen and accumulate in the media. Many metabolic products such as organic acids, while always present at low levels, may reach toxic levels as growth proceeds and they diffuse into the media. If tap water is used instead of distilled water, or if a sodium buffer solution is used, inhibiting cations may be present at levels sufficiently high to reduce growth because of their concentration or ability to modify pH (DENGLER and SCAMONI, 1939; SVOLBA, 1942). Germination of pollen at an improper pH may also give spurious results (SISA, 1930; HUREL-PY, 1934; COUPIN, 1936).

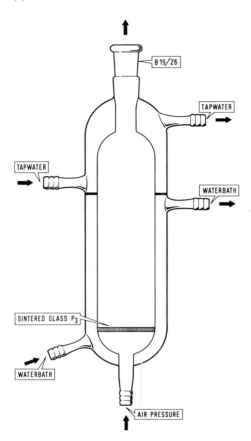

Fig. 6-2. Glass cylinder for mass culture of pollen (SCHRAUWEN and LINSKENS, 1967)

Reduced viability *in vitro* can result by excluding carbohydrate substrates or boron from the media of pollen that require these additives. Another factor which may result in a non-valid assay of viability is failure to recognize that the optimal osmotic concentrations of sugar substrates required for germination *in vitro* can vary for different pollen and even for pollen from a single plant at different times during the growing season (MOFFET, 1934).

Many nutrient factors can be added to pollen germination media to optimize tube extension (Table 6-1). A few pollens also require amino acids in the sugar media to germinate (SWADA, 1958), and many produce longer tubes in the presence of such additives. Before testing pollen samples stored over a desiccant or prepared by freeze drying (Chapter 5) rehydration of the pollen is generally necessary for obtaining a valid index of viability and maximum growth (SCEPOTJEV and PABEGAILO, 1954; DUFFIELD, 1954; WORSLEY, 1960). In viability assays we are primarily interested in germination of the pollen, not maximum tube growth.

Methods. Methods of recording germination vary from simple visual inspection and estimation of percent germination, to counting or photographing the sample under the microscope for future reference and accurate counting. In via-

Fig. 6-3. Lanometer projecting microscope with grid and thread for measuring length of pollen tubes

bility counts one generally considers a tube to have been formed when it is equal to, or greater than, the pollen grain diameter. The total number of grains observed in a field is compared to the number of grains producing tubes.

A simple procedure when viewing pollen under the microscope is to use a pair of hand tabulator counters either mounted on a rack or hand-held; one counter tabulates the total number of grains counted, the other tallies grains with tubes. It is best to keep one hand free for adjusting the microscope, since many grains may grow downward and require focusing to ascertain if a tube is produced. Some pollen may produce two very small tubes. One of the two must exceed the grain diameter to be considered a true tube and not a mere swelling. Some pollen bursts soon after germination begins. This pollen is often considered to have germinated for purposes of viability statistics (DEAN, 1964).

Standard 35 mm photomicrographic equipment can be used to record germination in hanging drop or microscope slide preparations. Two simple projection techniques, the Reichert Lanometer and the microscope projector, are often used for viewing germinating pollen in slide preparations (Fig. 6-3). A camera can be coupled to these viewers to produce a permanent photographic record for future analysis.

Statistical Analyses. The type of statistical treatment applied to the data obtained from pollen germination tests depends on whether one is seeking to select the most viable sample of pollen from a group of samples, or just to ascertain an index of viability of the pollen sample before using it. In all cases, there is a simple binomial response, i.e. the pollen does, or does not, germinate. Samples can be compared by a standard "F" or student's "T" test. The size of the sample counted for a valid T test will depend upon the mean range of germination and standard deviation of the mean. To place a high degree of significance, i.e. a 95 or 99% level of confidence, in samples counted with a mean of about 50% germination requires that a larger size sample be counted than when 20% or 80% of the pollen germinates (Table 6-2).

A variation of $\pm 8\%$ between replicate counts of 2×200 grains, with a germination of about 60%, is not unusual. An even distribution in the microscopic field

Table 6-2. Size of pollen sample count required for significant data

Percent Germination	No of grains to count (n)		
	0.01	0.05	0.10
	Confidence level = level of significance		
1 or 99	380	15	4
5 or 95	1825	73	18
10 or 90	3457	138	35
15 or 85	4898	196	49
20 or 80	6147	246	61
25 or 75	7203	288	72
30 or 70	8067	323	81
35 or 65	8740	350	87
40 or 60	9220	369	92
45 or 55	9507	380	95
50	9604	384	96

selected for counting generally produces the 8% variation between *in vitro* assays of pollen viability.

In vivo Assays

Pollen may not germinate *in vitro*, or may indicate a low viability, and yet yield a high percentage of seeds of fruits when used *in vivo* (JOHRI and VASIL, 1961). Conversely, pine pollen germinated after 15 years storage (STANLEY et al., 1960), but was unable to set seed (STANLEY, 1962). Since *in vitro* test media may be deficient or inhibit the pollen growth or may allow a tube to grow only a short distance, it is often suggested that the only valid test is application of pollen to a compatible stigma or micropyle and observation of pollen germination in the style or fruit development. In some cases, as was true of cotton until only recently, the only way to germinate the pollen is *in vivo* (MIRAVALLE, 1965). However, while *in vivo* growth criteria are not always valid, and the methods are slow and laborious, they do, in many instances, afford a more valid criteria of the quality of extracted or stored pollen than *in vitro* germination.

    Germination Procedures. Assuming that the primary purpose of a viability assay as distinct from the fertility assay is to determine if the pollen can germinate, then an *in vivo* test is often the most valid. This test requires that pollen be placed on a stigma; after a period the style is removed, and the number of tubes growing down into the style is compared to the number not penetrating and the number of grains not producing tubes.

    However, certain assumptions made in this test may not always be valid. If the stigmatic surface or style inhibits pollen tube penetration, the test is not valid. Examples of pollen germination being blocked by the surface cuticular layer in Cruciferaceae are known (KROH, 1964).

    If pollen tubes penetrate the style, then the problem is how to view and assay them. Staining techniques which differentiate between the pollen tubes and the stylar tissue have been developed (NEBEL, 1931; NAIR and NARASIMHAN, 1963).

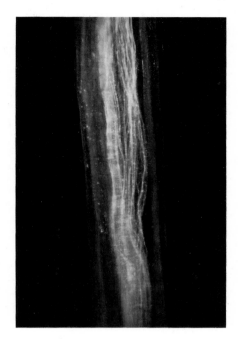

Fig. 6-4. *In situ* localization of pollen tubes after application of a fluorescent dye, which reacts with the callose

Such preparations, made by gently crushing open the styles, can be made into permanent mounts for future comparisons (ADAMS and MACKAY, 1953). Preparations must often be made of styles that cannot be readily crushed to view the pollen tubes inside. In such cases 5% pectinase can be used in preparing the stylar tissue for staining and mounting (PANDEY and HENRY, 1958).

Fluorescence Technique. A more sensitive and easily applied method of detecting germinated pollen tubes is based upon the fluorescence of callose in ultraviolet light. Readily squashed styles are stained with a flurochrome dye (LINSKENS and ESSER, 1957). The dye links to the callose occuring in germinating pollen tubes. When viewed with a U.V. microscope ($\gamma = 356$ mμ), callose appears light yellow-green. Although this method was initially applied to growing pollen tubes which contain callose, the florescent dye will also react with callose in dead pollen tubes (MARTIN, 1959). Also, the quantities of callose and the dye adsorbed by the pollen tubes in the styles can vary with the type and genetic origin of the stylar tissues through which the pollen must grow (LINSKENS and ESSER, 1957) (Fig. 6-4). The dye acridine organe stains pollen tube cytoplasm and does not depend on the presence of callose (SHELLHORN et al., 1964). This method may also be applicable to trace penetrating tubes and may aid in differentiating pollen grains on stigma tissues.

Difficulties can occur in attempting to use the *in vivo* germination technique to assay pollen viability. In *Gladiolus,* if water is present on the stigma when the pollen is dispersed, it will cause the pollen to rupture and not germinate (CAMPBELL, 1946). Genetic incompatibility barriers to germination and growth of the pollen must not be present. The pollen must be applied to the stigma at a time when the stigma is mature and receptive to pollen growth. Certain types of genetic incompatibilities between pollen and pistil tissue can be circumvented by injecting pollen suspensions in water directly into the ovary (BOSIO, 1940; KANTA and MAHESHWARI, 1963). Such interovarian pollination can avoid inhibiting reactions which block pollen growth in the stigma or style. The number of viable embryos developed by pollen injected into ovaries was lower than developed under normal pollination in five angiosperms tested (KANTA and MAHESHWARI, 1963). But this method does allow a crude test of pollen viability in certain angiosperms where incompatibility systems may inhibit tube growth.

Another factor which may distort the results of *in vivo* assays of pollen viability include application of a too high concentration of pollen. In such cases slow-growing pollen will be inhibited or arrested in germination (MODLIBOWSKA, 1942) and pollen tubes will not grow into the style. KAVETSKAYA and TOKAŘ (1963) showed that 10 to 18 grains deposited on each *Juglans* stigma produced maximum seed set. A marked lowering of air temperature after pollination can drastically modify pollen germination *in vivo*, and lead to an apparently low viability (STEINBERG, 1959).

Seed Formation. The capacity of pollen to give rise to viable seed is often used as a criterion of pollen viability. This test procedure can be the most valid test method, or conversely, because of interfering substances and other blocks to normal pollen growth, it may be an unreliable index of viability. Where a time delay is not important, and potential sources of interference are recognized, this method may be preferred in determining pollen viability.

In this test emasculated, female flowers are pollinated. The stigma or megastrobili must be protected from stray, noncontrolled pollinations; cloth or plastic bags coated with aluminum paint are frequently used to cover the pollinated flowers. When the fruit is ripe the seeds must be removed and counted. A comparison of seed set to flowers naturally pollinated or pollinated with fresh pollen by similar artificial techniques is generally made. Occasionally, tests with dead pollen or pollen diluted with a large amount of inert substances, such as talc powder, are used as additional controls. Since parthenocarpic setting of fruit can be induced by dead or poorly germinating pollen (GUSTAFSON, 1939; VISSER, 1955) (Chapter 16) this is an important control in this assay. Variables, such as the number of ovules available for fertilization, quantity of pollen applied to the stigmatic surface, and time of day of application are all statistically important in *Zea mays* pollen viability assays (WALDEN and EVERETT, 1961).

Two additional questions can be answered by the seed development index. Namely, not only can the pollen germinate, but are the stylar and ovarian tissues through which the tube must grow compatible with the pollen genotype, and does the pollen transmit a normal functioning set of chromosomes? Only cytological study can determine that germination was normal but syngamy was blocked. Assuming normal fertilization, if the pollen is old or modified by radiation, the germ plasm may give rise to a high percentage of abnormal mutants. The probability of growth defects can be evaluated by germinating the seeds and observing seedling growth characteristics. While the time required to complete such determinations separates this pollen viability assay test from others, this method does extend the analysis to the chromosome level. About 20 months are required to complete such determinations in pine; one to three months are required in many angiosperms. Cytological analysis of mitotic chromosomes in the embryonic tissue with accompanying idiograms have been used to shorten the time required to evaluate pollinations (SAYLOR and SMITH, 1966; WILMS et al., 1970).

Since a relatively large quantity of pollen is applied to each stigma, the chances of a viable grain setting seed are quite high even with pollen which germinates poorly during *in vitro* assays. VISSER (1955), in a study of pollen, compared an *in vitro* germination assay with fruit set in apples, pears and tomato plants. Table 6-3 summarizes his conclusions with pear pollen.

Table 6-3. Comparison of *in vitro* germination to fruit set (*Pyrus communis* var. Clapps Favorite)

% Germination in vitro	Fruit set
under 20	nil to poor
20—40	poor to moderate
40—60	moderate to normal
above 60	normal

Tests with tomato pollen which did not germinate *in vitro* showed that this pollen still set fruit and often with as many seeds as is set by fresh pollen (VISSER, 1955). But tomato is well known for its parthenocarpic fruit setting capacities (GUSTAFSON, 1939) (Chapter 17). Similar to VISSER's results in Table 6-3, apple pollen which yielded only 35% germination *in vitro* gave good fruit set (OVERLEY and BULLOCK, 1947). These results suggest that even though pollen may be relatively low in viability, i.e. 40% or less, it will still form fruit normally.

The genetics and nutritional status of the pollen parent plant may be more critical in *in vivo* studies than in *in vitro* viability assays if fruit set is used as the criteria of pollen quality. In genetic blocks to seed formation there are many different incompatible mechanisms. These can be merely physical, such as short pollen tubes unable to grow down long styles, the so-called "Pin-Thrum" incompatibility (LEWIS, 1955), or they may result from biochemical reactions in the style to the pollen (LINSKENS, 1955). Many variables, including plant age, nitrogen level and pre-treatment with low or high temperature can modify incompatibility reactions and the ability of the pollen to set seed. The genetic basis of the incompatibility reactions is well established (PANDEY, 1962), and studies on the mechanism, in particular as related to pollen proteins, continue as an active research area.

The genetic variations in a plant population are often overlooked as a source of differences in pollen growth. BARNES and CLEVELAND (1963) studied the genetic basis of differential pollen growth as an influence on seed set. They observed that in a natural population of diploid alfalfa (*Medicago sativa*), some plants produced long pollen tubes when germinated *in vitro*, and others produced only short pollen tubes *in vitro*; a difference of 100% existed between short and long tubes growing *in vitro*. This same growth pattern occurred in seed set. The pollen producing short tubes set few seeds, those of strains producing long tubes set many seeds. A mixed population of pollen placed on alfalfa stigmas induced fair to good seed set, but not as good as when only the long pollen tube strains were used.

Increasing *ploidy* will usually decrease pollen fertility, although it will not necessarily reduce viability as detected by *in vitro* germination. STAUDT and KASSRAWI (1972) studied the relationship between pollen viability and fertility in diploid and tetraploid grapes. In most of the 33 varieties, cultivars, and species that they studied, germination *in vitro* was reasonably similar (Table 6-4), but the tetraploid plants showed a reduction in pollen fertility of between 3% and 96%, as measured by fruit formation. A significant reduction occurred in most of the 33 cases of increased ploidy.

Table 6-4. *In vitro* pollen germination and seed set in diploid and tetraploid plants of *Vitis rupestris* (STAUDT and KASSRAWI, 1972)

Pollen ploidy	% Germination in vitro (20% sucrose + boron)	% Fruit developed
1 X	66.5 ± 2.9	54
2 X	54.3 ± 2.2	27

Extra precautions must be taken when collecting and testing trinucleate pollen, compared with methods of handling binucleate grains. For viability to be retained, trinucleate pollen must be tested soon after collection. Respiration and germination capacity of trinucleate pollen of *Cosmos*, *Triticum* and *Chrysanthemum* showed a 50% loss in about 6 hrs at 30° C as compared to an average 40 hrs active phase for binucleate pollen such as *Nicotiana* and *Corylus* (HOEKSTRA, 1972). Trinucleate pollen are particularly low in germination percent unless they are first resaturated with water before testing.

Mineral nutrient levels of stylar tissues can modify pollen growth and seed formation. When certain clones of potato flowers are removed and placed in nutrient solutions before the flower buds open, the percent of pollen germinating on the stigma and rate of tube growth is much greater than in flowers not given supplemental nutrients (BIENZ, 1959). VISSER (1955) found that in most pear species the percent of germination and length of pollen tubes formed *in vitro* in 10% sucrose solutions were related to the amount of boron supplied to the branch on or off the tree. Since the trees tested by VISSER (1955) were not boron deficient, he did not increase fruit set by spraying boron solutions directly onto flowering branches. KIBALENKO and SIDORSHINA (1971) increased size, rate and percent germination of pollen from boron deficient sugar beet (*Beta vulgaris*) by applying either a boron spray or fertilizer.

Other nutrients also influence pollen yield and growth rate. Nitrogen and phosphorus levels influence pollen production and seed formation. High soil salt concentrations during development can inhibit viability. *Petunia* was grown with up to 3 bars osmotic pressure concentrations of Na^+, Mg^{++} or Ca^{++} chlorides by REDDY and GOSS (1971). The pollen that was formed, assayed by *in vitro* germination, decreased in viability with increasing salt levels in the root environment. NaCl was most toxic; at some levels $MgCl_2$ relieved the extreme sodium toxicity. Obviously, nutritional status of the pollen source plants must be considered when seed or fruit formation is used as a criteria of pollen viability.

Many factors can affect the quantity of fruit and seeds produced after pollination. Since even poorly germinating pollen can generally yield a good or fair percent of seed formation (Table 6-3) and the viable seed assay is time consuming, most research workers prefer alternative methods to the *in vivo* or *in vitro* germination assays for measuring pollen viability. Procedures which do not require germination of the pollen have been devised.

Non-germination Assays

Specific Stains

Tests which stain pollen grains with a color dye are often used as indices of viability. In such tests, the chemicals are adsorbed to specific cell constituents present in mature pollen. Aniline blue in lactophenol stains callose with high specificity; potassium iodide is fairly specific for starch; acetocarmine preferentially stains chromosomes; and phloxin-methyl green stains both cytoplasm and cellulose (OWCZARZAK, 1952). The callose and starch stains have been the most widely used. A problem with this technique is that many immature or aborted pollen grains contain levels of constitutive chemicals sufficient to yield positive results in one or more of these stain tests. Non-viable pollen grains of hybrid wheat stain just as readily as normal mature grains (KIHARA, 1959); oak pollen may stain poorly for starch and yet be highly viable (JOVANCEVIĆ, 1962).

Most stains are not sufficiently accurate when compared to germination tests to give other than crude estimates of pollen viability (VAZHNITSKAYA 1960; NAGORAJAN et al., 1965). Flurochrome dyes can be adsorbed by protoplasmic organelles in germinating pollen (BHADURI and BHANJA, 1962). Acridine orange, when adsorbed by pine, birch and other tree pollens and viewed with a fluorescent microscope, shows a characteristic green fluorescence in the nucleus and orange-red fluorescence in the cytoplasm. This dye fluoresces green with DNA and orange with RNA. These fluorescent patterns are also characteristic of the normal germinating pollen grains (KOZUBOV, 1967). Dead cells fluoresce yellow-green.

Dye tests based on the presence of functional enzymes have been developed. These enzyme reactions (Chapter 14) are closely related to the pollen's metabolic capacity to respire and grow (LAPRIORE, 1928; BARTELS, 1960). They do not depend merely on the presence or absence of a substrate or chemical constituent.

Redox Dyes

The Mechanism. Viable pollen grains contain functioning enzymes, or enzymes capable of functioning when placed in germination conditions. The reduction-oxidation (redox) enzymes, such as the dehydrogenases and peroxidases, transfer protons and electrons to acceptor molecules. Certain dyes, on reduction by a proton, change from a colorless to a colored form. Active enzymes are usually necessary to transfer the protons, resulting in the color change of the dye. Thus, observing the change in dye color provides an indication of the redox enzyme activity present. The presence of these enzymes has been correlated with respiration and the ability of seeds to grow (LAKON, 1942; BROWN, 1954). Such redox dyes have also been used to determine pollen viability.

Topographical biochemical tests for pollen viability have been developed using several different dyes. The dyes and color characteristics of the reactions are given in Table 6-5.

Table 6-5. Redox dyes used to determine pollen viability

Reagent	Formula	Color change and enzyme source	Reference
2,3,5-Triphenyl-tetrazolium chloride (TTC)	$C_6H_5CN_2C_6H_5N_2$ C_6H_5Cl	colorless to red reductases	VIEITEZ, 1952
Triphenyl tetrazolium red		colorless to red reductases	SARVELLA, 1964
Na-biselenite	$NaHSeO_3$	colorless to red reductases	JACOPINI, 1955
Diaminobenzidine	$NH_2C_6H_4C_6H_4NH_2$	light yellow to blue peroxidase	SHARADAKOV, 1940
Guaiacol	$OHC_6H_4OCH_3$	colorless to brown-red peroxidase	OSTAPENKO, 1956

The dye most commonly used in pollen testing is 2,3,5-Triphenyl tetrazolium chloride (TTC), which is water soluble and easily applied. Adding catalytic amounts of thiazine derivatives increases the acceptance of H^+ ions from enzymes by TTC through the flavoprotein nucleotides, which are intermediate H^+ ion acceptors for dehydrogenase enzymes. The thiazine acts as an intermediate cofactor in moving the electrons to another redox-level (FARBER and LOUVIERE, 1956). Oxygen inhibits reduction of the dye. TTC also has the disadvantage of being easily reduced by chemical and photochemical agents (KUHN and JERCHEL, 1941). In measuring pollen viability by these dyes, precautions must be taken to avoid reduction by chemical elements of the nonliving system.

Applications. The first application of a redox dye to determine viability of pollen used benzidine to measure the peroxidase activity (SHARADAKOV, 1940). Viable pollen gave a red color which was detected by viewing the pollen grains under a microscope 3 or 5 minutes after application of the dye. Catalase activity released oxygen bubbles into the solution necessitating lifting the cover slip after about 10 minutes.

In the procedure of SHARADAKOV (1940), four solutions are prepared and stored separately in the cold in the dark: (a) 200 mg of benzidine, disolved in 100 ml of 50% ethanol; (b) 150 mg α-naphthol in 100 ml of 50% ethanol; (c) 250 mg sodium carbonate in 100 ml distilled H_2O; and (d) 0.3% hydrogen peroxide. Solutions a, b and c are pre-mixed in equal volumes and a drop is added to pollen grains sprinkled on a clean microscope slide. One drop of H_2O is then added and the reagents are mixed with the pollen. After 4 minutes a cover slip is applied and the pollen is viewed under the microscope. Dead pollen remains colorless or yellow; live pollen turns red. This reaction has been used on tree pollens but the results often indicate higher percent viability than shown by germination tests (MAURIN and KAUROV, 1956). Benzidine was reintroduced by KING (1960). However, because of nonspecificity (HAUSER and MORRISON, 1964), and other limitations, it generally has not been used for pollen viability assays.

LAKON (1942) used benzidine in seed testing. Because of the volume of chemical required, this technique for testing seeds was a hazard to humans. He therefore introduced triphenyl tetrazolium chloride (TTC) which has since been applied by many workers for determining pollen viability. Table 6-6 lists many of the pollen species to which the TTC test has been applied.

The TTC topographical enzyme reaction has the greatest potential of the methods available for determining pollen viability. We have found that a 1% TTC solution made up with 0.15 M tris-HCl buffer pH 7.8 is stable for about 3 months when stored in a dark bottle at 5° C. All pollens are not equally sensitive to the TTC assay (Table 6-6). Precautions should also be taken against those factors in the test which chemically reduce TTC, giving false positive results. Pre-soaking pollen in phosphate buffer increased uniformity of the test (DIACONU, 1968).

The search for possible alternative chemicals and methods of applying them in pollen viability assays should continue. Other indicator dyes have also been reported as useful in pollen viability tests. The most promising are those which react with specific enzymes present in redox pathways. HAUSER and MORRISON (1964) found nitroblue tetrazolium, which is specific for succinic dehydrogenase, to be a good index of viability in the 13 taxa of angiosperm pollen tests. SINKE et al. (1954) reported that the Nadi reaction, specific for cytochrome oxidase, is also a good index of pollen viability. Resazurin, a test for reductases, has been reported to give excellent correlation with germination in citrus seeds (PLAUT, 1957), and may be useful in pollen viability assays. Correlation of staining with neutral red at pH 4.7 to 7 was used to determine tree pollen viability (MANZOS, 1960), but the dye is too general in its reaction to provide a good assay tool.

A fluorescent reaction characteristic of living pollen employs a 10^{-6} M solution of fluorescein diacetate. After entering the pollen this molecule is probably hydrolyzed by esterase enzymes (Chapter 14) to yield fluorescein which can be easily detected (HESLOP-HARRISON and HESLOP-HARRISON, 1970). Assuming that the functional plasmalemma of the vegetative cell is a primary site for esterase activity and esterase is related to the pollen growth potential, then this microscopic fluorescent test can provide an index of viability. Extension of these methods to other pollen and the correlation of such techniques with germination and percent fertility remains to be done.

Adapting dye techniques for test paper application appears a logical step in the use of indicators for testing pollen viability. Such an enzyme test is used in testing of blood sugar levels and was suggested by BRUCKER (1948) for testing seed viability. In our laboratory we have, with moderate success, used filter paper pre-soaked in TTC, dried in the dark, and stored anaerobically in aluminium foil

Table 6-6. Viability tests of pollen by triphenyl-tetrazolium chloride

Species	Validity of test	Reference
Barbarea vulgaris	accurate	HAUSER and MORRISON (= HM), 1964
Beta spp.	accurate	HECKER, 1963
Capsicum spp.	accurate	KISIMOVA, 1966
Cichorium intybus	accurate	HM, 1964
Claytonia virginica	accurate	HM, 1964
Gossypium hirsutum	accurate	SARVELLA, 1964; ASLAM et al., 1964
Helianthus annus	accurate	HM, 1964
H. gigantes var.	accurate	HM, 1964
H. grosseserratus var.	accurate	HM, 1964
Lilium regale	accurate	HM, 1964
Lycopersicon spp.	accurate	KISIMOVA, 1966
Malus domestica	not accurate	PORLINGH, 1956; OBERLE and WATSON, 1953
Oenothera biennis	accurate	HM, 1964
Pelargonium hortorum	accurate	HM, 1964
Picea spp.	accurate	KOZUBOV, 1965
Pinus longifolia	accurate	COOK and STANLEY (= CS), 1960
P. nigra	accurate	CS, 1960
P. pinaster	accurate	GIORDANO and BONECHI, 1956
P. pinea	accurate	GIORDANO and BONECHI, 1956
P. ponderosa	accurate	CS, 1960
P. resinosa	accurate	CS, 1960
P. sabiniana	accurate	CS, 1960
P. sylvestris	accurate	KOZUBOV, 1965
Prunus. domestica (plum)	accurate	NORTON, 1966
P. domestica (plum)	not accurate	PORLINGH, 1956
P. persica (peach)	not accurate	PORLINGH, 1956; OBERLE and WATSON, 1953
P. amygdala (almond)	not accurate	PORLINGH, 1956
P. avium (cherry)	not accurate	PORLINGH, 1956
Pyrus spp.	not accurate	PORLINGH, 1956; OBERLE and WATSON, 1953
Solanum spp.	accurate	KISIMOVA, 1966
Verbascum blattaria	accurate	HM, 1964
Viburnum prunifolium	accurate	HM, 1964
V. recognitum	accurate	HM, 1964
Vitis spp.	not accurate	OBERLE and WATSON, 1953
Zea mays	accurate	DIACONU, 1961; PYLINEV and DIACONU, 1961; TOMOZEI and SCUMPU, 1964; VIEITEZ, 1952

for testing pollen viability. However, in such experiments it was difficult to prevent reduction of the TTC while impregnated in the paper. Possibly another enzyme-test paper which will afford a valid, convenient and rapid assay of pollen viability will be developed.

Inorganic Acids

Extrusion of a tube from a pollen grain, in the presence of acid, has been called "pseudo-germination" and has been suggested as an index of pollen viability (JOHRI and VASIL, 1961). The mechanism by which weak inorganic acids effect bursting and ejection of protoplasm by the ungerminated pollen grains has been

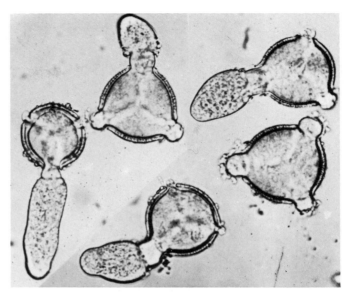

Fig. 6-5. "Instant pollen tubes" from *Petunia hybrida* after treatment with 4% sulfuric acid

attributed to either hydrogen ion absorption creating inward pressure, or rapid swelling of the membrane biocolloids in the acid media (LLOYD and ULEHLA, 1926).

Although SPRENGEL first reported this bursting phenomenon in 1817, it has only received careful study some 150 yrs later (LINSKENS and MULLENEERS, 1967). Because an intact pollen tube can be produced in seconds by addition of an acid, we called the reaction "*Instant pollen tubes*" (Fig. 6-5). KOUL and PALIWAL (1961) tested a series of monocot and dicot pollens and found many genera do not respond to the addition of HCl. LINSKENS and MULLENEERS (1967) showed that only apertured pollen grains respond by tube extrusion. These workers also showed that extrusion of the intine, and formation of the tube, is a function of the length of time which the pollens are exposed to acid at a given temperature. The acid concentration modifies the percent of grains extruding tubes, and the number of ruptured tubes and grains destroyed. A 14.4% concentration of H_2SO_4 produced a maximum number of extruded tubes, while 19.2% H_2SO_4 caused total destruction of the grains. Nitric acid was the most effective of 8 acids tested at a concentration of 0.8 M, while NaOH did not cause pollen tube extrusion.

The number of variables affecting the extrusion of intine and cytoplasm from the pollen, and the limitation of this phenomenon to pollen grains with apertures, obviates the widespread application of the inorganic acid test for determining pollen viability. It is an interesting phenomenon which has potential application in the study of pre-germination enzyme activity, and pollen wall composition.

Comparison of Viability Tests

As noted in Table 6-3, fair to good seed formation may occur even though the results of *in vitro* pollen germination assay indicate low viability. The same probably holds true for the results in *in vitro* nongermination assays. If the test indicates

40% or better viability, the pollen may be quite satisfactory for use in field pollinations. Inert materials and dead pollen are often used as diluents with fresh pollen to lower the percent viable grains in a sample. CALLAHAM and DUFFIELD (1961) tested seed set in pine with controlled pollination over a range from 2 to 100% viable pollen. They obtained maximum seed set with 10–30% viable pollen.

The capacity to form seed and seedlings is the ultimate criteria of pollen viability and is the usual purpose for assaying pollen viability. Thus, different assay methods must be judged for their relative merit and purpose in evaluating the pollen sample. If a rapid enzyme dye method gives satisfactory accuracy for the purpose for which the pollen is to be used, then it is a sufficient assay. If greater accuracy is needed, then alternative methods, up to and including actual counts of seeds formed may be employed.

The ability of pollen to grow, or to respond to a particular *in vivo* or *in vitro* assay is dependent upon the inherent chemistry of the pollen. As previously discussed (Chapter 5) changes in viability of pollen during storage can be related to carbohydrates and organic acids of the pollen. Studies of the chemistry of pollen (Chapter 8–16) provide not only insight into the mechanisms of testing pollen for viability, but also provide an understanding of the metabolic factors facilitating growth and seed formation and a basis for interpreting nutritional experiments involving pollen and animals.

Chapter 7. Nutritive Role

Although man often experiences discomfort on contact with pollen (Chapter 12), in other living organisms pollen is essential for their life cycle. Man has long been a consumer of pollen and pollen containing foods. Pollen products are now being marketed as human food supplements, with various nutritional and health benefits often claimed for the user. Since about 1960, research has shown that pollen is beneficial when incorporated in feed rations of certain farm animals. These newer uses of pollen are distinct from the interactions of insects with pollen. The adaptive and parallel evolution of insects with flowers is the key to survival and reproduction of many plant species. Patterns and mechanisms of insect pollination are well known (FREE, 1970), However, many aspects of how insects and other animals utilize pollen in their nutrition are unknown.

Bees

For many insects, and especially bees, pollen is the principal source of normal non-liquid food. Pollen contains most, if not all, the essential nutrients for production of royal jelly, which nourishes the larval queen and young worker larvae. Pollen is the ultimate source of protein and lipid for larvae and imagines of all species and genera of Apidae. The amount of protein and fat in nectar is insignificant. Older worker bees use protein directly from pollen; queen imagos, larval queens and the young larvae of both sexes receive protein in the royal jelly produced by nurse bees supplied with pollen. Thus, pollen is essential for normal growth and development of individual bees, as well as reproduction of colonies. Bee-collected pollen is stored in the hive in cells of the combs; changes during storage result in formation of bee-bread.

The Pollen Collecting Process

Nectar- and pollen-collecting bees were differentiated as early as 1623 (BUTLER, 1954). However, the importance of pollen for beekeeping was already recognized in ancient times. Virgil in *Georgica IV* recommended the establishment of pollen producing plants near the hive. Although early beekeepers recognized that bees collect pollen, the first detailed description of the methods was given by LANGSTROTH (1863). It was observed that the bee's body is covered with fine hairs to which pollen adheres when the bee visits flowers. The bee brushes pollen from its body with its legs and packs it onto the concave shaped femur known as the "pollen basket".

Packing Pollen Loads

Many observations on the collecting and packing of pollen by bees have been reported (SLADEN, 1912; 1913; CASTEEL, 1912a–b; BELING, 1931; HUBBE, 1954; FREE and WILLIAMS, 1972). HODGES (1952) summarized the whole process in her very readable book. The structure of the different flowers visited determines the way worker bees collect pollen (MEEUSE, 1961). In flowers of *Rosa* and *Tarax-acum* the gathering bee moves quickly among the stamens, often pulling them to itself with its legs and biting at them with the mandibles to dislodge the pollen grains. From other plants, such as sweet lupine, the bee only fortuitously gathers pollen while seeking nectar (BUTLER, 1954). In this case the hairy body is dusted with dry pollen. Scout bees show a high percentage of adhering pollen.

The quantity of pollen carried on the body hairs of honey bees is larger than that on other hairy insects (LUKOSCHUS, 1957). Some bee species, i.e. *Apis*, *Bombus* and *Colletes*, may carry a pollen load of 100–120 mg, equal to one-half their own body weight (CLEMENTS and LONG, 1923). There is a negative correlation between grain size and number of grains adhering. SKREBTZOVA (1957) found on the hairs of one bee between 250,000 and 6,000,000 pollen grains, depending on the pollen source. Distribution on different parts of the body is shown in Table 7-1. Statements of SKREBTZOVA (1957) on amounts of pollen grains on the different parts of the bee body seem improbably high. Three million *Prunus* grains have a volume of about 120 mm^3, which equals a 0.5 mm layer of closely packed pollen on the body. Such pollen layers are observed on the lower side of the abdomen of some Megachilae, but not in the genus *Apis*. Other hairy but solitary living species, such as *Osmia*, *Megachile*, *Anthidium*, *Eriades*, or *Bombus*, transport pollen only by body hairs, while *Apis mellifera*, *Dasypoda*, *Andrena* and *Halictus* are adapted to bring pollen loads home on the legs.

Table 7-1. Distribution of pollen on bees

Plant species visited	Number of grains per individual bee ($\times 10^6$)			
	on the thorax	on the abdomen	on the metathoraic legs	total on the whole body
Ribes grossularia (var. *Schwarzer Negus*)	1.7	0.67	0.44	2.81
Ribes nigrum (var. *Plodorodna*)	0.3	0.18	0.28	0.86
Pyrus malus (var. *Antowka*)	2.0	1.2	0.75	3.95
Prunus cerasus (var. *Wladimirska*)	3.0	1.2	1.25	5.45
Pyrus communis (var. *Ruska Moldawanka*)	1.5	1.0	1.25	3.75
Fragaria vesca (var. *Komsomolka*)	1.0	0.5	1.3	2.8
Rubus idaeus (var. *Gisrid 84/25 × 13/1*)	1.2	0.6	1.4	3.0
Fagopyrum esculentum	1.62	0.85	0.62	3.09

A bee usually performs a regular sequence of movements in gathering and placing pollen grains in the pollen basket. This occurs in two steps (PARKER, 1926) (Fig. 7-1). One: *sitting on a flower*—all pollen grains on the bee's face, appendages, and the first segments of the thorax are collected by means of brushes of stiff hairs on the front legs (Fig. 7-2). The hairs are moistened with a little regurgitated nectar or honey which facilitates packing of the pollen. The back of the head and the other two segments of the thorax are cleaned with pollen brushes on the middle pair of legs, which also take the pollen that has been collected on the brushes of the bee's front legs. The abdomen is cleaned with stiff combs on the hind-legs. Two: *during flight*—the mass on the pollen combs of one hind-leg is raked off by the pollen rake of the other hind-leg into a pollen-press formed by the joint between the tibia and metatarsus (Fig. 7-3). The press is closed and the pasty mass squeezed outwards and upwards, guided by guard hairs and spines, until it reaches the concavity forming the pollen basket. In the same way pollen from the other hind-leg is transferred into the pollen-press of its opposite number and thus to its pollen basket, the corbicula (Fig. 7-4).

A second, third, etc., load is prepared in the same way; each new pellet is pressed upwards from below, against the preceding pressing. During the packing process the bee remains on the wing, often suspended in the air without forward movement (BUTLER, 1954). In the corbicula a single spindle hair pierces through the middle of the pollen load and acts as holding pin (HODGES, 1967). As would be expected, pollen carrying capacity of mutant hairless bees is markedly diminished and absence of the spindle hair reduces the size of pollen load that can be carried in the corbicula (WITHERELL, 1972). Some bees, primarily nectar gatherers, will scrape the pollen from their bodies and discard it (FREE, 1971).

Methods of Packing and Gathering Pollen Vary for Different Plants. PARKER (1926) classified them into the following groups according to flower types:

Open Flowers. The worker bites the anthers with his mandibles and pulls them towards its body with its front-legs while it runs rapidly over the flowers, all the while packing pollen into the baskets. This type of gathering is observed in *Taraxacum, Pyrus, Papaver, Rosa, Fagopyrum, Spiraea, Malus, Ulmus, Sambucus, Acer* and *Ribes.*

Tubular Flowers. The bee alights on the corolla and inserts its proboscis into the tube in search of nectar. Collecting pollen is incidental to nectar gathering. The quantity of pollen obtained is small; it adheres to the mouth parts or fore-legs. In many flowers the corolla is inverted, and in these the bee is upside down when working. *Berberis, Syringa, Lonicera, Symphoricarpos, Medicago, Olea* and *Catalpa* are examples of the tubular type flower.

Closed Flowers. The bee alights on the wing of the flower and separates the petals by forcing its fore-legs between them on either side. The pollen is gathered on the mouth parts and fore-legs and packed in the usual way. Species in this group include *Trifolium, Robinia* and *Acacia.*

Spike or Catkin Flowers. In collecting pollen from spikes or catkins the bee may alight at the base of the lower part of the staminate flower, run up the catkin a short distance, then fly away to pack the gathered pollen, and return to gather more. It may repeat the process several times, all the while working up and around the catkin with the antennae extended toward the anthers. In many cases

Fig. 7-1. Bee collecting pollen. Close-up view on flower

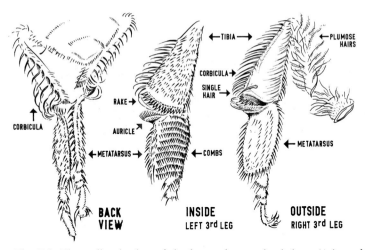

Fig. 7-2. The pollen basket of the honey bee on back legs. (Adapted from Hodges, 1952)

the catkin with the antenna extended toward the anthers. In many cases the bee does not actually touch the catkin, but suspends itself of darting toward the catkin and away from it. Flowers in this group include *Salix, Populus, Pinus, Zea mays, Juglans* and *Quercus.*

Presentation Flowers. Flowers of this type present free pollen to the visiting insects. *Apis bomus,* and many solitary bee species, press the abdomen against the inflorescence, causing a pollen mass to be pushed out of the disc flowers. Tubular

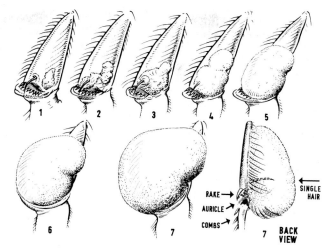

Fig. 7-3. Pollen load formation of the honey bee. From (1) to (7) the increasing amount of collected pollen fills up the bowl of the corbicula. (Adapted from HODGES, 1952)

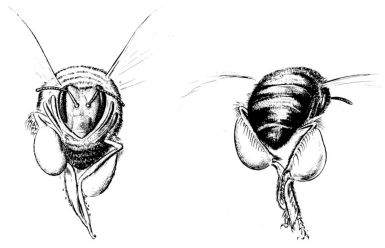

Fig. 7-4. Loading during flight. Left: front view, right: back view. (Adapted from HODGES, 1952)

flowers retract actively so that pollen is exposed. The collecting insect clamps to the corolla and the receptacle scale, and in this way presses its body into the corolla head. This way of collecting is observed on *Echinops* and other Compositae. In *Campanula* the flower also presents free pollen grains. Before the stigma opens, pollen is deposited on collecting hairs on the style surface. The collecting bee runs around the style columella and rakes it together with the hairs of the abdomen.

Application Flowers. This type includes all cases in which pollen is fastened on particular places of the insect body by special mechanisms. Flower construction guides the insect to the right place, resulting in the body contacting the jack, glue, pinch or sling mechanism. The classical example is the flower of *Salvia* where SPRENGEL (1793) described the mechanism. When the bumble bee contacts an arm or lever on the base of the fertile anthers the anther is pressed down on the upper side of the abdomen. Many orchid flowers belong in this category (VAN DER PIJL, 1966). When the right pollinator reaches the proper question in the flower, the pollinia attach themselves to the insect's head. This also occurs in *Scrophulariaceae*, *Labiateae* and *Borraginaceae*.

The *time* needed to collect pollen and return to the hive with a full load is about 10 minutes (PARK, 1928), but may be longer, i.e. up to 169 minutes was reported by KOBEL (1951). The area over which collecting takes place is generally limited to about 12 sq.m. (SINGH, 1950). An annual cycle of pollen storage by bees occurs (Fig. 7-6a). Well-marked peaks are observed in June, July and August. From September to April pollen storage remains relatively constant and low. Except for the rise, or high peak in the pollen storage curve in August, the shape of the curve of quantities of stored pollen is relatively similar to that of the brood reproduction curve. This indicates that the spring increase in brood-rearing is not entirely dependent on the quantity of pollen stored in the hive (JEFFREE and ALLEN, 1957).

Factors Influencing Pollen Collection

Available flora is only important after spring as a limiting factor to pollen collection. Temperature is the most important single factor (Fig. 7-5). No collecting activity can be observed by bees at temperatures below 10° C. However, above 10° C a correlation exists between temperature and the amount of pollen collected (LOUVEAUX, 1958a, b; 1959). Tropical bees collect pollen all year around (IBRAHIM and SELIM, 1962). Following the initial critical 10° C temperature, presence or absence of brood is next in importance in limiting pollen collecting activity. Above the critical temperature, light intenstiy may also be a limiting factor. In the temperature range of 13–21° C, light intensity seems to be most critical. Maximum bee activity occurs after solar radiation exceeds 3,600 calories per square centimeter per hr. On clear days bee flights begin at lower temperatures than on cloudy days.

Each colony shows its own particular behavior with regard to available flora. Some colonies follow complicated pollen collecting programs; others, simple ones. Bees prefer a source of pollen used previously. As expected, there is a correlation between flowering time and the main pollen species collected. This results in graphs which relate flowering time to the main collecting period of a pollen by bees, irrespective of the month of flowering (Fig. 7-6b). Analysis of pollen loads from heterostylic plants, e.g. *Fagopyrum*, *Lythrum* and *Primula* show that bees may visit both flower forms on one flight (DAVYDOVA, 1954). Bees can be selected and bred which have a preference for one flower (MARTIN and McGREGOR 1973).

The level of activity of bees, and when they discover a pollen source play large parts in establishing the collecting pattern. These factors are the basis of constancy

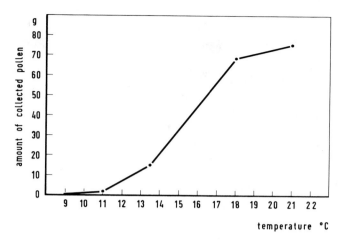

Fig. 7-5. Effect of temperature on pollen collection by bees. (From LOUVEAUX, 1958a)

in pollen collecting by bees (BETTS, 1935) and of the famous fragrance training of beekeepers by which an apiarist learns to detect the flower sources from the honey fragrance. The physical condition of the individual pollen grains and the nutritional value can also determine the selection of pollen by bees (LOUVEAUX, 1955; 1958a; b; 1959). Nitrogen content of pollen has been suggested as important in establishing preferences for certain pollen species. Oils, instead of nectar, are the bee attractants of some flowers. In *Calceolaria* the main oil component is a glyceride of acetoxypalmitic acid (VOGEL, 1971).

Specific chemicals in pollen may serve as attractants for bees. HÜGEL (1962) isolated a mixture of steroids, in particular 24-methylene cholesterol (Chapter 10), which was suggested as the bee attractant. Since bees apparently cannot synthesize this sterol, which constitutes 72% of the sterol fraction in bee larvae (BATTAGLINI et al., 1970), pollen would provide a logical source. A free fatty acid, octadecatrienoic acid, and an ester of the flavone pigment lutein have been isolated from mixed bee collected pollen and found to be specific attractants to honey bees (LEPAGE and BOCH, 1968; HOPKINS et al., 1969; STARRATT and BOCH, 1971). It would be fortuitous if the primary chemical attractant for bees was localized in pollen, not just in nectar odor or flower color, as is generally assumed. DOULL and STANDIFER (1970) suggested that some volatile chemicals from pollen act as activating "release" stimulants. Such compounds stimulate the hypopharyngeal glands in nurse bees resulting in further uptake of pollen to feed the larvae. Pollens differ in their attractiveness and capacity to supply nutrients. In all cases tested the bees selected mixtures of pollen plus sugar solutions over sugar solutions alone (DOULL and STANDIFER, 1970; STANDIFER et al., 1973). Field applications of specific chemical attractants have many limitations. When the flower-and bee-derived chemicals, geraniol and nerolic acid were sprayed on flowers to improve pollination, the bees spent their time collecting the applied attractant from foliage instead of gathering pollen (ANDERSON and ATKINS, 1968).

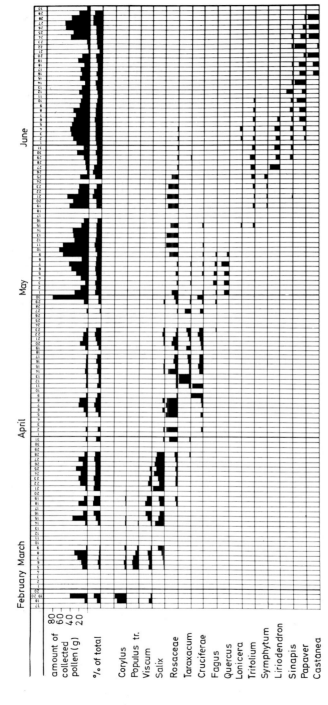

Fig. 7-6a. Annual cycle of pollen storage by bees (JEFFREE and ALLEN, 1957)

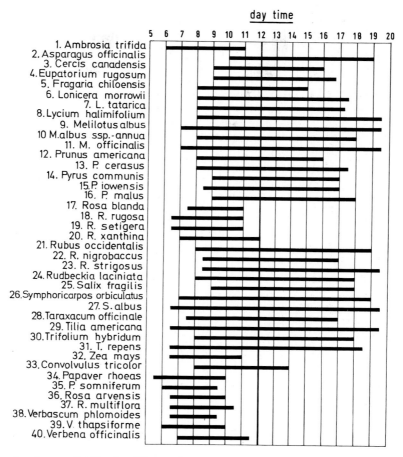

Fig. 7-6b. Time of pollen availability by different flowers. (1–32 from PARKER, 1926; 33–40 from KLEBER, 1935)

The Pollen Pellet

Size and weight of pollen basket loads collected by bees varies widely. The average weight of a load is about 7.5 mg (VON BERLEPSCH, 1864; HIRSCHFELDER, 1950; 1952; MAURIZIO, 1953) but may be as much as 15 mg (HIRSCHFELDER, 1952). Differences in the average weight of loads are related to the plant source. Large differences were found in the amount of pollen gathered from different plant sources (MAURIZIO, 1953) and also in the quantities collected by neighbouring colonies in the same apiary (SYNGE, 1947). Clover species generally supply small and light pollen loads. After a short time of collecting, a relatively constant weight is attained for the load of pollen collected by the bees. Specific gravity of entomophilous pollen is generally greater than one. Pollen pellets, in which hygroscopic pollen viscin cement is covered with secretions of the bee, sink in water.

A very large amount of pollen can be collected by a colony in one day. HIRSCHFELDER (1950; 1952) found that in July about 250 gm of pollen, requiring some 17,000 flights, were collected between 8 and 10 A.M. The total amount of pollen collected by one hive in a season varies between 15 and 40 kg (WEIPPL, 1928). This represents between 740 gm and 2,000 gm of pollen a day in the most active periods. Differences exist between individual colonies in regard to quantity and origin of pollen collected. ECKERT (1942) calculated the pollen demand of one colony to be about 50 kg a year; other apiarists consider this estimate too low. It has been computed that bees in the U.S. annually collect 80,000 tons of pollen, an amount comparable to the total weight of honey produced by these colonies (TODD and BRETHERICK, 1942; SAKATA et al., 1961).

It should be mentioned that pollen collection and honey production can be compared only if total honey production includes consumption by the colony plus the surplus taken away by the apiarist. Consumption of a normal hive can be calculated, in temperate climates, as about 35–60 kg honey per season.

Time of Collecting

The time pollen is collected varies with different species and depends primarily on when the flowers are open. Three categories of flower-opening are recognized: morning, afternoon and whole-day (MAURIZIO, 1953). PERCIVAL (1947; 1950; 1955) distinguished 8 time classes for flower opening which were correlated with bee activity. Significant differences occur among some plant groups and among the species comprising them. While the hourly distribution of pollen collected from the same plant species is approximately the same in different localities, it reflects the influences of weather on flowering (LOUVEAUX, 1955) (Fig. 7-6).

Daily flower opening and pollen collecting are correlated in some species. Generally, about a two-hour lag occurs in the maximum collecting activity of the bees and the time of maximum pollen availability (Fig. 7-7). The daily rhythm of pollen collecting by bees is reasonably well related to pollen availability of individual species (PERCIVAL, 1950; 1955).

Color of Pollen Loads

When bees visit many flowers of one species (monotropic harvesting) the color of the pollen load is uniform. Occasionally a bee collects pollen from more than one species (polytropic harvesting). This results in a "color mixed" pollen load (HODGES, 1952). Only about 0.1% of all pollen loads are mixed (SCHWAN and MARTINOVS, 1954). Two types of mixed loads are found, the "M" and the "S" types (BETTS, 1935). In the "M" type grains from two or more species are mixed in the corbicula and result in a uniform, and sometimes even a new, color. In the "S" or segregated loads, pollen is packed in different colors, resulting in striped loads. In mixed loads a distinction can be made between true mixtures of two or more species of pollen, and false mixtures of one pollen source with some other constituent, i.e. fungal spores, algae, or other foreign materials (MAURIZIO, 1953).

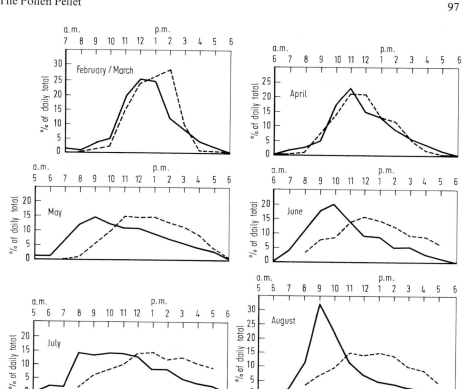

Fig. 7-7. Relation between availability of pollen, i.e. presentation time, and bee collection pattern in various months. ———— Flower opening. ----- Percent of pollen collected daily. (Adapted from PERCIVAL, 1955)

The plant source can often be identified from the color of the pollen loads (MAURIZIO, 1942; PERCIVAL, 1947; HODGES, 1952). Easy evaluation of pollen loads has been made possible by the development of traps which remove the pollen from the bees as it enters the hive. Figure 7-8 illustrates the arrangement of pollen collecting traps in relation to the hive entrance. The collecting trays where the pollen loads accumulate are removed for analysis of the pollen.

Excellent color charts of the species collected at the different seasons have been prepared by HODGES (1952). These charts suggest that many causes other than plant source can induce color variations. Dark loads may be expected in the early morning, after rain or frost, or at the start or end of the flowering period. Light colored loads prevail in sunny weather, when flowers have just opened, the pollen has just been released, or when bees work in large numbers on the same crop. Some color changes are characteristic of the pollen species (REITER, 1947); others can result from exposure to sunlight, moisture content of the pollen at dehiscence, and contamination with dust, soot or dark fungal spores. Variations can also be brought about by the kind and amount of liquid, sugar, or nectar mixed with pollen by the bee in making the pellet. These substances can also alter

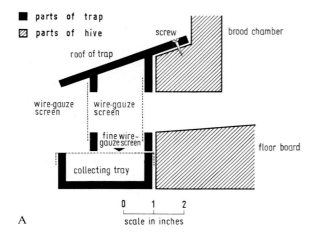

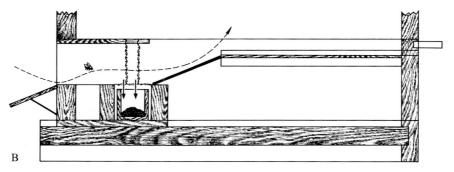

Fig. 7-8 A and B. Traps for removing bee pollen loads. (A) Cross section of pollen trap in position on national hive. (B) Cross section of pollen trap on the bottom of the hive

the chemical nature of the pollen cement. The color of pollen grains on the anthers of individual plants of the same species may also differ. A classical instance of the latter variation is described in *Lythrum salicaria* by SCHOCH-BODMER 1927; 1940).

Storage of Pollen in the Hive

Adult bees store pollen primarily to consume in their own diets and to feed the larvae. On returning to the hive the worker bee seeks out an empty or partly filled cell. The hind-legs are dangled into the selected cell and loads are stripped off by the middle-legs. The pollen-gathering worker-bee has now finished its duty. Another bee, generally a younger one, comes, breaks the loads and stamps them down firmly into the bottom of the cell. A small cover of honey is deposited on the pollen to prevent spoiling. This store of pollen is called "bee-bread" by the beekeeper. It requires about 18 loads to fill one cell (VON BERLEPSCH, 1864). Because different pollen loads are mixed in storing the pollen, the filling of the cells is stratified.

Changes in Pollen Brought into the Hive

Germination Capacity

Fresh pollen, collected directly from the corbicula of the returning bee, can generally germinate. In some cases the germination percentage remains the same as at
anther dehiscence; in other cases germination capacity decreases during the collecting process (FERNHOLZ and HINES, 1942; KREMER, 1949; JOHANSEN, 1956).
Stored pollen becomes nonviable in 1 to 8 days, depending on the species (MAURI-
ZIO, 1944; SINGH and BOYNTON, 1949; GOLUBINSKII, 1959). Pollen grains which
remain on the body of the bee longer than 12 hrs cannot germinate (KRAAL, 1962).
    A phytocide-like substance is probably secreted by bees to moisten the pollen
while packing it in the comb. Besides the sugar in the honey and nectar, an acid
material is present in the honeycomb cells during pollen packing; the acid is
probably responsible for the rapid loss of germination capacity (MAURIZIO, 1944;
1958a; 1959). Other chemical factors have also been suggested as responsible for
the loss in pollen viability. A specific inhibitory, thermostable factor is present in
the pharyngeal glands and in alcoholic extracts of bees, in larval foods of both
workers and queens, and in honey (CHAUVIN and LAVIE, 1956; LAVIE, 1958). The
chemical nature of the bacteriocidal and phytocidal substance is still unknown
(LAVIE, 1960; PAIN and MAUGENET, 1966).
    Experiments with honey bees, *Apis mellifera*, and the bumble bee, *Bombus
hypnorum*, (KEULARTS and LINSKENS, 1968) showed that the mandibular glands
produce a germination-inhibiting factor. This substance can be extracted from the
pellets or stored pollen by water, as well as by ethanol and diethylether; the
activity is not destroyed by heating the extract 1 hr at 100° C. Inhibition is irreversible and is also effective on germinating pollen. Comparative chromatographic
investigation suggested that the inhibition is linked to 10-hydroxy-2-decenoic acid
(*trans* isomer) which strongly inhibits pollen respiration. This inhibitory fatty
acid, present in the mandibular glands, after application to the stigma fluid of
Petunia, has no effect on pollen germination *in situ*. Therefore, it appears that the
inhibiting substance used by bees to preserve pollen in storage, is inactivated on
the stigma, or broken down or polymerized into glycerides of the stigmatic fluid.

Fermentation of Stored Pollen

Stored pollen undergoes further biochemical changes. LANGER, as early as 1915,
found that bee bread pollen increases in percent acidity from 0.26 to 1.78% and in
water-soluble protein from 2.9 to 5.6%. A lactic acid-type metabolism occurs,
indicating a lowering of oxygen tension; this probably contributes to the stability
and conservation of pollen as a food source (SVOBODA, 1935; PETERKA, 1939;
HEJTMANEK, 1943).
    During packing of the pollen loads sucrose inversion is initiated. Within
24 hrs, honey bees may invert 50% of the pollen sucrose (OKADA et al., 1968).
Stored pollen, therefore, has an increased content of reducing sugar. Stored pollen
also appears to have a high content of histamine, assumed to be produced from
histidine by bacteria present with the pollen (MARQUARDT and VOGG, 1952).

Stored pollen generally has a specific bacterial flora associated with (BURRI, 1947). Lactose digesting enzyme systems (HITCHCOCK, 1956) and a high content of vitamin K (HAYDAK and VIVINO, 1950; VIVINO and PALMER, 1944) are also present. This suggests that microorganisms are probably involved in the metabolism of stored pollen. Also, the enzyme-rich honey added by bees to stored pollen, directly influences anaerobic metabolism and fermentation before and during storage in the hive. Bee digestive tracts have a rich bacterial flora (GILLIAM and PREST, 1972). Further research on these subjects is of importance in evaluating the nutritional value of pollen to bees.

Some plant virus diseases are transmitted by pollen (GOLD et al., 1954; CAMARGO et al., 1969). Thus, bees collecting or ingesting pollen which contain viruses, presumably act as secondary vectors transmitting the virus. This has not yet been explored beyound the recognition that pollen can transmit plant virus diseases. Many interesting problems relating to when the virus particles multiply, and their affect on pollen or bee metabolism, must still be solved.

Digestion of Pollen

Relatively little is known about the mechanism of pollen digestion in bees. PARKER (1926) studied the digestion in larvae; DIETZ (1969) followed pollen movement in the digestive tract of young honey bees and WHITCOMB and WILSON (1929) concentrated on pollen metabolism in adult bees. Bees manipulate pollen masses, mixed with nectar, with their mandibles. Some authors infer this to be a grinding function to crack the hard coats of the pollen grains (SNODGRASS, 1925; PARKER, 1926). On the other hand, WHITCOMB and WILSON (1929) never found cracked pollen grains in the honey stomach; only 1% of the grains in the intestine showed signs of mechanical cracking. The true function of the mouth parts in relation to pollen feeding seems to be passive and practically the same as in the intake of liquids.

Material is mechanically carried from the mouth to the honey stomach by a slight peristaltic action along the esophagus. Pollen seldom stays in the honey stomach longer than 20 minutes; from there it is rapidly transferred by the honey stopper to the mid-intestine (Fig. 7-9).

It is surprising that in passing pollen from the honey stomach into the ventriculus (mid-intestine), very little of any other material, i.e. syrup or nectar, is removed from the honey stomach. The action of the honey stopper is one of straining rather than of crushing (BAILEY, 1954). It has 4 rapidly moving lips which open and close to collect pollen and material from the honey stomach into a bolus (BARKER and LEHNER, 1972).

The ventriculus is the most important part of the alimentary canal. Pollen passing into the ventriculus is directed by the pro-ventriculus valve. At this point the bolus of pollen rests well within the innermost of the peritrophic membranes. In the honey bee, which consumes only two types of food, pollen and honey, the peritropic membrane is important in pollen digestion. Electron microscopic observations show that the peritropic membrane has a fine network-like structure (MERCER and DAY, 1952). The form and function of this membrane changes with

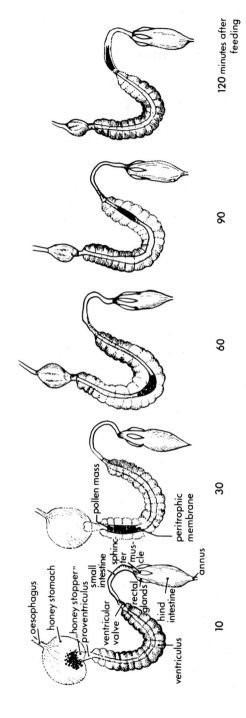

Fig. 7-9. Transport of pollen through bee digestive tract. (Adapted from WHITCOMB and WILSON, 1929)

the type of food (SCHREIBER, 1956). It tightens about the food mass, contracting mostly at the ends so that it forms a tube about the pollen mixture creating a sausage-like package. The surrounding peritrophic membrane also protects the bee digestive tract from the sharp spines and abrasive exine surface of pollen in the bolus (BARKER and LEHNER, 1972). At about 20° C pollen remains in the ventriculus 2 to 3 times longer than at 35° C (SCHREINER, 1952; HEJTMANEK, 1961). Usually 1 to 3 hrs are necessary for the passage of a pollen mass through the ventriculus. Food masses, the bolus surounded by the peritrophic membranes, enter the hind intestine about $2\frac{1}{2}$ hrs after ingestion by the bees. Here the pollen residue is stored until discharged in flight.

Pollen grains undergo physical changes during passage through the alimentary canal. The cellulose, pectic and sporopollenin coat is generally indigestible by insects. Therefore, extraction of nutrients must take place through the cell walls or through ruptures in those walls. Crushing of the grains is not necessary for their complete digestion. In the honey stomach few changes occur in the pollen, but there is a tendency toward a germination-like swelling of the apical caps in the pollen pore area. Digestion begins minutes after the pollen reaches the ventriculus. Within ten minutes after reaching the ventriculus the cell contents begin to pull away from the pollen wall leaving a transparent vacuole. Thirty minutes after ingestion the pollen shows a definite increase in vacuole size and some of the membrane stainable materials disappear.

Protein-degesting enzymes and lipases are found in the ventriculus (PAVLOVSKY and ZARIN, 1922). Digestion progresses rapidly and removes most of the cell contents. By the time the pollen grains are in the hind intestine they are almost devoid of cell content, colorless, and many have collapsed. The digestion process seems to be equally active in the adult and in the bee larvae (PARKER, 1926). Enzyme activity and digestive processes are highest in the front midgut, with absorption of metabolites highest in the lower end of the ventriculus (ZHEREBKIN, 1967).

These observations all confirm the concept that bees lack enzymes to break down pollen wall components. The digestion of proteins, lipids, and carbohydrates takes place through the pollen germination pores. MAURIZIO (Pers. communication) indicated that in many cases the external structure of the grains, especially the exine, remains unchanged.

Nutritive Value of Pollen for the Honey Bee

Pollen is the most important requirement for bee growth. It is particularly essential in feeding broods. During the early adult life of worker bees all nitrogen is derived from pollen protein. Physiologically young bees, nurse bees, are characterized by high percent protein content. Adding protein-containing food, pollen, to their diet substantially increases their longevity (DE GROOT, 1953). Bees fed a pollen free diet are unable to produce venom. Development of ovaries in bees is significantly enhanced by pollen in the diets (KROPACOVA et al., 1968). Egg laying capacity of the queen bee is initiated earlier and extended later by supplemental feeding of pollen to a colony (SHEESLEY and PODUSKA, 1969).

Stored quantities of bee-bread are available in the hive. Thus the suggestions of SIMPSON (1955) that larvae can develop normally without pollen and of JEFFREE

(1956) that bees overwinter practically without pollen are probably incorrect. Formation of fat bodies in the winter bees depends on pollen feeding during fall. The lipid fraction from pollen seems to serve as as source for buildup of reserve fat and glycogen for use during periods of food scarcity. The rich spectrum of fatty acids and the lipid constituents, which can comprise up to 20% of the bee weight, cannot be overlooked (STANDIFER, 1966).

While the royal jelly fed to larvae contains considerable amounts of carbohydrates, primarily fructose, glucose and sucrose derived by bees from nectar (CHRISTENSEN, 1962), royal jelly also contains considerable lipid material (PATEL et al., 1961) and the acid, 10-hydroxy-2-decenoic, synthesized from sucrose (BROWN et al., 1962). Presumably, pollen sucrose as well as nectar sucrose is metabolized by nurse bees to this acid and other nutritional substrates in the royal jelly.

Requirement for Pollen

Young adult bees require pollen within the first two weeks of life. Removal of pollen by pollen traps generally has no marked influence on the hive development, as long as all the protein supply is not disrupted. Setting pollen traps at a hive entrance (Fig. 7-8) induces a protein shortage in the hive which results in intensified pollen collecting activity by the bees at the expense of honey and nectar collecting (HIRSCHFELDER, 1950; VAN LOERE, 1971). Traps increased pollen collecting in Ukraine hives by about 70% (CARLILE, 1971). RYBAKOV (1961) observed an increase in the honey crop as a result of the pollen traps, increasing the number of bees in the hive collecting pollen. Pollen traps have long been advocated as a beneficial way to build up bee colony activity (SCHAEFER and FARRAR, 1941).

Feeding experiments show an average requirement of 145 mg of pollen for rearing one worker bee. More than 10,000 bees can develop normally on 1.5 kg of pollen (ALFONSUS, 1933). The nutritive value of artificially stored pollen depends on drying conditions (MAURIZIO, 1958b; HAYDAK, 1961), temperature and length of time of storage (HAYDAK, 1960; 1961), and the plant source of the pollen. These factors all indicate that chemical composition is the key which determines usefulness of pollen in bee nutrition (LOUVEAUX, 1963).

Pollen storage in the hive and colony size are inter-related because pollen supply and colony size independently influence brood rearing capacity (ALLEN and JEFFREE, 1956). During extended storage, pollen probably loses some of its nutritive value for growth of newly emerged bees. This may be correlated to diminishing digestible proteins in bee-bread with age. But vitamin content also declines; experiments with devitaminized food show that this can be a limiting factor to bee development (WAHL and BACK, 1955; SERIAN-BACK, 1961).

Nutritive Value of Different Pollens

All pollen species do not have the same nutritive value for bees. Three different pollen food types can be distinguished (MAURIZIO, 1950; 1955; WAHL, 1956; 1963):

a) Excellent: *Crocus, Salix, Papaver, Trifolium, Castanea, Raphanus, Sinapis, Erica* species and pollen from fruit trees are in this class.

b) Good: includes *Taraxacum*, *Ulmus* and some other anemophilous pollen. Very probably pollen of *Taraxacum* and some other Compositae are relatively high in nutritive value due to the high amount of pollen cement fastened to them.

c) Poor: this class includes pollen from *Corylus*, *Alnus*, *Betula* and *Populus*. The coniferous trees, *Pinus*, *Picea*, *Abies*, *Cedrus*, are especially bad.

The best feeding results are obtained with pollen mixtures which combine excellent with less nutritive pollen. Bee-collected pollen has a higher nutritive value than random hand-collected material.

The differences in nutritive value of pollen for bees is not well understood. The nitrogen content, particularly the amount of soluble protein (Chapter 11), may be the primary factor. The relation: entomophilic pollen = high nutritive value, anemophilic pollen = poor nutritive value is striking.

In comparing pollen chemistry with the nutritive value of pollen selected by bees, it is assumed that no differences exist in the capacity of bees to extract the chemical nutrients from the different pollen species. In fact, this may not be the case. Practically nothing is known about how bees extract the pollen contents. Conceivably, bees may have enzymes to break certain pollen grains selectively but not others, or internal solubilization of pollen contents and nutritive value may be related to the physical or chemical structure of the exine and the number of pores.

Toxic Substances in Pollen

Only rarely are substances toxic to bees found in pollen. Because of the content of anemonine, pollen from various *Ranunculus* species can be toxic for bees. Pollen of *Aesculus* and *Tilia* is toxic due to saponin content (MAURIZIO, 1945a; b). Furthermore, toxic effects are known from pollen of *Rhododendron*, *Andromeda*, *Corynocarpus*, *Scolypoda*, *Fagopyrum* (after drying only), *Polygonum bistorta* and *Hyoscyamus*. In *Hyoscyamus* the toxic substance is an alkaloid (SHAGINYAN, 1956). But the specific active chemical in this, as in most other pollens, is yet to be determined. There are many reports on highly toxic pollen, which are often immediately lethal to bees. Large scale deaths of bees foraging on species of *Astragalus* and *Veratrum* were observed (VANSELL and WATKINS, 1933; 1934). Pollen of *Asclepias* species contains highly toxic galitoxins (PRYCE-JONES, 1944).

One should carefully distinguish these pollen chemical effects from toxic honey dew secreted by certain aphids or poisonous nectar. Honey from *Arbutus unedo* has a toxic ingredient arbutin, a glucoside containing hydroquinone (SANNA, 1931). These are not related to the pollen constituents. Also, pesticide sprays can seriously impede pollen and nectar gathering and the survival of a bee colony (TODD and REED, 1969). Many dead bees around *Sophora japonica* flowers in Germany were due not to poisonous pollen as initially deduced, but to paralyzing attacks by the bee wolf, *Philanthus triangulum*, which robs the bee of nectar it carries (LØKEN, 1958).

Although pollen of specific plants may be highly poisonous to bees, there is only a slight chance of contamination of the stored honey harming the bees. But while bees may be immune, the honey stored may nevertheless be dangerous for human consumption. For example, in India *Lasiosiphon eriocephalus* provides a major pollen source for bees over a long period. Honey heavily contaminated with *Lasiosiphon* pollen will cause severe nausea and vomiting (CHAUBAL and DEODI-KAR, 1963).

Pollen Substitutes

Apiarists must frequently supply pollen or bee-bread substitutes artificially for the nourishment of their colonies. Natural pollen mixtures are more satisfactory than artificial protein supplements or foods. The less nutritious effect of inferior pollen is vastly compensated for in pollen mixtures. The biological effectiveness of artificially collected pollen is generally not substantially influenced by the drying conditions. But pollen prepared by freeze-drying has less value than fresh pollen as a life extender and for physiological conditioning of bees (MAURIZIO, 1958b). To some extent food nutrients from other sources can be used in place of pollen as a protein source for bees.

Mortality of bees fed various pollen substitutes varies greatly. HAYDAK (1933; 1939) found the highest mortality, 52%, when rye flour was substituted for pollen; the lowest, 15.5%, was in colonies fed dried yeast. This was confirmed by LEVIN and KNOWLTON (1951) and WAHL (1963). Including protein substances, even of poor value, increases the effectiveness of the pollen substitutes (WAHL, 1963). The use of fresh milk as a satisfactory substitute for pollen is recommended (HAYDAK, 1933). Both young and old bees respond by increased longevity to inclusion of proteincontaining foods in their diets. This suggests they need protein not only for growth and secretion of royal jelly, but also for maintenance of metabolism. It has been observed that wax production, resistance against *Nosema* infection, and successful wintering depend on provisions of pollen.

The bee requires protein primarily as a source of essential amino acids. The quantitative amino acid requirements for bees are in good agreement with those of other animals. The effective use of unnatural *d*-isomers by bees is less than those in other animals (DE GROOT, 1953). Yet a mixture of 19 amino acids compares poorly with casein and pollen in isonitrogenous amounts (Fig. 7-10). The pollen diet induces a higher growth rate than the amino acid and casein diets. The suggestion that free proline, present in a high level in some pollen (Chapter 11), is essential for bees has been fairly conclusively disproven by feeding experiments (BARKER, 1972). However, a possible source of error in the feeding experiments is that the average daily food consumption on the pollen diet was more than twice that on the amino acid diet (DE GROOT, 1953). Consumption is always stimulated by the inclusion of pollen in artificial diets (STANDIFER et al., 1973).

The question of whether or not bee larvae directly utilize pollen is still not clearly answered. The importance of the vitamin content of pollen with regard to normal larval development on bee-bread should not be overlooked (WAHL, 1954; 1963). Thus, vitamin content may explain why *Torula* yeast in mixture with soybean flour, despite a poor lysine and leucine content, has a good effect on bee growth and longevity (WAHL, 1963; SERIAN-BACK, 1961).

An artificial diet with gibberellic acid replacing all pollen in the diet allowed the rearing of bees from larvae through more than one generation (NATION and

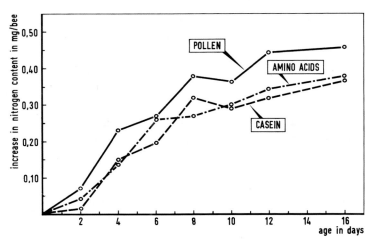

Fig. 7-10. Comparison of sources of nitrogen in diets for rearing bees. O ———— O A mixture of 19 amino acids. O ———— O Vitamin-free casein. O ———— O Pollen. (Adapted from DeGroot, 1953)

Robinson, 1966). When large amounts of inositol were substituted for gibberellic acid in an artificial diet, worker bees were able to rear the brood (Nation and Robinson, 1966; 1968). Inositol commonly occurs at fairly high levels in pollen (Chapter 8). Gibberellins are detected at trace levels (Chapter 16).

Pollen in Analysis of Honey

Delineation of the pollen species in honey is an important tool for control of honey production (Vorwohl, 1968). It determines what neutral components are present in the flora, i.e. bee pasture, from which the pollen material is collected by the bees (Martens et al., 1964). Information on the pollen content also provides guidelines for handling of the honey by the apiarist. The number of pollen grains in honey depends on the flower construction and especially on whether the pollen reaches the nectaries.

Methods of pollen analysis were defined in 1952 and 1964 by the International Commission of Bee Botany of the International Union of Biological Sciences. Standardization of the presentation of pollen analysis results is necessary (Maurizio and Louveaux, 1965). There are three principal methods of pollen analysis used in honey control.

1. *Determination of the Absolute Pollen Content* (Hammer et al., 1948; Maurizio, 1949; 1955; 1958a; Demianowicz and Demianowicz 1957). These extensive observations are summarized in Table 7-2.

2. *Absolute Pollen Frequency.* Large fluctuations in absolute pollen content, as well as in the percentage of certain pollen species in the honey, are caused by differences in the intensity of nectar coming into the colony and by secondary contamination during extraction (Demianowicz et al., 1966). During harvesting of honey, pollen can contaminate the honey. This is often observed when combs

Table 7-2. Quantity of pollen in honey relative to species

Pollen grains in 10 gm honey	Under 20,000	About 50,000	Above 100,000
Species	Lavandula	Citrus aurantium	Brassica napus
	Medicago	Erica carnea	Calluna
	Onobrychis	Filipendula	Castanea
	Robinia	Heracleum	Centaurea cyanus
	Salvia	Labiatae	Cynoglossum
	Sinapis	Prunus	Fagopyrum
	Tilia	Pyrus	Myosotis
	Trifolium	Rubus	Phacelia
		Salix	
		Taraxacum	
		Trifolium repens	

are pressed out. Nevertheless, the absolute number of pollen grains in samples may yield valuable information about the environment at collection time if assay errors are avoided (DAVIS, 1966).

3. *Pollen Spectrum.* The pollen analysis of honey provides qualitative information about the forage plants at the collecting site, and depends on the soil type and seasonal flowering pattern (PRITSCH, 1958). Adjacent colonies in a bee hive do not always produce honey with an identical spectrum (MAURIZIO, 1962). The difficulty in interpreting the observation that different amounts of pollen from different plants are found in honey can be overcome by use of an arithmetic correction, the Pollen Coefficient.

Pollen Coefficient. This concept was developed primarily by HAZSLINSZKY (1955), DEMIANOWICZ and JABLONSKI (1959) and DEMIANOWICZ (1961). It is based on the fact that nectar-gathering bees collect, with each droplet of nectar, a certain number of pollen grains from the flower. To determine the pollen coefficient, the average value of different pollen grains in four one gram samples of honey is ascertained. Pollen coefficients have been divided into 18 classes, so that each successive class has an average pollen content per gram of honey twice as high as the class below it.

In this method the coefficients of unifloral honey must first be determined. This coefficient can also be used to determine the origin of pollen poor honey. Most of the honey examined are in classes 2, 5 and 9, equal to about the mid-range of pollen in Table 7-2. Class 18 contained only *Myosotis* honey (DEMIANOWICZ, 1961; 1964). Errors may arise in determining the pollen coefficient. Pollen counts may not be accurate when made on pollen from reused combs from which all pollen is not removed between successive uses of the comb.

Other substances can also be used for identifying honey, e.g. crystals of calcium oxalate, which may become mixed with pollen as the anthers open, can be used. Such crystals have been used for characterizing *Tilia* honey (DEMIANOWICZ, 1963).

Plants Supplying Pollen. Generally plants supply bees with both pollen and nectar for honey production at the same time. This agrees with the fact that most of the working bees collect both plant products during the same flight. But within

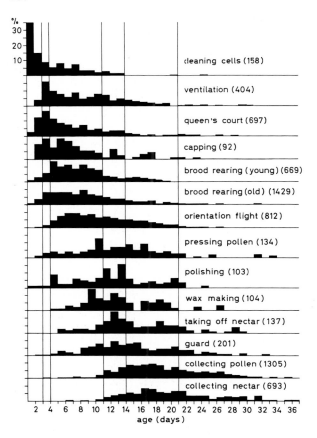

Fig. 7-11. Work distribution of pollen collection by bees in relation to the hive population

the hive there is a work distribution (Fig. 7-11): some bees collect only pollen or visit plants in the afternoon when no nectar is presented. These are generally older individuals, while nectar collecting bees are commonly younger ones. PARKER (1926) reports that of the collecting bees about 58% collect only nectar, 25% only pollen and 17% both pollen and nectar.

Plants supplying only pollen are generally those which are specialized for wind pollination, i.e. *Rosa canina* and other rose species, *Cirsium lanceolatum*. Other plants supply bees only with nectar, i.e. *Marrubium vulgare, Nepeta cataria, Salvia officinalis, Ranunculus repens, Rhamnus cathartica,* and *Veronica spicata*. The seasonal distribution of plants observed to be main pollen suppliers in one area are listed in Table 7-3 (WEDMORE, 1932; ZANDER, 1935).

Other Animals

While bees are the principal and most thoroughly studied collectors, consumers and transmitters of pollen, other animals are known to ingest pollen. Insects, birds and bats frequently have a role in pollination and some consume pollen.

Table 7-3. Main bee pollen supplying plants in Central Europe

Species	Flower-Period
Corylus avellana	February-March
Crocus vernus	February-March
Salix repens	March-April
Salix viminalis	March-April
Salix capraea	March-May
Taraxacum officinale	March-September
Prunus cerasus	April-May
Pyrus malus	April-May
Taxus baccata	April-May
Brassica oleracea	May
Picea excelsa	May
Pinus sylvestris	May
Acer pseudoplatanus	May-June
Genista tinctoria	May-June
Robinia pseudoacacia	May-July
Centaurea cyanus	June-July
Fagopyrum esculentum	June-July
Ambrosia artemisifolia	June-September
Reseda odorata	July-October
Solidago virgaurea	July-October
Aster spp.	August-November
Hedera helix	August-November

Insects

Bees are the most common pollination vectors in the north temperate areas of Europe. Early studies of floral pollination mechanisms were extensively summarized by KNUTH (1906). Other insects, in particular, beetles, flies and moths may transmit pollen incidentally during their quest for nectar of floral tissues. The patterns of pollen dispersal mechanisms and the parallel evolution of some independent plant and insect species are given in detail by MEEUSE (1961), FAEGRI and VAN DER PIJL (1966) and BAKER and HURD (1968). These patterns may be particularly important for plant breeders attempting to introduce a crop species into an area which may lack the natural pollinator for the introduced species. However, our emphasis will be upon a few of the well documented cases where pollen is utilized in insect diets.

Beetles. Some beetles collect and consume pollen of *Macrozamia tridentata, Zamia* and *Encephalartis* (FAEGRI and VAN DER PIJL, 1966). Beetles and some primitive Lepidoptera utilize pollen by chewing; beetles crush the pollen in their jaws and mix it with nectar. Beetle pollination (cantharophily) and pollen ingestion occurs primarily in flowers of plants growing in semiarid regions, commonly found in the southwestern United States. Beetles and other anthophilous insects usually feed on flower parts by eating the soft tissue of the flower at the base of the anthers and pistil where pollen may be lying, or chewing and ingesting intact stamens with anthers. The flower food is used to feed beetle larvae of the species *Heloides* and *Meligethes*. In the Phlox *(Polemoniaceae)* family, Melyridae beetles were among 6 other insects active as pollinators in some of the 122 species observed (GRANT and GRANT, 1965). The beetles did not appear to harm these flowers, nor were they particularly active pollen eaters. Many mature beetles seen on flowers, e.g. of Cucurbitae, are often more destructive to the flowers than beneficial as pollinators (MICHELBACHER et al., 1964). In *Victoria amazonica* the beetles crawl in the open flower at night, then the flower closes, the anthers dehisce and the beetles are covered with pollen. They consume some pollen along with the succulent floral tissue and then move on to another flower the next night.

Whether the eating of pollen is obligatory or facultative, and at what stage pollen is consumed are the primary factors determining the degree of damage done by the insect. Pollen beetles *(Meligethes aeneus)* may be numerous in Cruciferae flowers but it is the larvae of the cabbage seed weevil *Ceuthonhynchus assimilis* which feeds on the anthers and flowers of many Cruciferous plants (EDWARDS and HEATH, 1964).

About 23 families of beetles are known to feed on pollen. Some, as *Oxacis subfusca*, are highly specific for the pollen source they ingest. This latter species is restricted to ecotypes of poppy flowers in which they develop, and the female appears to require the pollen to produce eggs (ARNETT, 1963). All *Oedemerid* beetles are obligatory pollen feeders and have a special intestine for storing pollen (ARNETT Pers. communication). The larvae of *Celetes* and *Phytotribus* feed on the inside of spathes in the tropical palms *Asterogyne*, while the adults feed on pollen (SCHMID, 1970).

Mosquitoes in the Alaska-Canada region often feed on pollen (HØPLA, 1965). These mosquitoes frequently go through their life cycle without a blood meal and it has been suggested that pollen protein may substitute for mammalian blood protein. Several species of mosquitoes, which actively search and consume nectar, are known to transmit pollen (HOCKING et al., 1950; THEIN, 1969). However, a definitive role for pollen in the mosquito diets is still not proven.

When **insect larvae** are obligatory feeders on pollen, then the nutritional role is easily adduced. The slash pine sawflies and related species of *Xyelidae* isolated on pine catkins, in particular those of *Pinus palustris*, are highly damaging to male cones, from which they consume developing microspores (EBEL, 1966). *Leptoglossus occidentalis*, a seed cone insect in western North American pines, will also consume developing pollen in staminate cones (KRUGMAN, Pers. communication). Thrips *(Gnophothrips fucus)* will eat pollen in catkins non-specifically along with pine female strobili (DEBARR, 1969). The larvae of the apple blossom weevil, *Anthonomous pomorum*, destructively feed on anthers and other parts of the flower blossom (KNUTH, 1906).

Collembola. Consumption by some insects may provide clues to pollen chemical composition. In the Collembola, *Onychiurus pseudofimetarus*, a semi-transparent exoskeleton permits one to follow the digestion path of juniper pollen after ingestion. SCOTT and STOJANOVICH (1963) followed the pollen until the exine disappeared, i.e. just before it entered the pyloric valve (Fig. 7-12). Isolation of an intestinal enzyme, tentatively called exinase, may provide a good tool for studying pollen wall constituents. Bees, beetles, and most other insects excrete the exine after consuming the pollen protoplasm since they lack exine degradative capacity. The evolution of an exine degrading enzyme has apparently occurred only rarely in the animal kingdom. The almost total absence of such an exinase in the plant

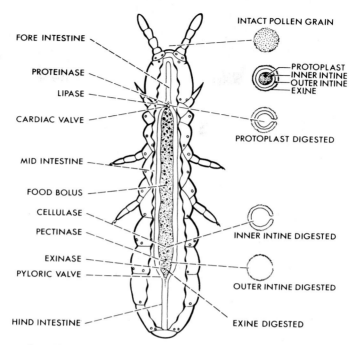

Fig. 7-12. Diagram illustrating digestion of juniper *(Juniperus pachyphloea)* pollen by Collembola (in Hypogastruridae). (From SCOTT and STOJANOVICH, 1963)

kingdom, with the possible exception of some members of the *Chytridiales* (SPAR-ROW, 1960) (Chapter 9) raises many interesting questions.

In the case of the chytrid fungi, e.g. *Rhizophydium sphaerotheca*, growth is only on and in the pollen grain, probably absorbing exudates and internal protoplasm as substrates. However, the obligative specificity of such fungi for a pollen germi-nation substrate, and the fact that deep pollen deposits do not accumulate from year to year in most places where large quantities of pollen are dispersed suggest that, as in the case of the insect Collembola, exine degradative enzymes have occasionally evolved in soil microflora.

Butterflies and **moths** are frequent visitors to flowers. Mature Lepidotera generally have only sucking mouth parts so that they primarily seek out nectar with their proboscis and only incidentally act as pollinators. However, 14 species of the neo-tropical butterfly *Heliconius* were observed collecting and consuming pollen (GILBERT, 1972). The butterfly collects pollen by scraping its proboscis tip over the anther. Based on rapid protein-amino acid diffusion rates from pollen (STANLEY and LINSKENS, 1965) and ^{14}C-amino acid incorporation experiments, GILBERT (1972) suggested that this butterfly uses pollen amino acids for the nutri-tion of its developing eggs.

The female yucca moth, *Pronuba yuccasella*, has specially modified mouth parts for collecting pollen into a ball which she then applies to the stigma, thus assuring pollination (BRUES, 1946). She oviposits at the base of the pistil; the fertilized ovules yield a seed supply for the larvae to consume (RILEY, 1892). Larvae of some moths consume pollen and are quite injurious to male cones of many gymnosperms. The spruce and pine cone worms, larvae of the moths *Dioryctria abietella* and *D. amentella*, are probably the most extensively reported. However, reports of damage by these insects are merely general surveys characterizing the types of damage, with no effort made to discern the nutritional role of the pollen. They do not appear to be specific for male cones but may, as observed for larvae of *Holcocera lepidophaga*, move from male cones to vegetative buds depending on what is available (EBEL, 1965). In the case of spruce and balsam fir budworms, pollen is the preferred food of the young larvae (MORRIS, 1951).

Birds and Bats

It has been estimated that beetles or birds pollinate 93% of plant families producing flowers with inferior ovaries (BAKER and HURD, 1968). Bird and bat pollinations are particularly prevalent in the tropics and in cacti plants (VOGEL, 1968). It has been suggested that birds who visit flowers first evolved as consumers of insects concentrating around flowers, and that such birds only secondarily began to respond to flower secreted nectar (BENT, 1940). These birds all have acute vision and very poor sense of smell, compared to the insects which seek out flowers. The insects generally have poor vision and a highly developed sensitivity to odors. In many cases, it is more logical to reason evolutionarily on energetic concepts than on morphological adaptations. The energy gain, i.e. nutrient yield, from a plant which a forager visits and from which it collects pollen, must be sufficient to warrant the energy expenditure (HEINRICH and RAVEN, 1972). Thus, birds or bats tend to dominate in collecting nectar and pollen under conditions of the tropical rain forest with widely separated individuals in the plant community, or at high elevations where low temperatures or rains prevail (CRUDEN, 1972). Bees and other insects energetically cannot gain or survive by such a collecting effort.

About 2,000 species of birds in 50 families regularly visit flowers. Families with the largest number of flower visiting, pollen ingesting species are: the honey eaters of Australia *(Meliphagidae)*, the nectar birds of Africa and Asia *(Nectariniidae)*, and the hummingbirds *(Trochilidae)*. Analysis of stomachs of Arizona blue-throated hummingbirds indicated that about 15% of the food in 2 out of 3 birds was pollen (BENT, 1940).

In bats, pollen ingestion also appears incidental to nectar consumption. One tropical plant, the sausage tree *(Kigelia aethiopica)*, supplies nectar and pollen to a bat species which is entirely dependent upon this tree for its food (MEEUSE, 1961).

No role for pollen has been established in the nutrition of either birds or bats. It is possible that when insect protein sources are scarce, pollen may provide some nutrition. However, the fact that many flower associated birds and bats are found without pollen in their stomachs indicates pollen is probably only incidentally ingested along with other primary foods.

Domestic Animals

Experiments with laboratory rats established that pollen and pollen protein extracts can serve as biostimulants in feeding rations (CHAUVIN, 1957). This work suggested to agronomists associated with large hybrid *Zea mays* breeding programs in Romania a possible use for the 50 kg/ha pollen removed in detasseling corn plants. Through by-product recovery and active collection, about 10,000 kg of pollen were harvested at Clij in 1970. Feeding experiments suggested that under some circumstances it may be feasible to harvest and use such pollen in animal feeds.

In piglet diets pollen had a nutritive stimulant value (SĂLĂJAN, 1970) (Table 7-4). Control diets were presumed deficient in the amino acid lysine; however, the biostimulant nature of pollen may also have been due to other factors present in the total spectrum of pollen trace elements and vitamins. The feed level of pollen to sustain the weight gain and improvement in food conversion could be reduced from 2% to 0.5% in older animals.

Significant weight gains were realized by adding 40 gm of *Zea mays* pollen to the daily milk solid-liquid diet of calves (POPA et al., 1970). At age 2 yrs, controls had a mean weight of 444 kg vs. 495 kg for animals with pollen in their diets. The improved food conversion was attributed to enhanced digestion. A better control than diet with minus pollen might be to substitute an equal weight of soya meal. Ethanol extracts of pollen have been used as beneficial additives in calf diets (BESLIN et al., 1972).

Probably the most significant animal feeding benefits demonstrated with *Zea mays* pollen involve egg production in hens. When the feed ration included pollen, egg laying increased 17% in the first 60 days (SĂLĂJAN, 1972). The egg yolks were a more intense and desirable yellow color. At the histological level added pollen increased the primordial follicular diameter and level of gonadotropins in the pullet capillaries. In these experiments control hens received balanced diets complete in vitamins and micronutrients.

Broiler chickens studies by COSTANTINI and D'ALBORE (1971) showed weight gains in both sexes when pollen was added to their diets; females consistently gained more than males. Pollen in the diets increased the percent hemoglobin and decreased glycohaemia.

Table 7-4. Effect of *Zea mays* pollen in piglet diet, age 41–150 days (SĂLĂJAN,1972)

	Group of 8 piglets				
	A	B	C	D	E
% Pollen added	0	1	2	3	4
Wt.-kg initial	13.35	13.37	13.43	13.81	13.50
final	32.82	34.58	38.22	36.18	36.35
Mean daily gain %	100	109.09	127.27	115.05[a]	117.55[a]
Feed consumption/kg gain Digest. prot. %					88.58
Digest. prot. %	100	91.11	77.52	88.05	

[a] Highly significant.

Human Consumption

Pollen is being actively marketed (Table 4-1) for alleviating certain health afflictions and as a beneficial dietary supplement. It has been adopted as a highly concentrated energy source for athletes. Although these claims have a certain amount of supporting evidence, the benefits of pollen to humans is an area still subject to question.

Chronic prostatitis appears to be one medical problem for which pollen has been found to have a favorable influence (ASK-UPMARK, 1963; 1967; DENIS, 1966; SAITO, 1967). The chief source of pollen medication used in clinical practice in 1970 was Cernilton, manufactured by A.B. Cernelle, Sweden (Table 4-1). The pollen tablets are listed as containing a mixture of 4 or 5 different pollen extracts; they are rich in B vitamins and contain, in 63 mg of extract, about 0.58 mg of steroids. The therapy varies from 3 to 6 tablets a day. About 55% of all patients treated reported benefits. Although only limited placebo-double blind experiments have been published, the consistent positive clinical response under supervision of urology specialists make this one of the better-established medical benefits attributed to pollen. The mechanism of the beneficial effect induced by pollen extracts is still unknown.

Respiratory infections and allergen reactions may be reduced by oral ingestion of pollen or pollen extracts (Chapter 12). Apparently, the esophagus and stomach acids quickly inactivate the allergens, or the sites of sensitive cells are limited primarily to the mucous nasal and eye tissues and are not reached by the ingested pollen tablets. This means that pollen tablets can be ingested by most people, even those that normally have hay fever. In fact, a well known way of desensitizing a person is to build up his immune response by oral hay fever tablets (HELANDER, 1960).

A tablet mixture of pollen extracts combined with acetylsalicylic acid (aspirin), called Fluaxin by A.B. Cernelle, Sweden, is marketed to afford relief from the common cold. Whether the pollen additive is more effective than the analgesic action of aspirin and vitamin C alone, or anything else now recommended to relieve the discomfort of colds, will have to await further clinical assessment.

Bleeding stomach ulcers have been reported as responding favorably in a study of 40 patients ingesting two tablespoons (250 mg) per day of pollen (GEORGIEVA and VASILEV, 1971). It would be very meaningful if the favorable effect of pollen on bleeding ulcers can be confirmed by others. In such studies pollen is derived from bee hive traps, used reasonably fresh and contains additives mixed in by the bee.

Nutritive Supplements

Pollen, as tablets or mixed with honey, is often sold as a "natural food" (BINDING, 1971). It has been praised by some athletes as being of benefit in their conditioning (BERTUGLIA, 1970). Many Finnish athletes use tablets of pollen extracts in their training program to assist in weight gains and ward off respiratory infections (KVANTA, 1972). But, whether pollen is superior to any high protein, nutritionally balanced food for such dietary purposes must still be established. While there is probably little doubt in the minds of such athletes and their trainers that pollen pills or extracts are beneficial, medical evidence supporting this claim is still scanty.

One reasonably valid observation is that pollen ingested in even large amounts has not yet been reported as harmful to humans. Many athletes and nutritionally concious people relate the true stories of many Ukrainians living vigorously at ages around 120 yrs to the honey with pollen these older Russians generally include in their diets. Based on such observations, people often decide to employ pollen to advance their efforts for superior physical performance or prolongation of life, conveniently overlooking the many other possible contributing factors in such long-lived peoples. Pollen probably is truly beneficial, although the benefits, we suspect, can be equalled by many other less expensive, more readily available foods.

III. Biochemistry

Chapter 8. General Chemistry

Many chemical analyses have been made on pollen loads collected by bees (TODD and BRETHERICK, 1942; VIVINO and PALMER, 1944). Such loads include flower nectar secretions and bee packing cement (Chapter 7). Data discussed in this chapter and those which follow will be, wherever possible, from chemical analyses of pure pollen, i.e. that collected free of anther residues and insect secretions, rather than from analyses of mixed pollen batches and bee-derived pollen.

The wind-borne, anemophilous pollen, such as pine and palm, is the type most frequently chemically analyzed. This pollen is easier to collect in large quantities than are insect-borne types. Techniques for collecting relatively pure samples of entomophilous types, e.g. *Pyrus and Lilium*, have been developed (Chapter 4) so that, with care, pure samples of this pollen is also available for chemical analysis.

Gross Chemical Analysis

One of the earliest chemical analyses of pollen was reported by BRACONNOT (1829). Many other analyses have since been reported. To provide a general overview of the chemical components, a few values are summarized in Table 8-1. At anthesis, water content of *Zea mays* and other grass pollens will generally be above 50%; in *Typha, Pinus* and longer surviving pollen the water content is usually 20% or less. Moisture level is normally determined by drying pollen in an oven at about 90° C for 24 hrs. The iodometric titration assay is also accurate and offers rapid measurement of pollen water content (GRINKEVICH, 1968).

Minerals

Mineral analyses are usually made by ashing the pollen in concentrated acid. By early 1800, analyses had established that about 4% of pollen was ash mineral. THOMSON (1838) summarized most of this early work.

Variation with Species. Studies of mineral composition by v. PLANTA (1886) showed that potassium, phosphorus, calcium and iron are the most commonly occurring minerals in pine pollen. Analysis of *Zea mays* added chlorine and magnesium to the list of elements present (ANDERSON and KULP, 1922). However, mineral levels vary with the species, and sources of error may arise because different colorimetric or qualitative analytical methods are used, or because some elements such as boron or chlorine may be volatilized by certain ashing conditions and not by others. Results are sometimes reported as percent fresh weight,

Table 8-1. Gross chemical analysis of pollen dry matter (%)

Species	Ash	Carbohydrates	Fiber (residue)	Protein	Lipid	Reference
Cryptomeria japonica	2.14	33.79	5.34	5.89	1.85	Mizuno, 1958
Zea mays	1.79	36.59	5.32	14.33	1.55	Miyake, 1922
Zea mays	2.55	34.26		20.32	3.67	Todd and Bretherick, 1942
Zea mays	3.46	17.78	5.35	28.30	1.48	Anderson and Kulp, 1922
Typha latifolia	3.70	13.15		18.90	1.16	Watanabe et al., 1961
Pinus sabiniana	2.59			11.36	2.73	Todd and Bretherick, 1942
Pinus radiata	2.35	13.92		13.45	1.80	Todd and Bretherick, 1942

Table 8-2. Percent major mineral composition of pollen ash

Species	Total Ash	K	Na	Ca	Mg	P	S	H₂O	Reference
Pinus radiata	2.35	0.88		0.03	0.11	0.30		11.25	Todd and Bretherick, 1942
P. sabiniana	2.59	0.87		0.04	0.09	0.36		14.08	Todd and Bretherick, 1942
P. montana	3.00					0.30	0.18		Nielsen et al., 1955
P. sylvestris	3.04	0.59		0.12					Kiesel, 1922
P. pinea	4.70	0.33	0.23	0.33	0.24	0.61		3.91	Anelli and Lotti, 1971
Juglans nigra	3.07	0.55		0.13	0.18	0.49		6.43	Todd and Bretherick, 1942
Typha latifolia	3.82	0.97	0.13	0.30	0.24	0.44			Todd and Bretherick, 1942
T. latifolia	3.80	1.24		0.10	0.28	0.71	0.24		Tischer and Antoni, 1938
Phoenix dactylifera	6.36	1.14		1.18	0.38	0.26		17.14	Todd and Bretherick, 1942
Zea mays	2.55	0.67		0.10	0.21			5.53	Todd and Bretherick, 1942
Z. mays 1953	4.90					0.58	0.43		Nielsen, et al., 1955
Z. mays 1954	4.90					0.75	0.30		Nielsen, et al., 1955
Z. mays yellow dust	3.46	1.24	0.19			0.63	0.34		Anderson and Kulp, 1922

but generally they are recorded as percent of dry weight or percent of total pollen ash (Andronescu, 1915). A comparison is made in Table 8-2 of values for the major mineral elements in pollen ash.

Total ash generally lies between 2.5 to 6.5% of pollen dry weight. The relative order of concentration for the principal elements is: potassium, phosphorus, and sulfur or calcium. In *Pinus sylvestris*, relative percents of K, P and S in pollen ash were 35.0, 28.6 and 14.8%, respectively (Kressling, 1891). Three different pine pollens contained about one-half the amount of phosphorus found in seven angiosperm pollens (Togasawa et al., 1967).

Date palm pollen *(Phoenix dactylifera)* is particularly high in mineral content, especially calcium (Table 8-2). This may be related to the high mineral content of the soil horizons in dry areas where this species grows. But, the capacity of the parent plant to accumulate salts in the pollen is also related to the species. The 26.8% ash in *Atriplex patula* pollen (Table 8-3) is a good example of this relationship. *Atriplex*, salt bush plants, are well known accumulators of minerals in their leaves and stems.

Gymnosperm pollen commonly contains less potassium and phosphorus than angiosperm species. Analyses of 10 angiosperms and 9 gymnosperms gave mean values of 0.98% P and 0.72% P, respectively (SOSA-BOURDOUIL, 1943). The values ranged from 0.34% in *Taxus* to 1.57% in *Lupinus*. Data in Table 8-2, while from many different sources, indicate that plants of the same family accumulate comparable levels of any particular element.

Sulfur is present at about 0.2 to 1% levels (LUNDÉN, 1954). Some workers have reported common elements, such as magnesium and chlorine, to be absent from pollen (ELSER and GANZMÜLLER, 1930). More careful analyses have established that magnesium is probably present in all pollen in microgram trace quantities along with many other common and uncommon elements (BERTRAND, 1940). These earlier results (Table 8-2), and magnesium concentration data are supported by studies of KNIGHT et al. (1972) on the mineral composition of 58 species (Table 8-3) from Sweden and England. The completeness of analyses, milliequivalents per 100 gms dry pollen, affords improved opportunities for mineral content comparisons, even though the purpose of the analyses was to compare the potential exchange capacities of pollen with other plant organs (Chapter 9). Nitrogen values are included for comparisons in Table 8-3. The main bulk of pollen nitrogen is in the protein fraction, the second most abundant group of compounds, after the carbohydrates (Table 8-1).

Trace Elements. Along with the major elements found in plant tissue, pollen also contains many minerals in trace and micro quantities. Iron, on the basis of percent dry weight, occurs at levels of 0.0013% in pine and corn *(Zea mays)* pollen (TODD and BRETHERICK, 1942). This corresponds to 130 µg/gm dry weight.

Iron assay by KNIGHT et al. (1972) of 58 species listed in Table 8-3 showed a considerably greater range than the limited values given in Table 8-4. The values were distributed between 68 and 9559 ppm. A few of these extreme ranges were observed within closely related species (Table 8-5). Why this element varies over such a broad range in different pollens is not yet understood. It may be related to function or it may primarily reflect different levels dependent on plant growth location.

Seven elements present in pollen at micro levels, determined by DEDIC and KOCH (1957) are summarized in Table 8-4. Iodine occurs in pollen at levels of about 4 to 10 ng/gm dry weight (PORTYANKO et al., 1971), about 2 and 3 orders of magnitude less than other trace elements commonly assayed in pollen (Table 8-4). Since iodine has been shown to stimulate pollen germination (PORTYANKO and KUDRYA, 1966) it is important to determine its endogenous level and possible function. Demonstrating an essential requirement for iodine is very difficult and such a requirement has not yet been conclusively shown for higher plants.

Table 8-3. Mineral composition of pollen (mequiv./100 gms dry wt.) (KNIGHT et al., 1972)

Species		Total ash (%)	K	Na	Ca	Mg	N	P	S
Dicots									
Cruciferae	Brassica napus L.	6.0	42	6	26	20	364	24	38
Chenopodiaceae	Atriplex patula L.	26.8	193	18	58	48	195	10	16
	Beta vulgaris L.	9.6	45	73	21	23	276	16	26
Hippocastanaceae	Aesculus hippocastanum L.	9.4	32	12	15	18	305	20	30
Papilionaceae	Trifolium hybridum L.	4.8	14	3	27	18	157	10	22
	Trifolium pratense L.	5.9	30	1	18	15	368	27	22
Rosaceae	Prunus padus L.	8.6	32	4	26	16	326	17	18
	Sorbus aucuparia L.	10.4	23	12	23	20	398	22	26
Umbelliferae	Anthriscus sylvestris (L.) Bernh.	8.4	26	1	15	16	331	23	30
Polygonaceae	Rumex acetosa L.	4.9	24	1	12	12	192	16	12
Urticaceae	Urtica dioica L.	11.2	49	1	88	32	200	20	30
Ulmaceae	Ulmus glabra Huds.	3.1	21	2	9	11	302	22	18
Myricaceae	Myrica gale L.	2.2	18	3	7	6	242	18	20
Betulaceae	Alnus glutinosa (L.) Gaertn.	2.0	17	2	10	5	276	13	14
	Alnus incana (L.) Moench.	2.4	17	2	12	6	269	13	14
	Betula pendula Roth.	2.9	19	2	9	7	329	19	18
Fagaceae	Fagus sylvatica L.	2.1	18	3	2	11	199	15	18
	Quercus robur L.	4.4	24	3	7	23	300	26	18
Salicaceae	Populus tremula L.	6.4	44	1	36	18	365	32	18
	Salix caprea L.	7.3	31	4	21	19	421	39	26
	Salix repens L.	5.9	29	1	19	23	442	39	24
Oleaceae	Fraxinus excelsior L.	4.3	34	2	8	18	381	29	22
	Syringa vulgaris L.	7.9	19	3	15	17	196	14	26
Caprifoliaceae	Sambucus nigra L.	5.2	30	1	11	23	428	37	22
Compositae	Anthemis arvensis L.	16.9	11	6	14	8	200	12	18
	Artemisia campestris L.	6.2	35	10	23	15	198	10	16
	Artemisia vulgaris L.	7.0	34	6	18	14	223	13	18
	Centaurea cyanus L.	9.2	20	3	25	35	299	17	20
	Chrysanthemum segetum L.	15.6	39	41	20	16	208	14	20

Boron in Pollen and Floral Organs

One trace element, boron, has been assayed in many flowers and in pollen (BER-
TRAND and SILBERSTEIN, 1938; GLENK and WAGNER, 1960). *Oenothera* and *Lilium*
pollen is lower in boron than most vegetative tissues of the flower; ovaries and
stigma are relatively high in boron compared to amounts in other flower tissues.
Even anthers with pollen usually contain less boron than stigma tissue. In 66
flowers of *Lilium regale* anthers contained 16 µg/gm boron; this was 46% of the
level found in the pistil (THOMAS, 1952). However, in rye plants the maximum

Compositae								
Tanacetum vulgare L.	4.7	40	2	20	18	134	8	10
Taraxacum officinale Web.	5.9	38	1	19	14	214	10	20
Tripleurospermum maritimum spp. inodorum (L.)	13.7	44	8	26	22	171	16	16
Monocots								
Typhaceae								
Typha angustifolia L.	4.4	39	2	8	14	251	20	14
Typha latifola L.	6.3	51	2	15	23	273	29	18
Cyperaceae								
Eriophorum vaginatum L.	3.7	25	6	6	1	227	19	14
Gramineae								
Agrostis tenuis Sibth.	4.5	32	6	12	21	222	16	16
Calamagrostis arenaria L.	6.7	43	56	9	21	251	16	22
Dactylis glomerata L.	7.0	33	2	17	20	251	15	22
Elymus arenarius L.	7.4	54	27	8	18	305	19	24
Festuca pratensis Huds.	5.0	28	8	12	15	281	17	20
Festuca rubra L.	3.0	23	5	8	16	250	18	14
Helictotrichon pubescens (Huds.) Pilger	6.4	36	8	10	20	296	15	20
Holcus lanatus L.	5.5	35	6	8	21	296	21	16
Lolium perenne L.	4.6	24	1	10	13	221	21	18
Phalaris arundinacea L.	7.0	35	8	13	21	314	16	18
Phleum pratense L.	5.3	39	1	19	14	226	21	18
Poa nemoralis L.	4.9	38	2	11	17	314	25	6
Poa pratensis L.	4.0	36	1	11	16	236	17	14
Poa trivialis L.	11.2	29	3	18	19	239	18	20
Secale cereale L.	4.5	32	1	9	14	236	17	16
Trisetum flavescens (L.) Beauw.	11.0	24	12	19	15	281	12	22
Triticum sativum Lam.	5.3	25	2	10	22	225	16	18
Zea mays	3.5	32	1	3	13	264	21	18
Conifers								
Juniperus communis L.	4.8	25	1	10	12	100	8	6
Picea abies (L.) Karst.	5.5	56	1	2	14	239	21	20
Pinus contorta Dougl.	2.5	25	1	1	8	143	9	12
Pinus mugo Turra	2.2	24	1	1	7	157	10	12
Pinus sylvestris L.	2.4	24	1	4	7	160	10	10

boron accumulation occurs in anthers and pollen not in the female tissues (LOHNIS, 1940).

Boron levels in pollen are influenced by the amount of boron available to the plant during development (VISSER, 1956). When developing floral organs are placed in solution containing a high boron concentration the resulting pollen germinates *in vitro* better than pollen having developed with little or no boron available. Boron levels in flowers and pollen can be increased by externally added sources (VISSER, 1956). This procedure is often used in orchard management to assure good fruit production. However, the inherent levels of boron usually vary

Table 8-4. Trace elements in pollen

Species	μg/gm dry wt. Element									Reference
	Al	Cu	Fe	Mn	Ni	Ti	Zn	Cl	Si	
Corylus avellana	0.3	1.5	120	37	0	0.3	30			DEDIC and KOCH, 1957
Tulipa suaveolena	49	10	250	36	0	2.0	251			DEDIC and KOCH, 1957
Paradisia liliastrum	8	11	133	17	0	0.8	120			DEDIC and KOCH, 1957
Caltha palustris	8	11	184	44	20	0.8	160			DEDIC and KOCH, 1957
Anemone nemorosa	27	14	286	112	75	1.5	150			DEDIC and KOCH, 1957
Prunus avium	80	18	363	40	0	4	140			DEDIC and KOCH, 1957
Prunus spinosa	1.6	20	350	25	0	1.6	80			DEDIC and KOCH, 1957
Typha latifolia				70				50	10	TISCHER and ANTONI, 1938
Pinus sylvestris		0	281	27			88			D'ALCONTRES and TROZZI, 1957

Table 8-5. Range of iron found in 58 different species (KNIGHT et al., 1972)

Coniferales	ppm (μg/gm dry wt.)
Pinus mugo	60
Juniperus communis	748
Monocots — all Gramineae	
Lolium perenne	67
Trisetum flavescens	4251
Poa pratensis	108
Poa trivialis	1431
Dicots — from the Papilionaceae (Leguminosae)	
Trifolium hybridum	9559
Trifolium pratense	504

within a limited characteristic range for a genus and within a single species (Fig. 8-1 a and b). Pollens which require added boron to germinate in viability assays (Table 6-1), e.g. *Pyrus* and *Sequoia*, are usually high in endogenous boron; pollen low in endogenous boron, e.g. *Pinus*, is generally that does not require boron to germinate *in vitro*.

Role and Distribution

Most essential elements, with the exception of boron and possibly calcium, are present in mature pollen at levels sufficient to facilitate normal growth and fertilization. The female tissue through which the pollen grows probably supplies boron and other elements required by the pollen. Presumably, mineral elements in mature pollen are associated with the parts of the cell in which they are required for physiological function, e.g. iron in enzymes (Chapter 14) and calcium in cell wall pectin. The Mg^{++} bound in ribosomes and polysomes becomes more tightly bound as germination begins (MASCARENHAS and BELL, 1970).

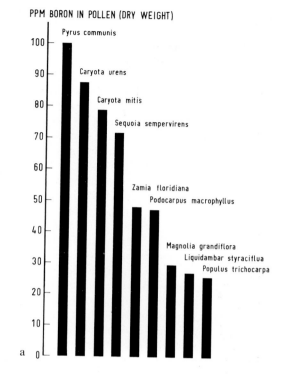

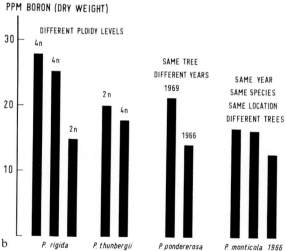

Fig. 8-1 a and b. Boron concentrations, ppm (dry weight). (a) In different pollen species; (b) within one species

Not all elements in pollen have recognized roles in growth or reproduction. Silica and aluminum are commonly reported in the residues of pollen. In *Zea mays* 3.7% and 0.2% of the ash was silica and aluminum, respectively (ANDERSON and KULP, 1922). ICHIKAWA (1936) found values of 23% silica and 2% aluminum in the ash of *Carpinus laxifolia* pollen. ANELLI and LOTTI (1971) found between 22.5 and 42.4% silica in the mineral ash of 15 hand-collected pollen species. The analyses, by atomic absorption spectra, showed unusually high levels of this non-essential element. Assuming that the presence of these elements is not due to dust contamination, these observations suggest that pollen accumulates extraneous, non-essential elements from tissues which supply nutrients to the developing microspores.

Methods used to assay mineral content afford little, if any, specific information about how the elements are localized in particular parts of the pollen. Most analyses are total assays. Pollen is extracted for water soluble minerals, and the water solution and residue are assayed. Several variations in methods of analyzing pollen are worth considering in terms of the potential information they may afford. If the function of the different elements in pollen is to be ascertained, then it is particularly important to determine the cytoplasmic constituents with which the element is associated.

One way to analyze the organelles with which an element is associated is to isolate specific cytoplasmic units and determine the level of element associated with the cell fraction. This assumes that the element will not diffuse from the fraction during isolation. Providing optimum levels of an element during microsporogenesis or during pollen tube growth *in vitro*, assuming growth *in vitro* simulates growth *in vivo*, can provide material for such studies. When iodine was added to the germinating media of Rosaceae pollen the percent distribution was relatively similar throughout the 3 separate parts of the tube analyzed (Table 8-6). When pear pollen was grown in the presence of boron, washed free of boron, and cell fractions separated and their boron content measured, the distribution was as noted in Table 8-7. On the basis of relative distribution the most boron was found in the vesicle and wall fractions of the pollen (Table 8-7) (STANLEY et al., 1974). Tips of pollen tubes contain high concentrations of vesicles.

Table 8-6. Percent distribution of iodine in germinated pollen (PORTYANKO et al., 1971)

Species	Tube tip %	Tube % (minus tip)	Exine %
Pyrus communis (pear)	38.9	37.9	23.2
Cydonia oblonga (quince)	63.1	23.1	13.8
Armeniaca vulgaris (apricot)	42.4	34.5	23.1
Ribes aureum (currant)	47.5	35.3	17.2

Table 8-7. Boron distribution in germinated *Pyrus communis* pollen (STANLEY et al., 1974)

Cell fraction	Concentration ppm
Whole-ungerminated	120
Cell wall	194
Mitochondria	76
Vesicles	73
Supernatant + ribosomes	10

Ether extraction of pollen generally removes mineral elements associated with lipid components. Analyses of residues after extraction with ethanol or ether may supply information on carbohydrate or pigment-bound elements. Such extractive treatments can provide information about the localization of elements and components in different pollen organelles. Efforts have been made to detect low levels of elements, and to relate them to specific cellular sites in pollen. However, the different effects of each organic solvent on viability (Table 5-3) and the differences in responses of pollen to rapid elution by buffers and water (STANLEY and SEARCH, 1971) suggest that short term differential extractive methods are of limited use in studying localization of bound elements.

Radioisotope Analyses

The element calcium has been localized in the tube membrane of germinating pollen by means of ^{45}Ca (STEFFENSEN and BERGERON, 1959; KWACK, 1967). Developing microspores and mature pollen have been infiltrated with ^{32}P to follow pollination patterns (COLWELL, 1951), but the technique involving radioisotope labeling with inorganic salts during microsporogenesis has not been coupled to autoradiographic studies of microspores or pollen differentially extracted by solvents or enzymes. Neutron activation analysis has been used to assay trace elements in pollen; elements with a molecular weight above 12 can be detected with this technique (FAWCETT et al., 1971). This neutron activation technique provides

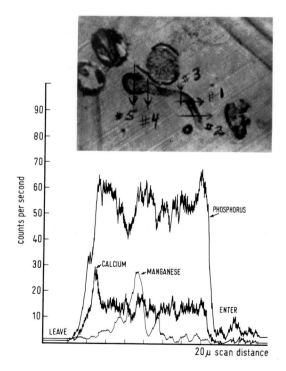

Fig. 8-2. Elemental analysis across the germinated pollen tube of *Pyrus communis* by electron microprobe. Arrows 1, 2 and 3 indicate direction of beam scan for calcium, phosphorus and manganese

a sensitive gross quantitative assay, rather than a selective localization of the element.

The electron X-ray microprobe method of analysis (TOUSIMIS, 1964) has been applied to pollen grains (STANLEY, 1971). By this technique a single pollen grain can be scanned in a known pattern with a 1μ beam of X-rays. As the beam passes over the pollen it strikes elements which radiate a detectable fluorescence at a wave length characteristic for each element. The assay for several elements including calcium and phosphorus are given as the probe moves across the grain (Fig. 8-2).

The highest levels of calcium were present in the walls of the pollen grains. The highest levels of phosphorus were in the cytoplasm. By this technique it may be possible to localize elements in actual subcellular organelles of pollen. This type of analysis requires that the X-ray microprobe be coordinated with an electron microscope. Elements as low as atomic number 12, magnesium, are easily detected *in situ* by this technique. Such localization could help to identify further the distribution and role of the mineral elements in pollen.

Chapter 9. Carbohydrates and Cell Walls

In most pollen, carbohydrates constitute the major dry matter fraction (Table 8-1). Polysaccharides, primarily starch and cell wall consituents, may comprise up to 50% of the dry weight. Low molecular weight carbohydrates average 4 to 10% of the pollen dry weight. Carbohydrate levels vary with the species (NIELSEN et al., 1955). Gymnosperm pollen, e.g. *Pinus*, is generally low in total carbohydrates (Table 9-1). In angiosperms, the trinucleate pollen of *Zea mays* is high in total carbohydrates, but the trinucleate pollengrains of *Beta* and *Ambrosia* are reported to be low (LUNDÉN, 1954). Binucleate angiosperm pollen may also be either high or low. The number of nuclei is apparently not directly correlated with carbohydrate content. Table 9-1 gives a few examples of between species variation in carbohydrate content of pollen collected directly from the plant.

Angiosperm pollen collected from bees is higher in reducing sugars and generally lower in non-reducing sugars than pollen isolated directly from the plant (TODD and BRETHERICK, 1942). The increase in reducing sugars in bee-collected pollen is probably due to the nectar used to cement the mass together (Fig. 7-3). The decrease in non-reducing sugars probably reflects a high rate of metabolism in bee-collected pollen (LUNDÉN, 1954).

Total extractable, soluble sugars in pollen vary with the conditions of handling and storage, and are related to germinating ability. For example, glucose and oligosaccharides decreased more significantly in pine pollen stored for 15 yrs at $5°$ C and 25% relative humidity (R.H.) than when stored at 10% R.H. The polysaccharides did not differ significantly after 15 yrs. storage at the two different humidities (STANLEY and POOSTCHI, 1962). The decrease in endogenous glucose and oligosaccharides in stored pine pollen was accompanied by a decrease in germination capacity. Non-viable grass pollen shows a loss of fructose and glucose and appearance of an unknown aldohexose when compared to the viable pollen (SHELLARD and JOLLIFEE, 1969). Correlation between viability and soluble carbohydrates was also observed when high and low germinating varieties of quince, *Cydonia vulgaris*, were compared (DŽAMIĆ and PEJKIĆ, 1970). The via-

Table 9-1. Carbohydrates in pollen. (Adapted from TODD and BRETHERICK, 1942)

Species	Carbohydrates as % total dry wt.	Reducing sugars	Non-reducing sugars	Starch
Pinus radiata	13.92	0.05	11.45	2.42
Pinus sabiniana	13.15	7.50	3.47	2.18
Typha latifolia	31.93	0.04	18.88	13.01
Phoenix dactylifera	1.20	1.07	0.13	0.00
Zea mays	36.59	6.88	7.31	22.40

Table 9-2. Sugars in *Cydonia vulgaris* pollen (DŽAMIĆ and PEJKIĆ, 1970)

| | Quince variety | |
	Vranjska (%)	Leskovaćka (%)
Germination	40	20
Total soluble sugars dry wt.	33.75	21.37
Sucrose	11.10	8.60

bility differences in these plants (Table 9-2) are due to genitically induced sterile grains.

Comparisons of carbohydrates in pollen of different species and varieties are valid as long as they recognize any potentially modifying environmental or genetic factors which occurred during development. Total carbohydrates from a variety of corn grown at the same site for two successive years were found to be about the same from one year to the next (NIELSEN et al., 1955). Since the predominate carbohydrates in pollen occur as polysaccharides in the wall and cytoplasm, little variation in total carbohydrate level would be expected between mature grains on the same plant. Greatest seasonal and species variation would be anticipated, and does occur, in the soluble carbohydrates.

Low Molecular Weight Sugars and Related Compounds

Fructose, glucose and sucrose are the free sugars found in highest concentration in ethanol extracts of pollen. The levels vary with species (Table 9-3), and as already described, with harvesting and storing conditions. Most data summarized here are based on studies where the pollen was either handled so as to assure minimum changes, or received comparable pre-treatment prior to chemical analysis. Intergeneric differences in relative levels of the low molecular weight sugars are usually greater than interspecific differences (CHIRA and BERTA, 1965). As noted in *Pinus thunbergii* (Table 9-3), a high percent of sucrose occurs, relative to glucose or fructose; this proportion was also found in *Pinus ponderosa* pollen (HELLMERS and MACHLIS, 1956). Angiosperm pollen generally contains more nearly equivalent proportions of dissaccharides and monosaccharides. For example, the ethanol extractable sugars of *Rosa damascena* contained about 26% fructose, 31% glucose, 33% sucrose and about 10% chromatographing as rhamnose (ZOLOTOVITCH and SÉCENSKÁ, 1963).

Raffinose and stachyose are usually present as free sugars in pollen. Raffinose was found in all 15 conifers analyzed by UENO (1954) and stachyose in 10 of the 15 species. However, while 7 species of *Abies* pollen analyzed by KANTOR and CHIRA (1971) all contained stachyose, only *Abies grandis* contained raffinose. Galactose and rhamnose are also occasionally identified in pollen extracts (MIZUNO, 1958). Rhamnose completely disappeared from three species of *Rosa* pollen stored 50 and 80 days, and there was a general decrease in the total free sugars (ZOLOTOVITCH et al., 1964). This, again, substantiates the observation that detec-

Table 9-3. Distribution of simple sugars (MOTOMURA et al., 1962)

Pollen species	Fructose (%)	Glucose (%)	Sucrose (%)
Pinus thunbergii	3.19	3.27	93.54
Typha latifolia	43.84	34.62	21.54
Lilium lancifolium	21.52	24.81	53.68
Lilium auratum	25.96	23.84	50.21
Cucurbita moschata	42.21	21.11	54.04
Oenothera lamarckiana	27.14	20.29	54.04

tion of sugars present at low concentrations is a function of both the assay method and of the time and method of storing the pollen.

Sugars in the water-ethanol extracts of Golgi-derived vesicles and tube membranes of germinating pollen reflect the sugar pattern found in mature pollen grains. In *Lilium longiflorum*, the pollen tube vesicles and membranes contain the following sugars: rhamnose, fucose, arabinose, xylose, mannose, galactose and glucose (VAN DER WOUDE et al., 1969). The mechanism of linkage, how they are incorporated, and role of these specific sugars in the pollen cell wall must still be discerned. In the Golgi vesicles as well as in the tube walls of *Petunia* the presence of cellulose components was demonstrated (ENGELS, 1974a). The resemblance in monosaccharide composition of the polysaccharides of the Golgi vesicles and the wall, as well as the fusion of the Golgi vesicles with the plasmalemma (Chapter 14) strongly support the idea that Golgi vesicles are involved in the synthesis of the cell wall material (ENGELS, 1974). But sugars may be brought to the cell wall not only by Golgi vesicles but perhaps also by other systems. The particles derived from the endoplasmic reticulum, which were observed by VAN DER WOUDE et al. (1971) in lily represent such a system.

WATANABE et al. (1961) reported that rhamnose, xylose, arabinose, glucose and fructose constitute 97% of the soluble sugars in *Typha latifolia* pollen. The remaining 3% of the soluble sugars consisted of several exotic trisaccharides such as maltotriose and kijibiose, along with raffinose, and the uncommon dissaccharides isomaltose, nigerose, turanose and leucrose, along with sucrose. About 0.12% of the sugar produced by hydrolysis in the Vranjska variety of quince was melezitose. This triose is totally absent from the hydrolytic product of the less viable (Table 9-2) Leskovacká quince. In extracts of the latter pollen, maltose occurs at about an equal concentration to that of melezitose in the other variety. Maltose is lacking in extracts of the viable variety. Splitting out the fructofuranosyl moiety and recombination of the 2 residual glucose units to yield maltose, or a differential hydrolytic reaction pattern in pollen extracts of one variety from that in the other, could account for such a difference. Fructose, in fact, is 10 times higher in extracts of the less viable quince pollen (0.569%) where maltose is present, than in the more viable (0.059%) without maltose (DŽAMIĆ and PEJKIĆ, 1970). Several of these exotic sugars are known to occur as breakdown products of starch and other polysaccharides.

Lactose has been reported in *Forsythia* pollen (KUHN and LÖW, 1949), but this could not be confirmed by others (BREWBAKER, 1959). Nevertheless, occasional reports of lactose in pollen extracts, based on chromatographic evidence, have been published (NAVARA and POSPÍSILOVÁ, 1962). Lactose can also occur as a breakdown product of polysaccharides; this may account for its occasional isolation. Several unidentified, noncommon sugars have been reported in extracts of pine (CHIRA and BERTA, 1965) and corn pollen (HIURA and HIURA, 1964).

Free myo-inositol, at levels up to 45 mg/gm dry weight, is frequently isolated, in particular from grass pollen (AUGUSTIN and NIXON, 1957). This level is several-fold higher than that normally encountered in other plant tissues. Inositol is incorporated into pollen pectin (STANLEY and LOEWUS, 1964; KROH and LOEWUS, 1968) and is not merely an enzyme cofactor (Chapter 14), the usual role assigned to this compound. Related cyclitols, the methyl derivatives sequoyitol and pinitol, were isolated from *Pinus montana* by NILSSON (1956).

The pentose sugars, ribose and deoxyribose, when isolated from pollen are probably breakdown products or precursors of nucleic acids (LUNDÉN, 1954). However, a pentose has been isolated in association with an allergenic protein extract from *Ambrosia* (ABRAMSON, 1947). Arabinose and galactose also were found associated with protein and pigments in *Ambrosia* pollen (LEA and SEHON, 1962). A glycoprotein containing hydroxyproline and arabinose was isolated from lily pollen tubes (DASHEK et al., 1970). Two mucoproteins from *Zea mays* pollen yielded mannose, glucosamine and glucose or galactose on acid hydrolysis (GLADYSHEV, 1962). The pollen-derived, heat stable inhibitor, M.W. about 850, of the hemolytic enzyme streptolysin, may be a polysaccharide-peptide complex (KVANTA, 1970).

Starch Content

Starch content of pollen is highly variable, comprising as much as 12.4% of the dry weight of *Typha latifolia* down to 2.6% in *Pinus thunbergii*, 1.4% in *Lilium auratum* and 3.6% in *Lilium lamfolium* (HÜGEL, 1965). The high starch content of *Typha elephantina* pollen has led to its use as flour (PATON, 1921). TODD and BRETHERICK (1942) also found 13% starch in *Typha*, they reported starch completely absent in only two of 34 species analyzed, i.e. *Ceanothus integerrimus* and *Phoenix dactylifera*. Of the 34 species, 70% contained less than 3% starch, and three species contained over 10%. *Zea mays* contained 22.4% (Table 9-1).

In a heterozygous line of corn described by MANGELSDORF (1932), the sugary and starchy pollen could be separated mechanically by a sieve, the smaller grains being the starchy gametes. The question of whether or not the starch content and form of carbohydrate in corn endosperm is under the primary control of the paternal parent has been long debated by geneticists (BRINK et al., 1926). ANDERSON and KULP (1922) reported that pollen of yellow dent corn contained 11% starch while that of white flint contained 19%. Variety names refer to the appearance of the seed endosperm. This work followed PARNELL'S (1921) study in rice, where pollen and endosperm starch of plants homozygous for the waxy-glutinous recessive gene (wx) stained red with iodine, but normal Wx rice plants produced pollen and endosperm starch which stained blue. So, it was initially proposed that a simple starch iodine blue color test for total pollen starch could be correlated with the corn endosperm starch amylose content. KARPER (1933) applied the iodine test for starch in sorghum pollen and endosperm to detect a 1:1 ratio of segregation in the heterozygous plants of the waxy starchy gene. In sorghum, unlike corn (MANGELSDORF, 1932), pollen dimorphism cannot be detected by size.

Detailed analysis of seven cultivars of corn, ranging in endosperm starch amylose content from 0 to 75% (ZUBER et al., 1960), showed no apparent correlation between the percent amylose in endosperm starch and the percent in pollen starch, except that the waxy endosperm genotype consistently yielded pollen starch low in amylose. This, if correct, would mean that the earlier work is in error and that haploid pollen, the gametophytic generation, has a different starch amylose content than the endosperm and diploid sporophytic generation. BANKS et al. (1971) checked to determine if the pollen starch in two different cultivars of Amylomaize corn variety were similar. They concluded that the starch of the sporophytic and gametophytic generations are similar, and that the results of ZUBER et al. (1960) might be the result of differences in the procedures they used in preparing starch from pollen and endosperm. BANKS et al. (1971) obtained amylose with only slightly different iodine binding capacity, but yielding grossly different color reactions depending upon the temperature during measurement of the color reaction. A small change in glucose chain length does cause a great difference in iodine color reaction.

Under some conditions, the environment can induce greater changes in pollen starch level than can be attributed to genetic factors. Mature pollen of *Corylus avellana*, *Cornus mas* and *Poa annua* on flowers developed in high humidity contained less starch, or even no starch, compared to pollen on flowers reaching anthesis in low relative humidity (KAUFMAN, 1920). Flowers maturing in high humidity would not release their pollen as early as those where low humidity prevailed at the time of flower maturity. Delayed dehiscence would reduce certain low molecular weight sugars and the starch level, due to on-going respiration.

Temperature during maturation also affects carbohydrate content. *Pelargonium* pollen, matured at 25° C, contained less starch than that maturing at 15° C (SEARS and METCALF, 1926). In *Moraceae*, as pollen reaches maturity starch content decreases (SEKI, 1954). Induced premature dehiscence prior to full grain maturity may yield grains with a different level of starch than would occur under normal flower development.

CALVINO (1952) attributed ecological significance to the presence or absence of starch in pollen. She concluded, as had earlier workers (TISCHLER, 1910; QUADRIO, 1928), that starchy pollen is generally produced by anemophilous plants, while most entomophilous plants produce pollen lower in starch and higher in sugar and fat. She also attempted to relate starchy pollen grains to plant species where growth of the pollen tube to the egg cell is other than via the micropyle (aporogamy). However, JOHRI and VASIL (1961) pointed out that in some plants with starchy pollen, the tube does in fact enter through the micropyle, i.e. porogamous growth.

Callose

Callose, an amorphous, colorless substance, is a β-1,3-polyglucan composed of β-D glucopyranose residues. It is insoluble in water, ethanol, solutions of basic carbonates and cuprammonium; it is soluble in cold concentrated H_2SO_4, cold dilute KOH, and solutions of concentrated $CaCl_2$ and $SnCl_2$. The solubility properties of callose differ from those of cellulose. The latter, a β-1,4-polyglucan formed in microfibrillar chains, is considered insoluble in dilute KOH.

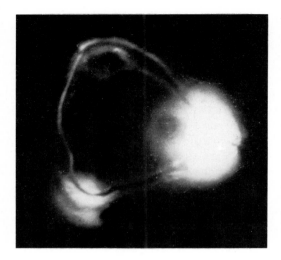

Fig. 9-1. Callose in ungerminated pollen grains of *Pyrus communis*. Viewed as translucent white under ultraviolet light

Callose, first described in 1864 as associated with the sieve plate of *Cucurbita*, was recognized by MANGIN (1890) in pollen mother cell walls (Chapter 2) and then in mature pollen grains (CURRIER, 1957).

Aqueous staining methods are generally used to detect callose in plant cells (ESCHRICH, 1956). Aniline blue stains callose a bright blue. Other stains used to detect callose include resorcin blue, coralline or resolic acid (CURRIER, 1957). The most widely applied method in recent years employs the U.V. fluorescent micro-scope to observe cells stained with an ammoniacal water solution of aniline blue. The fluorescent method has been adapted to study callose in germinating pollen (LINSKENS and ESSER, 1957).

In ungerminated pollen, callose shows up yellow-green under the U.V. fluores-cent microscope and is a light or white color in black and white prints (Fig. 9-1). In gymnosperms, during maturation (Chapter 2), callose is detected as a very thin layer around the prothallial cells (GÓRSKA-BRYLASS, 1968); at maturity it is gener-ally concentrated inside the intine with maximum localization in the furrow region. Callose increases in angiosperm pollen pore areas and probably under furrows of gymnosperm grains placed in a non-germinating condition, or when it ceases to be viable (ROGGEN and STANLEY, 1969). However, KNOX and HESLOP-HARRISON (1970) did not detect callose in the intine pore region of mature, fresh angiosperm pollen. Yet callose is generally reported in all developing microspores (Chapter 2) and during pollen germination.

A callose layer was also localized around the disintegrated prothallial and antheridial cells in mature pine pollen. In maturing pollen of *Chlorophytum* and *Tradescantia*, the generative cell is separated from the vegetative cell by a callose layer (GÓRSKA-BRYLASS, 1967). After moving away from the pollen wall, the gener-ative cell changes shape and loses the callose envelope. Callose at these times probably serves as a diffusion barrier surrounding or separating cell organelles or parts. Development of the cytoplasmic dimorphism observed in mature pollen

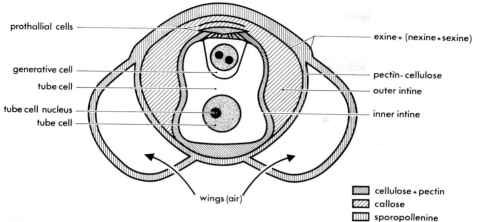

Fig. 9-2. Illustration of callose layering in pine pollen *(Pinus sylvestris)*. (Drawn after MARTENS and WATERKEYN, 1962)

(LARSON, 1963) would be facilitated by such a complete separation of the generative cell at some stages of its development.

MARTENS and WATERKEYN (1962) studied the pollen walls in 10 species of gymnosperms. In *Pinus sylvestris* pollen, they found that the outer intine is composed primarily of callose free of pectin and lipids, although it may contain cellulose. The callose layer (Fig. 9-2) is hydrated swollen pollen intine is isotropic, but when the intine was dehydrated and shrunken, the callose was anisotropic. However, the intine of the mature *Picea glauca*, unlike *Pinus* pollen, was lacking in callose (MARTENS and WATERKEYN, 1962). It was also not detected in the intine of EM preparations which ROLAND (1971) made of several *Ranunculaceae*, possibly due to the pectinase treatment he applied to the sections. Such an enzyme treatment may have loosened the bonds, or association between the two amorphous polysaccharides, and solubilized the callose along with the pectin. CRANG and HEIN (1971) failed to detect callose using fluorescent stains in *in vitro* grown *Lychinus alba* tube walls or plugs, although they did find callose in *Amaryllis* tubes. SOUTHWORTH (1973) found callose to be prominent in intine of 3 Compositae she studied with the histochemical stain, aniline blue.

In summary: Three functions can be attributed to callose. During pollen development and initial germination, callose layers, particularly those subtending pore areas, are readily mobilized as reserve material (ESCHRICH, 1964). In developing pollen, the callose wall which begins forming in the meiotic prophase (Chapter 2), functions as a molecular sieve to enable the autonomous development of the haploid pollen nuclei, independently segregated to their own cytoplasm (HESLOP-HARRISON, 1966; HESLOP-HARRISON and MACKENZIE, 1967). By forming along the inner tube membrane during germination, and rapidly forming at small puncture holes in the tube membrane, callose acts to add wall strength and delimit, restrict tube cytoplasm. Callose plugs, which form in the back of extending pollen tube, thus limit and contain the path of the cytoplasmic stream inside the tube.

Pectin and Other Polysaccharides

Other polysaccharides extractable from pollen include the hemicelluloses and pectin. In the mature grain these components are localized primarily in the intine (ROLAND, 1971; SOUTHWORTH, 1973). On hydrolysis they yield arabinose, galactose, glucose, xylose, galacturonic acid and rhamnose. In some pollens, such as *Cryptomeria japonica*, the products of hydrolysis suggest that the principle extractable polysaccharide in the mature pollen is pectin (MIZUNO, 1958).

The amount of pectin extracted depends upon the method used as well as the species. OHTANI (1955) compared extraction procedures using hot water and ammonium oxalate in 15 species of pollen. The ammonium oxalate, but not the hot water, yielded larger amounts of pectin from deciduous than from conifer pollens. This could be an artifact. The basis of comparison, i.e. percent dry weight, may contribute to the differences, since the proportion of crude fiber residue, sporopollenin, is higher in most gymnosperm than in the thinner-exined angiosperm pollens; or, the ammonium solvent may more effectively penetrate the multipore regions common in the angiosperm pollens.

Hemicelluloses of *Pinus mugo* pollen were characterized by extracting the intine layer with monoethanolamine (BOUVENG, 1965). He isolated several polysaccharde constituents, including an arabinogalactan containing primarily chains of D-galactose, L-rhamnose and D-glucuronic acid with L-arabinose at branching positions and D-galactose terminal groups; also a polysaccharide containing D-xylose and D-galacturonic acid in the ratio 1:2 combined with a major component of linear β-1,3-glucan. The xylose was limited to the terminal position in the xylogalacturonan side chains. The arabinogalactan of pine pollen was chemically very similar to the arabinogalactan-type gums from gum tragacanth and the *Acacia* tree gums (BOUVENG and LUNDSTRÖM, 1965).

Table 9-4. Pollen total and free uronic acid content (KNIGHT et al., 1972)

Species	mequiv./100 g dry matter		
	UA	Methyl content	% UA which is methylated
Prunus padus L.	36	3.7	10
Sorbus aucuparia L.	24	2.0	8
Betula pendula Roth.	21	1.5	7
Populus tremula L.	33	3.3	10
Salix caprea L.	24	3.3	14
Salix repens L.	25	3.3	13
Centaurea cyanus L.	26	2.3	9
Dactylis glomerata L.	30	1.9	6
Festuca sp.	32	2.8	9
Holcus lanatus L.	22	2.2	10
Phalaris arundinacea L.	27	2.1	8
Phleum pratense L.	33	2.1	6
Poa pratensis L.	27	1.6	6
Zea mays L.	17	3.2	19
Picea abies (L.) Karst.	25	2.8	11
Pinus mugo Turra	25	3.6	14

The free carboxyl groups in pectin, generally in the uronic acid (UA) fraction, provide a cation exchange capacity (CEC), and negative charge for the pollen wall (KNIGHT et al., 1972). The UA value is fairly constant for a species, i.e. between 17 (*Fagus sylvatica, Zea mays*) and 52 meq. UA/100 gm dry weight (*Urtica dioica*). Pines and most conifer species have values between 22–30. About 10% of the UA is commonly methylated, although in *Zea mays* it is 19% (Table 9-4). While deep cytoplasmic evaginations across the intine may have facilitated the movement of lanthanum nitrate into pollen of *Nuphar* and *Epilobium* (ROWLEY and FLYNN, 1971), UA binding sites along the membrane extensions and in the intine, must not be overlooked as possible agents in such cation translocation. Variation in CEC capacity of the intine in the pore area may also help explain certain uptake diffusion rates, although exchange undoubtedly also involves many other wall and cytoplasmic factors besides those associated with the UA of pectin.

Cellulose and Sporopollenin

Pollen exine does not react with most chemicals. The inert, nearly non-destructible nature of the exine permits survival of pollen grains from past geological periods with intact, recognizable surface layers. Chemical analysis of the exine is difficult because the products are a function of the hydrolytic method used, and may not necessarily represent natural molecular components of the wall. We will review the common methods of analysis, the deduced composition of the exine based on the different analytical methods, and lastly, discuss something of the destruction and survival of pollen exine in nature.

Complete dissolution of the exine will generally result from heating pollen at 97° C in monoethanolamine for 3 hrs. Formalinacetic acid (FAA) fixation does not dissolve pollen. Using a short-period extraction with monoethanolamine, BAILEY (1960) removed the exine of fresh *Liriodendron* pollen, leaving the protoplast surrounded by intine with oil droplets visibly undisturbed. However, MARTENS and WATERKEYN (1962) failed to remove the exine selectively from pine pollen by the same monoethanolamine treatment.

The standard chemical treatments for extracting cellulose, i.e. cuprammonium, 75% H_2SO_4 and prolonged acetylation, permit preparation of samples with varying quantities of cellulose microfibrils localized in the exine, particularly in the nexine and inside of the intine (Fig. 9-3). Removal of all cellulose still leaves an exine residue. Electron micrographs and staining (MARTENS and WATERKEYN, 1962) have also been applied to confirm the sites of cellulose and other chemical constituents in and adjacent to the exine.

The chemical analysis of the pollen exine requires successive removal of all pollen wall constituents not resistant to a particular treatment. A typical extractive sequence employs an alcohol wash to remove surface coatings, then solutions of potassium hydroxide to break down proteins, fats and waxes, followed by ether-alcohol extraction to remove fats and oils (Chapter 10). Carbohydrates, except for cellulose, are removed with solutions of concentrated NaOH and water; the cellulose is removed by 72% sulfuric acid. The residue remaining is considered the chemically resistant exine. Variations in time and temperature of each extractive step provide a somewhat different final residue.

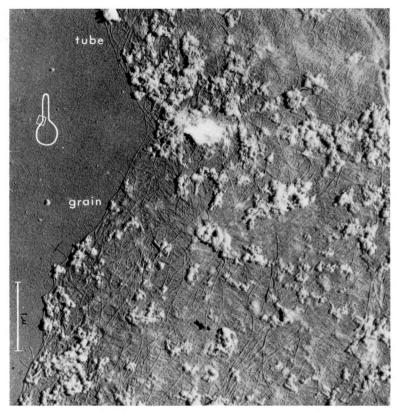

Fig. 9-3. Localization of cellulose microfibrils in pollen grain wall. (Courtesy M. M. A. SASSEN)

BRACONNOT (1829) followed the 1814 work of JOHN and called the pollen residue "pollenin" (KWIATKOWSKI and LUBLINER-MIANOWSKA, 1957). Early studies (HERAPATH, 1848) reported pollen to contain about 40% pollenin. VON PLANTA (1886) called the 21.9% dry weight, residue exine of *Pinus sylvestris* "Cuticula"; KRESSLING (1891) also obtained about a 20% residue of pine pollen after exhaustive chemical extraction, but considered the residue a modified cellulose. However, BIOURGE (1892) found the outer exine layers more resistant to chemical degradation than were cutinized membranes. CZAPEK (1920), summarizing the known exine chemistry, concluded that many pollen species contained a cutinlike layer on the exine. ZETZSCHE and his colleagues worked in Switzerland for a period of about ten years, starting in the 1920's, analyzing *Lycopodium* spores and pollen walls. At first they termed the residue sporonin. However, because of the similarity of spore wall sporonin to pollen wall pollenin they subsequently used the word "*sporopollenin.*" A symposium reviewing all aspects of sporopollenin was held in London in 1971 (BROOKS et al., 1971).

ZETZSCHE and HUGGLER (1928) found that sporopollenin had a C/H ratio of 1/1.6, similar to that of terpenes. However, the exine chemical residue included oxygen, thus they proposed the formula $(C_{10}H_{16}O_3)X$. Further quantitative elemental analysis (ZETZSCHE and VICARI, 1931) suggested a basic C_{90} unit; ozonation suggested a product similar to caoutchouc, the polymeric isoprene $(C_5H_8)_n$ units in rubber. The formulas varied with the method of isolation and the species. A typical proposal, for example, for sporopollenin of rye pollen is $C_{90}H_{134}O_{31}$, for pine it is $C_{90}H_{158}O_{44}$.

FREY-WYSSLING and MÜHLETHALER (1965) considered sporopollenin to be a polymerized series of hydrocarboxylic acids with similar properties as cutin and suberin, but differing in resistance to saponification in alkali. SHAW and YEADON (1966), on the basis of their chemical analysis, concluded that sporopollenin is a fatty acid-lignin-like material. They isolated a series of mono- and dicarboxylic acids as the breakdown product of the exine. A lignin-like fraction of 10 to 15% was found in the sporopollenin. Since this lignin is not detected in mature pollen grains by the usual histochemical tests, they suggested that the lignin is masked by lipids or other wall constituents. In addition to the lignin and fatty acid fraction, SHAW and YEADON (1966) found pine exine to contain about 10% hemicellulose and about 15% cellulose. The ratio of exine sporopollenin to intine cellulose to total pollen dry weight varies between species in the same genera (Table 9-5). However, the amounts of these wall components, in a given genus, are relatively similar. Considering the constancy of exine pattern, similar chemical characteristics are to be expected. As plants evolved from the gymnosperms to the angiosperms, the exine thickness decreased, probably representing a physical and chemical adaptation to the less severe climate, or the shorter time the pollen had to survive between dehiscence and fertilization (BROOKS, 1971).

Lignin. Survival of a totally masked lignin is not likely after differential solvent extraction and cross-sectioning of pollen grain walls. SOUTHWORTH (1969) compared absorption spectra in lignified plant cell walls and several angiosperm pollens and concluded it was not a significant fraction of the sporopollenin. Interestingly, a protosporopollenin deposited in the early stages of exine formation has the characteristic resistance to acid hydrolysis of sporopollenin, but stains differently (HESLOP-HARRISON, 1968a). Thus, a modified phenolic component with some lignin-like properties may survive to maturity as an exine compo-

Table 9-5. Levels of sporopollenin and cellulose in pollen (BROOKS, 1971)

	% Dry wt.		
	Sporopollenin	Cellulose	S/C
Gymnosperms			
Pinus sylvestris	23.8	6.0	4.0
Pinus montana	27.7	7.1	3.9
Pinus contorta	20.7	5.5	3.8
Pinus radiata	24.4	6.6	3.7
Angiosperms			
Lilium henryi	5.3	3.1	1.7
Lilium longiflorum	5.1	3.3	1.5
Rumex thyrsiflorus	5.7	4.2	1.3
Rumex acetocella	6.3	6.7	0.9
Rumex acetosa	4.2	7.0	0.6

Table 9-6. Comparison of ozonation products of pollen wall

Material	Mol. formula (adapted to C_{90} base)	% acids produced		
		Branched chain mono.	Straight chain mono.	Dicarboxylic
Lilium henryi				32.9
pollen wall	$C_{90}H_{134}O_{36}$	59.3	8.4	
carotenoid polymer	$C_{90}H_{148}O_{38}$	64.1	3.4	33.4
β-carotene polymer	$C_{90}H_{130}O_{30}$	61.3	1.9	35.0

nent. The degree of polymerization and crosslinking of the components provides a basis for recognizing sporopollenin as a chemically variable group or class of substances.

Subsequent analyses by BROOKS and SHAW (1968; 1971) of *Lilium henryi* indicated that a lignin-like material can also arise in hydrolysis of carotenoid polymers. The products are similar to those obtained by ozonation of β-carotenoid polymer (Table 9-6). In the closely related *Lilium longiflorum*, formation of carotenoids for deposition on the exine does not occur in the tapetum until after exine formation is nearly complete (HESLOP-HARRISON, 1968b, c). The carotenoid products detected by BROOKS and SHAW (1968) may be odorless derivatives, or may be from carotene infused and trapped in the exine after the exine is formed. They were led to conclude that sporopollenin is an oxidative co-polymer of carotenoids and carotenoid esters chemically bound into a matrix.

Carotenoids are common constituents of most pollen species, predominantly the *"Pollenkitt,"* the surface adhesive-like materials which facilitates pollen dispersal and water uptake. However, carotenes have never been recovered by extraction or hydrolysis of pine pollen, a species devoid of Pollenkitt, but with an exine very yellow in color (Chapter 15). Just as the plant cell wall varies from one species to another, and within different parts of the plant, so too one would not expect all exine to be chemically the same throughout the plant kingdom. One possible way to resolve the problems of the exine composition would be to follow the biosynthesis of the exine in the developing anthers infused with suspected precursor compounds labeled with radioactivity (SOUTHWORTH, 1971). It will be important, in testing the hypothesis of "sporopollenin arising as a carotenoid derivative," to ascertain if non-polymerized carotenes present in developing pine microspores are incorporated into the exine matrix, and hence never found free in the pollen.

Concepts of the structural units of sporopollenin are primarily derived from indirect evidence of product analysis. The exine constituents proposed by SHAW and YEADON (1964) and BROOKS and SHAW (1968; 1971) are polymerized, oxygenated products of carotenoids, derived from β-carotene (Chapter 15) and carotene ester as antheraranthin dipalmitate (Fig. 9-4 A, B). Many methyl and hydroxy and keto groups (Fig. 9-4 C) would be free to polymerize along with the ester terminal groups or facilitate ring closurer and aromatization along the unsaturated aliphatic part of the carotene molecule. POTONIÉ and REHNELT (1971), basing primarily on studies with fossil exine, suggest that the keto groups separated in the

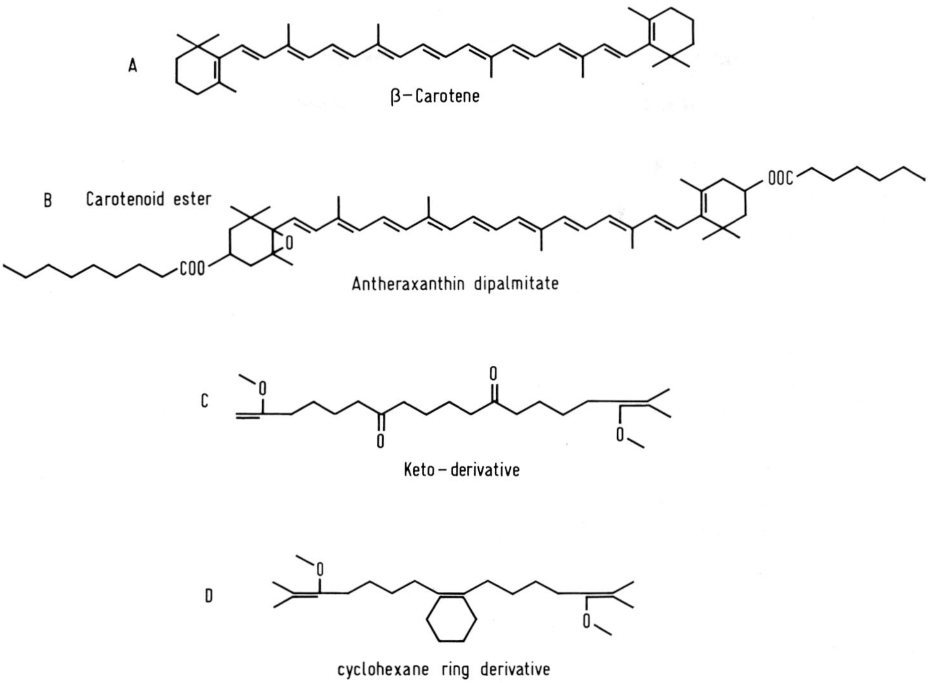

Fig. 9-4 A-D. Proposed structural units of sporopollenin. Carotenoid derivatives A, B and C: After SHAW and YEADON, 1964; BROOKS and SHAW, 1968. D: After POTONIÉ and REHNELT, 1971

chain by 4-methyl radicals readily form a cyclohexene ring (Fig. 9-4D). Such a cyclization is quite possible and would be expected to occur under the thermal and pressure characteristics of the environment in which the pollen or spores are fossilized.

Some of the U.V. fluorescence microphotometry differences recognized by VAN GIJZEL (1971) in fresh and fossil exines may be due to such environmentally induced cyclization, as well as selective erosion and removal of part of the exine. Artificially simulated pollen fossils may, when the fluorescence or U.V. absorption spectra (SOUTHWORTH, 1969) are related to recognized chemical changes in the wall, help resolve the problem of the nature of the chemical bonding and crosslinking in the exine structural moieties.

The chemically inert components of the exine are distributed as indicated in Fig. 9-3. TSINGERS and PETROVSKAYA-BARANOVA (1961) concluded that the inert exine wall of freshly shed mature pollen also contains proteins and enzymes and that the wall is thus, a physiologically active structure. Some pollen devoid of pores rapidly swells and excretes a drop on its exines when landing in the germination media. This lends support to the conclusion that some easily mobilized, physiologically active and diffusible constituents are localized at the pollen surface.

Using the histochemical stains chlorozinc-iodine for cellulose, ruthenium red for pectins and sudan III for lipids, TSINGER and PETROVSKAYA-BARANOVA (1961) found cellulose in both the intine and exine. It was difficult to detect in the exine because of the hindrance of dye penetration by the sporopollenin; pectins accompanied the cellulose and were in high concentration in the intine. Using the protein stain bromophenol blue, they localized protein in *Amaryllis* pollen wall areas which they designated intexine and mesine (Chapter 2). Using biuret and Millon's reaction, they also characterized proteins in the cell wall of pollen from peony, narcissus and sweet pea. Removal of fats and proteins by alkaline hydrolysis eliminated the protein in the pollen wall. They also characterized wall protein as including enzymes such as dehydrogenases, cytochrome oxidase and phosphatases (Chapter 14).

The above results and other work (ROWLEY et al., 1959; STANLEY and LINSKENS, 1964; SKVARLA and LARSON, 1966; KNOX and HESLOP-HARRISON, 1969; STANLEY and SEARCH, 1970) show that while the exine is highly chemical-resistant, it is not impenetrable and it, or the intine, contains protein as protoplasmic strands or inclusions which can be readily solubilized in water (Chapter 11). Relative levels of ribonuclease and acid phosphatase were assayed in the walls of 50 species of angiosperm and one gymnosperm pollen (KNOX and HESLOP-HARRISON, 1970). The intine, particularly in the pore aperture regions, were high in enzyme activity. These and other enzymes readily diffuse off the pollen (STANLEY and LINSKENS, 1965). Radial channels through the exine of grass pollen which are not cytoplasmic strands (SKVARLA and LARSON, 1966) were suggested as transfer channels through the exine (ROWLEY, 1959). The movement inward of water and solvent may be facilitated by such transfer channels.

It is interesting to note that the spaces within the outer sculptured portions of the exine, the sexine, are often filled with oily, flavnoid or pigmented viscous substances which give pollens such as *Liliaceae* and *Oenothera*, their characteristic odor, stickiness and some of their color (HESLOP-HARRISON, 1968). These rapidly solubilized fractions presumably become crustations on the exine—in either the nexine or exine (Fig. 9-3)—and may be sources of protein, pigment and other components readily solubilized from the pollen wall. Proportionately, the level of surface-localized proteins is low compared to enzymes concentrated in the intine pore area (KNOX and HESLOP-HARRISON, 1970).

Pollen Destruction in Nature

Studies of pollen corrosion were related to the determination of sporopollenin composition (ZETZSCHE and KÄLIN, 1931; KIRCHHEIMER, 1933a, b; 1935). From these, and studies of HAVINGA (1963, 1964, 1971), pollen has been characterized on the basis of differential susceptibility to corrosion (Table 9-7). Pollen resistance to experimental oxidation is correlated with resistance to natural corrosion. KWIATKOWSKI and LUBLINER-MIANOWSKA (1957) related the stability of pollen in geological sediments to the quantitative relationship between sporopollenin and membrane mass.

Table 9-7. Relative corrosionability of pollen

Species	Susceptibility to corrosion[a]	Sporopollenin	% Content[b] Cellulose
Pinus sylvestris	relatively slight	19.6	3.6
Tilia sp.	relatively slight	14.9	8.1
Alnus incana	intermediate	8.8	1.8
Corylus avellana	intermediate	8.5	2.7
Betula verrucosa	intermediate	8.2	3.2
Carpinus betulus	intermediate	8.2	3.5
Ulmus sp.	relatively great	7.5	5.8
Acer negundo	relatively great	7.4	2.4
Quercus sessiliflora	relatively great	5.9	1.9
Populus alba	relatively great	5.1	4.2

[a] from HAVINGA, 1964
[b] from KWIATKOWSKI and LUBLINER-MIANOWSKI, 1957.

Species differences in wall corrosion observed in soil residues are related primarily to the percent composition and variation in chemicals of the sporopollenin and cannot be correlated with cellulose (Table 9-7). River clay soils and leaf mulch are biologically very active; these mixes accelerate corrosion compared to peat bog soils or to the rates observed for pollen exposed to non-biological, chemical oxidation (HAVINGA, 1971). A more gradual thinning of the wall occurs in the slower corrosion processes.

Soil pH, exchange capacity and climate, all modify the microflora and rate of pollen degradation. Oxidized pollen wall is rapidly attacked by bacteria, yeast and other fungi. Among the bacteria the *Actinomycetes* often colonize pollen, while members of the aquatic phycomycetes, especially the *Chytridiaceae* are particularly destructive to pollen (GOLDSTEIN, 1960; ELSIK, 1971). One species, *Rhizophidium pollinis*, was recognized as a saprophytic fungi regularly associated with pollen (BRAUN, 1885). *R. sphaerotheca* and *R. racemosum* also commonly occur on pollen (SPARROW, 1960). Fertile soil with high levels of biological activity are, therefore, poor sites for pollen preservation (HAVINGA, 1967). Autooxidation of *Pinus* and *Picea* pollen occurs when the pollen is exposed to air for long periods (ZETZSCHE and KÄLIN, 1931). Pollen in a well aerated soil would be rapidly oxidized, particularly if the soil solution is in a pH range of 6.5–7.5. Such soils and most tropical soils tend to be free of pollen (DIMBLEBY, 1957).

In storage experiments, *Pinus sylvestris* and *Tilia* pollen were partially decomposed after 20 months in river clay and leaf mulch. In these controlled environments, the microflora varies just as it does in different soils. In the soil, oxidative and biodegradative activities remove the pollen cutin and other surface coatings, and then the cellulose. The perforation and intrusion generally leads to a specific scar pattern, corresponding to exine areas removed during breakdown by the invading organism (ELSIK, 1971). The presence of different microflora provides a range of destructive potential which complicates the effort to define corrosion susceptibility of pollen in the soil as due merely to the percentage or structural differences in exine sporopollenin (HAVINGA, 1967).

The process of pollen exine degradation is quite similar to the mechanism of enzymatic degradation of surface cuticular compounds. Chemically, the cutin is first oxidized; then, following the break down of peroxide complexes the cutin material becomes alkali-soluble (HEINEN and LINSKENS, 1960; HEINEN, 1963). *Pharbitis* was reported to contain an enzyme(s) in its intine which is secreted into the exine after pollination and dissolves the exine (GHERARDINI and HEALY, 1969). Biologically, the penetrating rhizoids of the fungi secrete enzymes which dissolve rather than puncture the exine. The availability of specific microorganisms, insects (Fig. 7-12) and pollen with enzymes able to degrade the exine, may afford a strong biochemical tool for the study of sporopollenin in a much less destructive, and more revealing manner than has hitherto been possible.

Chapter 10. Organic Acids, Lipids and Sterols

The organic acids are a poorly assayed group of pollen constituents. These key metabolic intermediates vary with respect to development, species, and prior handling of the samples. Initial assays of acid content were made by titrating alcohol extracts with base. KRESSLING (1891) reported that an extract of 5 gms of *Pinus sylvestris* pollen contained 1.76 equivalents of acid (H^+). By salt formation and a specific color reaction he determined that the principle acid in the extract was tartaric, with malic and acetic acids also present. Assay of an alcohol extract of 100 mg samples of *Pinus ponderosa*, *P. echinata*, *P. lambertiana* and *P. radiata* showed that viable pollen stored 15 years contained about 1.65 equivalents of acid per 5 gms. Non-viable pollen, stored the same length of time but at higher relative humidity than the viable samples, contained 20% less organic acids (STANLEY and POOSTCHI, 1962). Quince pollen with 20% viability also contained 20% less total organic acid than a variety with 40% viability (DŽAMIĆ and PEJKIĆ, 1970). This suggests that high levels of organic acids, as well as levels of complex carbohydrates, are associated with germination capacity of pollen.

Ragweed (*Ambrosia*) pollen contained the following free organic and fatty acids: formic, acetic, valeric, lauric, oleic, linoleic, palmitic and myristic (HEYL, 1923). The free ascorbic acid content of pollen collected from bees (WEYGAND and HOFMANN, 1950) varied from 7 mg/g pollen (dry wt.) in *Taraxacum officinale* to about 15 mg/g in *Brassica napus olifera*. Total vitamin C content in the pollen was several times higher than the free ascorbic acid, i.e. 36 and 59 mg/g respectively. This high content of ascorbic acid has led to the suggestion that pollen is a potential source of therapeutic benefit to men and animals.

Pine pollen was also tested for polyphenolic acids and their flavanol and coumaric derivatives. The phenolic acids in pine pollen (STROHL and SEIKEL, 1965) are listed in Table 10-1. Few clear species differences were recognized in the phenolic acid assays. Lack of vanillic acid in *P. ponderosa* and *P. echinata*, and the large amounts of the cinnamic acid derivatives, p-coumaric and ferulic acids, in *P. strobus* were suggested as having possible taxonomic significance (STROHL and SEIKEL, 1965). However, variations in handling and lack of a broad survey of other pollen species for these constituents limits conclusions at this time.

When pollen is stained and viewed by light microscope, or prepared and viewed in the electron microscope, lipids appear as droplets dispersed throughout the cytoplasm (SASSEN, 1964). The lipids in pollen cytoplasm occur primarily in spherosomes (TSINGER and PETROVSKAYA-BARANOVA, 1967). GÓRSKA-BRYLASS (1962) established that the fats in pollen of *Campanula* are primarily in elaioplasts. Both terms, spherosome or elaioplast, have been applied to the fat-containing particulates in pollen cytoplasm. Spherosomes have a more extensive biochemical

Table 10-1. Known phenolic acids in *Pinus* pollen

Species	p-hydroxy-benzoic	p-coumaric	Vanillic	Proto-catechuic	Gallic	Ferulic
P. banksiana	4	3	3	2	2	2
P. echinata	3	3	0	2	2	2
P. elliottii	4	3	3	3	0	2
P. palustris	4	3	3	3	3	1
P. ponderosa	4	4	0	3	2	1
P. resinosa	4	3	3	3	2	2
P. strobus	3	4	1	2	0	3
P. taeda	4	3	3	2	2	2

Relative levels: 0 (absent) to 4 (very high).

role and enzyme complement than the elaioplastids which are primarily fat reservoirs.

Ether extracts may include lipids, fatty acids, sterols, branched and straight chain hydrocarbons and their alcohol derivatives. The nonsaponifiable portion of the extract contains the sterols, hydrocarbons and higher molecular weight alcohols.

Total Lipids

Ether extracts of pollen vary from 1% to 20% of the dry weight. KERNER (1891) found 400 of 520 species of pollen he examined contained a layer of fatty-viscous oil on their exine. This fatty oil was usually yellow, but occasionally it was colorless or a different color from the exine pigment. This exine-contained lipid material is probably the major source of variation in ether-extractable materials. Air dried, bee-collected pollen contains about 5% ether-extractable materials (TODD and BRETHERICK, 1942). Generally, the percent lipids will run 5% or less (Table 8-1). Some pollen, i.e. *Taraxacum officinale* and *Brassica* spp., contains as much as 19% (TODD and BRETHERICK, 1942; STANDIFER, 1966a). *Corylus avellana* pollen contained 15% ether extractables, including 3% non-saponifiable material (SOSA and SOSA-BOURDOUIL, 1952). Percent distribution of the non-saponifiable: saponifiable fractions in an ether extract of pollen generally runs about 60:40 (KWIATKOWSKI, 1964). Gross fractionation of the polar and neutral lipids extracted from *Zea mays* and *Typha latifolia* (cattail) is given in Table 10-2. The major fractions of the polar lipids in these two species are lecithin, lysolecithin, phosphoinositol and phosphatidylcholine (GUNASEKARAN and ANDERSEN, 1973).

The comparison of fat and starch contents of different pollen was the basis for the classification by CALVINO (1952) of 1,170 species. She correlated the storage energy source with the dispersal mechanism, i.e. species high in fat are distributed by bees, while pollens high in starch are wind-dispersed.

Roles other than as a storage source of energy have been ascribed to lipids and related compounds in pollen. Esters of fatty acids in corn pollen were suggested as growth substances (FUKUI et al., 1958; FATHIPOUR et al., 1967). Pollen high in fat

Table 10-2. Distribution of different classes of lipids in pollen (GUNASEKARAN and ANDERSEN, 1973)

Lipids	Typha latifolia	Zea mays
	% of total lipids	
Polar lipids	39.7	36.6
Neutral lipids		
Monoglycerides	3.2	4.9
Diglycerides	1.6	4.7
Triglycerides	41.3	19.5
Free fatty acids	3.1	0.0
Sterols	3.1	2.5
Hydrocarbons	7.9	31.7
Total in pollen, % dry wt.	7.6	3.9

have been considered to be selectively sought by bees as a nutritive source. Doubt is cast upon this hypothesis by the work of STANDIFER (1966a) which showed large variations in the fat contents of pollen selected by bees. The lipid content of hand-collected samples of pollen from similar sources of *Populus fremontii* in Tucson, Arizona, was three times higher than the content in bee-collected pollen (STANDIFER, 1966a). Such a magnitude of variation places definite limitations upon the uses of different sources of pollen for chemical analytical data. Some pollen at maturity contain predominantly fat or starch, but the teleological relations of such differences are not as obvious as the physiological functions in pollen growth.

Bees undoubtedly metabolize pollen fats along with proteins and the other constituents in the pollen (Chapter 7). Selection of pollen by bees for fat content is unlikely, although some evidence suggests that a free fatty acid in pollen, octadeca-trans-2, cis-9-12-trienoic acid, is an attractant for the honey bee (HOPKINS et al., 1969). The volatility of this fatty acid was not established. In contrast, however, selection of pollen with high fat content for corn breeding and correlation of the pollen fat level with kernel oil content is a valid application of the index of lipid content.

Fatty Acids

Fatty acids are primarily recovered in the saponified portion of the lipid extract. As demonstrated with corn pollen (BARR et al., 1959), extraction and separation by gas chromatographic analyses of the fatty acids as their methyl esters is rapid and especially useful. To avoid errors in comparative analysis of specific fatty acids, they should be compared on the basis of the quantity per gram pollen dry weight, rather than amount recovered in the saponified fraction alone (STANDIFER, 1966). Three species of *Pseudotsuga* contained 0.76–0.89% dry weight fatty acids compared to 1.25–1.33% in two species of freshly harvested *Pinus* pollen (CHING and CHING 1962). Extracts of bee collected *Taraxacum officinale* pollen contained 7.3% fatty acids, while the 15% dry weight of crude lipid extract in *Corylus avellana* contained 5% fatty acids (SOSA and SOSA-BOURDOUIL, 1952).

Unsaturated fatty acids occur quite commonly; in *Papaver rhoeas* pollen they comprise 91% of the total fatty acids. Analysis of 15 different species of bee-collected pollen showed linoleic acid present in all the pollens; 11 of the 15 contained myristic, stearic, palmitic, palmitoleic, oleic and lauric, arachidic was found in 8, and 2 species contained eicosanoic as a free fatty acid (BATTAGLINI and BOSI, 1968). The unsaturated free fatty acids are easily oxidized by activated atmospheric oxidants such as singlet oxygen. When *Pinus echinata* pollen was exposed, in a water solution, most of the $C_{18:1}$ was converted to $C_{18:0}$ (DOWTY et al., 1973). However, the conversion occurred after a two-hour exposure with an abnormally high concentration of activated singlet oxygen being supplied to the pollen. The surface-localized, or diffusible fatty acid peroxide derivatives formed in such a reaction were suggested as possible sources of adverse pulmonary reactions in man. Epidemiological analyses may help evaluate this hypothesis, which appears at this time, rather untenable as an important environmental factor. However, obviously under certain experimental conditions, unsaturated fatty acids in pollen can be saturated and oxidized to potentially harmful chemicals.

Pine pollen acid-fast lipids, along with the uncommon amino acid α,ε-diamino pimelic acid were suggested as possible causative agents in human sarcoidosis (CUMMINGS et al., 1956; CUMMINGS and HUDGINS, 1958; WALLGREN, 1958). This conclusion was based upon the pattern of disease incidence in relation to pine forest distribution. Subsequently, CUMMINGS (1959; 1964) cautioned against accepting the unproven epidemiological suggestions as valid evidence for pine pollen as the etiological and pathological source of this lung disease. Granulomatous lesions can be induced in animals by injection of abnormally high levels of pine pollen, but not by inhalation of even finely ground pollen (BRIEGER et al., 1962; ARIENZO et al., 1969; ISHIKAWA and WYATT, 1970). Thus, while a few cases of natural pine pollen allergy (Chapter 12) have been reported (ROWE, 1939; PANZANI, 1962; NEWMARK and ITKIN, 1967), there is as yet no valid clinical evidence supporting the hypothesis, based on epidemiological comparisons, that pine pollen and the lipid fraction therein, is the source of sarcoidosis. But neither has definitive negating clinical evidence been produced. Fortunately, forest situations have both a limited pine shedding period and also afford only a very limited source of potential oxygenating, peroxide forming atmospheric factors to interact with the pollen as per the DOWTY et al., (1973) hypothesis discussed above.

Early chemical analysis of *Pinus sylvestris* pollen by KRESSLING (1891) revealed only palmitic acid, but the C_{26} hexacosenoic (cerotic) acid was found a few years later (HENRIQUES, 1897). More recent studies by CHING and CHING (1962), applying gas chromatographic analysis, detected 16 different fatty acids in 5 species of *Pinaceae*. Palmitic, linolenic, and oleic acids were the major components, e.g. 14 to 31% of the fatty acids extracted from *P. ponderosa* and *P. contorta* pollen. Compared to pine, pollen of *Pseudotsuga* species contained relatively little of the C_{18}-unsaturated fatty acid linolenic, but were several-fold higher in the C_{18}-saturated fatty acid oleic. The range of the principal fatty acids are listed in Table 10-3.

Petunia contained four times the fatty acid content present in pine pollen (HOEBERICHTS and LINSKENS, 1968). Palmitic acid was the most abundant fatty

Table 10-3. Distribution of principle fatty acid components in saponifiable fractions of pollen fat[a]

Species	Fatty acids (% by weight of methyl esters)				
	Palmitic $C_{16:0}$	Stearic $C_{18:0}$	Oleic $C_{18:1}$	Linoleic $C_{18:2}$	Linolenic $C_{18:3}$
Taraxacum officinale	20.1	9.3	4.2	14.3	25.4
Pseudotsuga menziesii	20.9	2.7	62.2	11.9	0.9
P. wilsoniana	26.5	2.5	52.9	16.4	0.9
P. macrocarpa	26.4	15.6	39.0	8.0	4.5
Pinus ponderosa	17.6	10.9	23.1	5.4	24.1
P. contorta	13.4	12.2	16.5	4.4	31.5

[a] From data in CHING and CHING, 1962; STANDIFER, 1966.

acid (42%) in *Petunia* pollen, as in pine and hazel nut *Corylus avellana* (SOSA-BOURDOUIL and SOSA, 1954). In *Papaver rhoeas*, palmitate comprised 68% of the fatty acid fraction (KWIATKOWSKI, 1964). *Petunia* pollen also contained a fatty acid with an uneven number of carbons; this may be similar to the branched C_{23} fatty acid found by SOSA and SOSA-BOURDOUIL (1952) in *Corylus avellana* pollen. Such uneven numbered carbon fatty acids have not been found in conifer pollens (Table 10-3). About 77% of the fatty acids in *Petunia* pollen is bound primarily as phospholipid.

Phospholipids in *Oenothera missouriensis* contain 87% of their fatty acids as palmitic, 13% as linolenic, while the glycolipids in this pollen contain 42% palmitic and 58% linolenic (CARON, 1972). Similar analyses should be made of lipid fractions in other pollens to facilitate an understanding of the role and probable sites of activity of these compounds in pollen. The amount of phospholipids in *Oenothera missouriensis*, in particular proportions of phosphatidyl-ethanolamines, -cholines and -inositols were related to pollen genotype (CARON, 1972).

The phospholipid fraction of *Petunia* pollen contained cephalin, lecithin and inositides (HOEBERICHTS and LINSKENS, 1968). Cephalin can contain either an esterified long chain alcohol or a sterol. MCILWAIN and BALLOU (1966) found the saponifiable phospholipid fraction of *Pinus ponderosa* pollen consisted of phosphotidyl derivatives of choline, ethanolamine, inositol, glycerol, serine and bis-phosphatidyl glycerol. Gas chromatographic analysis of the methyl esters of fatty acids obtained from each phosphatide from the pine pollen showed that palmitic, oleic and linoleic are the major fatty acids of each pollen phospholipid. Less sensitive analytical procedures of MICHEL-DURAND (1938) failed to find phytic acid in pine pollen, but did find the hexaphosphate derivatives in several species of angiosperm pollen.

Alcohols and Long Chain Hydrocarbons

The neutral fractions of the lipid extracts of pollen often contain higher alcohols plus saturated and unsaturated straight chain hydrocarbons. *Secale cereale* pollen contained 1.3% of its air dry weight as hydrocarbons. The C_{25}, C_{27} saturated and monoenic hydrocarbons were the most common, although a high percentage, 12.4

and 16.6%, occurred as the C_{29} and C_{31} unsaturated compounds respectively. With increasing chain length, the degree of unsaturation also increased (HALL-GREN and LARSSON, 1963).

Other pollen from which long chain hydrocarbons have been isolated include *Zea mays*: pentacosane, C_{25}; haptacosane, C_{27} (NILSSON et al., 1957); and nonaco-sane, C_{29} (ANDERSON, 1923); *Corylus avellana*: tricosane, C_{23} (SOSA-BOURDOUIL and SOSA, 1954); and *Alnus glutinosa*: mainly heptacosene and nonacosane (NILS-SON et al., 1957). Many of these odd-numbered carbon compounds, presumably derived by decarboxylation of the even chained hydrocarbons, occur as waxy layers on or in the pollen exine.

Long chain alcohols, closely related to the hydrocarbons, generally occur in the same waxy-like fraction. *Pinus mugo* pollen yielded a series of diols, i.e. tetra-cosanol-l, hexacosanol-l, and octacosanol (NILSSON, 1956); *Cedrus deodara* pollen was high in heptacosanol, while pentacosanol was the dominant long chained alcohol isolated from *C. atlantica* (SPADA et al., 1958). The presence of octade-canol-l, stearyl alcohol, in *Ambrosia* pollen was at one time suggested as a factor influencing the allergic response to this pollen (HEYL, 1923), a hypothesis which finds no current support. The alcohol, 31-norcycloartenol, was isolated from the unsaponified fraction of cactus, *Carnegiea gigantea*. An interrelationship between the lower alcohols and hydrocarbons was clearly indicated in studies of six pollen species (KWIATKOWSKI, 1964). The sequence of organic acids and diols in these pollen are closely related chemically, although differences in the components present occur even between pollens in the same family.

Lipid materials are localized in various parts of pollen. Exine and intine are considered the primary sites of long chain fatty acids, alcohols and waxy esters. Lipid fractions in pollen cytoplasm are primarily a triglyceride mixture of linoleic, oleic and palmitic acids (SCOTT and STROHL, 1962). The isoprene derived terpenes and sterols, on the other hand, are probably associated with the cell wall and membranes. However, the ease of extraction of sterols from pollen suggest they are also contained in the cytoplasm. Intracellular sites of sterols or terpenes in pollen have not as yet been determined.

Terpenes. Terpenes are among the essential oils and organic acids contribut-ing to the distinct flavors of some wines. The pollen source may thus be important in grape breeding and viticulture. EGOROV and EGOFAROVA (1971) reported the following essential oils, terpenoid components, in two varieties of Muscat grape pollen: farnesol, geraniol, nerol, linalool and β-ionone. The three monoterpenes assayed quantitatively varied greatly between the two varieties (Table 10-4).

Sterols. The amounts and types of sterols vary between different pollen species. In early studies of *Ambrosia* pollen, KOESSLER (1918) reported 0.34% of the organic constituents were sterols. *Zea mays* pollen was shown to contain sterols by MI-YAKE (1922) and ANDERSONS (1923). The corn sterols, present at about 0.1%, are primarily cholesterol with some stigmasterol (GOSS, 1968); androgenic activity was not found (BUTZ and FRAPS, 1945). Pollen of all 26 species analyzed, repre-senting a wide range of families in both gymnosperms and angiosperms, contained steroids or triterpenoids (HISAMICHI and ABE, 1967).

The animal hormone β-estradiol was first reported in pollen by SKARZINSKY (1933). EULER et al. (1945) confirmed the presence of this steroidal hormone

Table 10-4. Composition of monoterpenes in *Vitis domestica* (EGOROV
and EGOFAROVA, 1971)

Terpene	Muscat pollen variety	
	Alexandria (µg/gm)	Aligote (µg/gm)
Linalool	0.66	1.08
Nerol	3.49	0.84
Geraniol	1.46	0.52

reaction in pollen. Considerable work on the isolation and characterization of this female hormone from date palm pollen was carried out by WAFA and his colleagues (RIDI and WAFA, 1947; HASSAN and WAFA, 1947; SOLIMAN and SOLIMAN, 1957). The definitive isolation of estrone and cholesterol from date palm, *Phoenix dactylifera* L., pollen was completed by BENNETT et al. (1966). Estrone is also present as a relatively high percent of the sterols in *Hyphaene thebaica* pollen (AMIN and PALEOLOGOU, 1973).

An acetone-powder extract of pollen sold commercially by the A. B. Cernelle Company (Table 4-1) was found to be high in lipids (HALLGREN and LARSSON, 1963) and sterols of the 3-β sterol complex, identified as stigmasterol and estrone (KVANTA, 1968). Pollen of six species: *Zea mays, Secale cereale, Pinus montana, Phleum pratense, Alnus glutinosa* and *Dactylis glomerata* are used in preparing the extract. The extractable, bound or combined forms of sterols vary in different pollen species. Free and esterified cholesterol occurs in the extract of ungerminated *Petunia* pollen (HOEBERICHTS and LINSKENS, 1968). Sterol glucosides and esterified derivatives probably occur along with the free sterols. *Pinus pinaster* pollen extracts contained about 15% more bound sterols and alcohols than were free (DUNGWORTH et al., 1971).

Using the Libermann-Buchard color reactions of sterols, efforts were made in early survey studies to establish phylogenetic relationships based on color reaction in pollen of different species and genera (BERGER, 1933). Further studies have shown that grouping and classifying plants taxonomically on the basis of their pollen sterols is not valid (BARBIER, 1971). Analysis of unsaponifiable lipid extracts of pollen by gas chromatography and mass spectrophotometry has extended our knowledge of pollen sterols and their interrelationships (HÜGEL et al., 1964; BARBIER, 1966; 1971; DEVYS and BARBIER, 1967). The general conclusion of STANDIFER et al. (1968) based on data from 15 pollen species, is that no simple phylogenetic relationship is discernible from the sterol data. Distribution of the main pollen sterols and approximate percentages are listed in Table 10-5. Although cholesterol is present in many pollen species, in others the 24-methyl or 24-ethyl derivative predominates (KNIGHTS, 1968). Unlike most pollen which contain a mixture of up to 7 different sterols, *Quercus robur* contains 100% β-sitosterol as its sterol (BATTAGLINI et al., 1970).

Corrections are necessary for purity of the pollen sample analyzed, i.e. bee-versus hand-collected pollen. Some of the seasonal variation reported in pollen sterol content, particularly in the fucosterol and β-sitosterol fractions (BATTAG-

Table 10-5. Pollen sterols

Plant family	Species	Collection method	Principal sterols (% in total sterols)
Compositae	*Helianthus annus*	Hand	β-sitosterol, 42
	Taraxacum officinale	Bee	β-sitosterol, 38
	Hypochoeris radicata	Bee	cholesterol, 90
Rosaceae	*Malus sylvestris*	Bee	24-M-ch[a], 50
	Pyrus malus	Hand	24-M-ch, 60
Betulaceae	*Corylus avellana*	Hand	β-sitosterol, 75
	Alnus glutinosa	Hand	β-sitosterol, 64
			C_{29}-di-unsat, 17
Salicaceae	*Populus fremontii*	Hand	cholesterol, 59
	Salix spp.	Hand	24-M-ch, 50
			β-sitosterol, 25
Gramineae	*Zea mays*	Hand	24-M-ch, 59
			β-sitosterol, 17
			campesterols, 12
			stigmasterol, 12
	Secale cereale	Hand	24-M-ch, 49
			β-sitosterol, 13
	Phleum pratense	Hand	24-M-ch, 62
			β-sitosterol, 13
Pinaceae	*Pinus sylvestris*	Hand	β-sitosterol, 54
			24-M-ch, 9
	Pinus montana	Hand	β-sitosterol, 65
			24-M-ch (trace)
			campesterol, 17
			cholesterol, 8
	Pinus mugo	Hand	β-sitosterol, 65

[a] 24 Methylene-cholesterol.

LINI et al., 1970) may be related to the metabolism of the collecting bee, or to the redox levels in the pollen tissues, where conversion of fucosterol to sitosterol can also probably occur (HÜGEL, 1965). Bees, as many insects, seem to require an exogenous source of sterols (HÜGEL, 1962) for development (Chapter 7). The chemical relationship of a few common pollen sterols are illustrated in Fig. 10-1. Fucosterol and isofucosterol probably intermediates between β-sitosterol and stigmasterol or cholesterol occur at high levels in oil palm, *Elaeis guinensis* (RICHERT, 1971) and in lower percentages in many other pollen species.

Nothing is known about changes in pollen sterols with aging or germination. The triterpenoid squalene is present in corn pollen at levels of 0.02% (DEVYS and BARBIER, 1965); other sterols are also present. The C_{29} sterols are the most prevalent component in the palm pollen sterol fraction (RICHERT, 1971), and in many other pollens; the C_{28} sterol occurs at about 1/2 the level and C_{27} at 1/10 that of the C_{29}. This suggests that C_{29} sterols are the terminal or dominant metabolic pathway, which accordingly suggests that sterols in pollen are derived through the acetate mevalonate pathway first recognized as the mechanism of animal sterol biosynthesis. Pollinastanol, a 14-methyl sterol, was isolated from *Taraxacum dens-leonis* pollen, tritiated in the 3-carbon position and infiltrated into

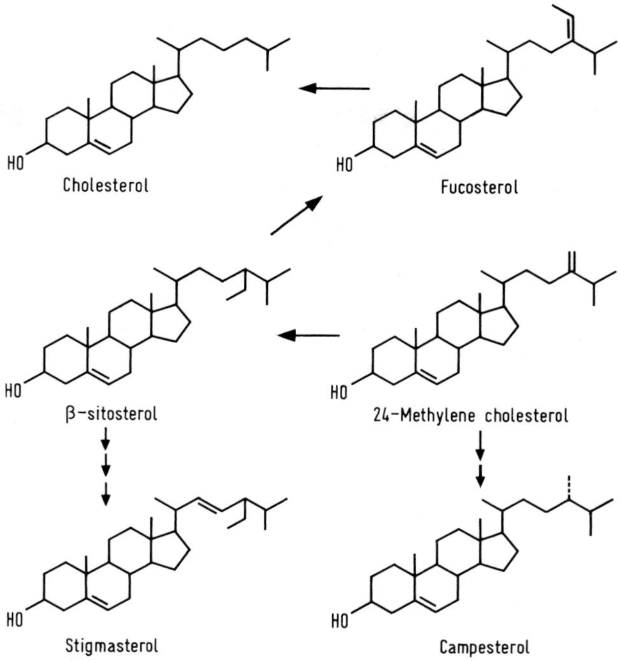

Fig. 10-1. Sterols commonly isolated from pollen, arrows indicate possible relationships

leaves of *Nicotiana tabacum* plants, where it was converted into cholesterol (DEVYS et al., 1969). The same pathway may possibly function in pollen and may include or combine compounds containing the related triterpenes, cycloartenol and 31-norcycloartenol which, along with the 31-nordihydrolanosterol, have also been isolated from *Taraxacum* pollen (DEVYS et al., 1970; ATALLAH and NICHOLAS, 1971). The cyclic triterpenic alcohol 31-norcycloartenol, a possible precursor of pollen sterols, was isolated from the unsaponified fraction of cactus *Carnegiae gigantea*. This latter pollen contains 94% of its sterol as 24-methylene cholesterol (BARBIER, 1971), a common dominant sterol in pollen (Table 10-5). Further clues as to the metabolic role and importance of these compounds in plants may be gained by radioisotopic studies on pollen.

Interestingly, the male sex-organ-inducing sterol, antheridiol, produced by the male mycelium of *Achlya bisexualis* (MCMORRIS and BARKSDALE, 1967), can be chemically derived from sterols common to pollen. No specific male sterol has as yet been isolated from pollen or pistil tissues although after hydrolysis with *β*-glucosidase (Chapter 14), low levels of testosterone, epitestosterone and androstenedione were found in *Pinus sylvestris* (SADEN-KREHULA et al., 1971), suggesting that these male hormones may be present as glucuronides. Some sterols can stimulate pollen tube growth when added to the *in vitro* growth medium (MATSUBARA, 1971; STANLEY, 1971).

Chapter 11. Amino Acids and Proteins

Analyses have shown that all the essential amino acids are present in pollen. The free amino acids do not necessarily reflect the mole ratio of amino acids in pollen protein; polypeptides may also be conjugated with glucosides and pigments. The amino acid content and total nitrogen can vary with the climatic and nutritional conditions of the plants on which the pollen matures. Thus, growing *Zea mays* under optimum nutrition and light conditions increased the levels of alanine, proline, arginine and protein in its pollen (TSELUIKO, 1968). Mature sterile corn pollen had lower levels of proline and increased levels of alanine than viable, fertile pollen (KHOO and STINSON, 1957). BAUMANE et al. (1968) correlated viability and subspecies of *Prunus* with amino acid differences during pollen development. Storage and handling methods also modify the free amino acids; extended storage decreased, in particular, glutamine and glutamic acid (KATSUMATA et al., 1963). Such changes may be very critical for the survival of grass and other ephemeral pollen and may account for some of the quantitative variation in amino acids reported among species (BIEBERDORF et al., 1961).

Because of the importance of pollen protein in allergy responses and bee nutrition, most reports of pollen amino acids represent data from hydrolysates of total protein fractions. We will review information relating to studies of the free amino acids before comparing data from protein hydrolysates.

Free Amino Acids

Total free amino acids and amides are usually higher in pollen than in leaves or other plant tissues (VIRTANEN and KARI, 1955). Calculations with *Petunia* indicated that 6% of the pollen dry weight is composed of free amino acids, while 25% is accounted for as amino acids after hydrolysis, peptide- and protein-bound amino acids constituting 6% and 13%, respectively (LINSKENS and SCHRAUWEN, 1969).

Proline is one of the most abundant free amino acids found in pollen; it comprises 1.65% of the dry weight in some grasses (BATHURST, 1954), and as high as 2.2% in apple (TUPÝ, 1963). In corn, free proline may constitute up to 1% of the pollen dry weight (ANDERSON and KULP, 1922; BRITIKOV and MUSATOVA, 1964). Proline is not incorporated into pollen protein in the mole proportion in which it occurs as a free amino acid in extracts (Goss, 1968). Only traces of proline were found in 50 analyses of fresh pollen of *Pinus nigra* and *P. sylvestris* made in each of 5 successive years (DJURBABIĆ et al., 1967). However, a similar alcoholic extract of *P. thunbergii* indicated proline in the free amino acid pools (VAN BUIJTENEN, 1952). Since the extraction and chromatographic procedures were reasonably similar, one might attribute these differences in reports for *P. sylvestris* as due to the fact that *P. sylvestris* pollen used by VAN BUIJTENEN (1952) was harvested from trees growing outside the normal range and therefore reflected differences in climatic and nutritional conditions prevailing during development.

Unbound hydroxyproline does not occur in any substantial amount. In *Picea glauca* the percent soluble nitrogen found as hydroxyproline was 0.4%, as compared to 56.6% occurring as free proline (DURZAN, 1964). The general incorporation pattern of oxygen into the proline *in situ* to form hydroxyproline in the wall probably prevails (LAMPORT, 1963); the predominant amount of hydroxyproline in walls occurs in a glycopeptide (DASHEK et al. 1970).

On a dry weight basis, relative values for extractable proline from pine pollen were 16.9 n mols/gm in *Pinus halepensis*, but only 1.0 n mol/gm in *P. echinata* (BIEBERDORF et al., 1961). Proline from three mutant strains of corn averaged 186.3 n mols/gm and ranged from 161.8 to 217.4 n mol/gm pollen (PFAHLER and LINSKENS, 1970). A potential metabolic role for proline in pollen tube growth has been suggested (BRITIKOV et al., 1964; TUPÝ, 1964; DASHEK et al., 1970). By following L-proline-C^{14} movement into the developing microspores, BRITIKOV et al. (1964) showed it was translocated from the vegetative anther tissues and was fairly evenly distributed throughout the pollen protoplast. It was not correlated with taxonomic or other species characteristics such as monoecious-dioeciousness of the plants or entomophilic-anemophilic mechanisms of pollination or with the expression on incompatibility. BRITIKOV et al. (1964) concluded proline is used by pollen and pistil tissues during growth, possibly in tube or wall protein formation, or in a more fundamental metabolic reaction associated with the sexual process. DASHEK et al. (1972) found evidence that the 0.14% soluble proline in the cytoplasm of ungerminated *Lilium longiflorum* pollen serves as a precursor for the wall-bound proline and hydroxyproline in the germinating pollen tube.

An unusual free amino acid-like compound found in pollen is pipecolic acid (VIRTANEN and KARI, 1955). *Rosa damascena* pollen contained pipecolic acid as a free amino acid; it was not present in protein hydrolysates (ZOLOTOVITCH and SECENSKA, 1962). It is possible that pipecolic acid, commonly derived from lysine, may serve as a storage precursor for amino compounds (FOWDEN, 1965). Taurine, a sulfur-containing amino acid, has been found in five species of bee-collected angiosperm pollen (MARQUARDT and VOGG, 1952). Taurine may involved in translocation or regulation of cation influx (JACOBSEN and SMITH, 1968).

Some values for free amino acid content extracted from different pollen (BIEBERDORF et al., 1961) are compiled in Table 11-1. The qualitative range, in 107 different species analyzed, was from a low of only 4 amino acids in one grass, *Agrostis alba* to a high of 19 in western cottonwood, *Populus sargentia*. About 50% of the 107 species analyzed had 10 or more free amino acids. Pollen in certain taxonomic groups, such as Salicaceae and Fagaceae, yields a large number of different free amino acids; others, such as Liliaceae and Gramineae, generally have fewer amino acids. Amino acids detected in trace levels of chromatograms (Table 11-1) were not included in numerical qualitative tabulations. NIELSEN et al. (1955) and BELLARTZ (1956) routinely detected not only the free amino acids listed in Table 11-1, but also good levels of α- or γ-amino butyric acid and serine. The amides, asparagine and glutamine, and trace levels of cysteine and threonine are

Table 11-1. Free amino acids detected in pollen[a] (µ moles/g pollen)

| | Species | | | | | |
	Ambrosia aptera	Solidago speciosa	Plantago lanceolata	Pinus halepensis	Pinus echinata	Cynodom dactylon
α-β-alanine	16.0	4.8	T	8.9	18.0	30.0
arginine	19.1	1.5	1.5	10.2	1.7	36.5
aspartic acid	T	3.2	T	13.5	10.7	17.7
glutamic acid	0	0	T	T	2.7	25.4
glycine	5.7	0.9	4.1	5.2	8.7	20.1
histine	18.4		0	3.9	0	7.3
leucine/isoleucine	4.8	3.1	T	6.7	13.5	34.9
lysine	0	T	10.0	11.9	3.6	17.3
methionine	T	0	0	T	0.4	T
phenylalanine	0	0.7	0	T	0.7	3.7
proline/hydroxyproline	5.9	2.5	T	16.9	1.0	32.3
tyrosine	T	T	0	T	1.4	2.6
valine	2.6	2.9	T	5.3	11.5	20.3

[a] From BIEBERDORF et al., 1961.
T = trace amount detected.

also often found in alcohol-water extracts. The oligoamine, spermine, which occurs bound to fungal and animal nuclei acids is absent in some pollen species (NISHIDA, 1935). Spermine and spermidine could be found in *Petunia* pollen. On the basis of uridine-5-T and thymidine-T incorporation, which are stimulated by spermine and spermidine in the germination medium of pollen, it is suggested, that polyamine concentration may be a factor in the regulation of the nucleic acid synthesis (LINSKENS et al., 1968).

SHELLARD and JOLLIFFE (1968) failed to detect any qualitative differences in the free amino acids extracted from pollen of 11 different grass species. Killing grass pollen by drying 4 days over silica gel did not change the kind of extractable free amino acids, although it modified the quantity released. *Zea mays* plants exposed to low levels (3 ppm) of ozone produced pollen with a 50% increase in free amino acids (MUMFORD et al., 1972). Glutamic acid, α-alanine and proline showed the greatest changes; suggesting that amino acid synthesis or protein hydrolysis was enhanced in plants exposed to ozone. Non-viable pollen resulted on corn plants grown at high concentrations of ozone. When the amino acid contents in pollen of two quince (*Cydonia vulgaris*) varieties, one with 20% the other with 40% viability were compared, the total and free amino acid content were greater. In the less viable pollen, a 30 to 50% increase in free lysine, arginine and valine occurred with decreasing viability (DŽAMIĆ and PEJKIĆ, 1970). Proteins in the less viable pollen were generally lower in the amino acids found at increased levels in the free amino acid pool. This suggests that in *Cydonia* and probably other pollen, if certain paths to protein synthesis are interrupted, the percent sterile grains will increase, along with the increase in unincorporated free amino acids in those less viable grains. Quantitatively, the amino acids extracted from dead and living *Petunia* pollen increased to 0.32 in dead vs. 0.21 n mol/gm in live pollen (LINSKENS and SCHRAUWEN, 1969). Storage of *Zea mays* pollen for 5 days, resulting in non-viable pollen, produced changes in all 16 amino acids

Table 11-2. Changes in extractable amino acids during storage of
Zea mays pollen (LINSKENS and PFAHLER, 1973)

Amino acid	Storage days at 2° C		
	0	1	5
	μ moles/mg dry wt.		
Aspartic acid	4.12	5.86	10.20
Threonine-serine	33.19	34.36	35.67
Glutamic acid	7.36	6.04	3.95
Proline	208.31	198.26	191.08
Glycine	1.22	1.23	1.06
Alanine	12.57	12.58	8.77
Valine	1.19	1.13	1.21
Isoleucine	0.14	0.21	0.33
Leucine	0.41	0.47	0.64
Phenylalanine	0.59	0.69	0.83
Ethanolanine	2.60	3.16	8.61
α-Amino butyric acid	2.66	4.43	6.08
NH_3	5.81	5.87	7.30
Lysine	0.78	0.78	0.83
Histidine	1.23	1.21	1.27
Unknown	5.90	5.84	5.87

assayed by LINSKENS and PFAHLER (1973); the greatest increase in extractable amino acids occurred in aspartic acid, with glutamic acid showing the greatest percent decrease (Table 11-2). Possibly the increase in α-amino butyric acid during storage reflects a decarboxylation of the glutamic acid. The total quantity of free amino acids in 7 pine pollen species, harvested in an approximately similar manner, depended on the species (VAN BUITJENEN, 1952).

The variation in amino acids resulting from differences in handling and storage was avoided in a comparative study of *Zea mays* pollen derived from known endosperm mutants. All pollen was harvested and rapidly dried over silica gel at 30° C (PFAHLER and LINSKENS, 1970). The waxy mutant produced pollen with increased aspartic acid, valine and histidine and decreased α-amino butyric acid. The allele for shrunken endosperm increased the glutamic acid, proline, lysine and histidine as much as 100%, but decreased α-amino butyric acid. The sugary endosperm mutant produced pollen in which the change in free amino acids reflected the initial genotype as well as the mutant. Genotypic differences have also been related to the free amino acids in *Oenothera* pollen (LINDER and COUSTAUT, 1966); however, in the latter species gene markers are not as well recognized as in corn.

The presence and activity of glutamic dehydrogenase would influence the level of amino butyric acid formed, while other reductases could influence the level of proline formed from glutamate. Thus modified levels of free amino acids, particularly those occurring in low amounts, may offer a tool for detecting the expression of gene-controlled enzyme synthesis in non-lethal plant mutations.

Free amino acids were compared in pollen from fertile 2 n apple varieties with that extracted from the poorly germinating, sterile 3 n varieties (TUPÝ, 1963). The

Table 11-3. The relation of free amino acids and apple pollen fertility

Variety	μg/mg fresh weight		
	Proline	Histidine	P/H
2n-Golden Winter Pearmain	17.9	0.6	29.4
James Grive	14.9	0.3	45.9
Ontario	14.0	0.5	27.8
3n-Boskoop red	8.6	1.6	5.2
Pepping	7.8	1.7	4.7
Habert's	5.4	1.5	3.8

(From: TUPÝ, 1963; 2n derived pollen are fertile, 3n functionally sterile)

main differences occurred in levels of proline and histidine and were suggested as an index (proline/histidine) for differentiating between fertile and sterile apple varieties (Table 11-3). While the level of glutamic dehydrogenase may account for the increased proline formation in fertile pollen, the number of complex enzyme steps leading to formation of histidine suggest that the mutation is accompanied by multiple gene changes, or, the increased levels of this amino acid might be accounted for by a gene block of purine metabolism, with a rechannelling through imidazole glycerophosphate or a failure to incorporate histidine into protein in the sterile variety.

It will be interesting to unravel the mechanism behind these gene-linked biochemical differences. Comparison of the proteins and their hydrolysates will help construct the balance sheet for understanding gene control of phenotypic expressions and incompatibility in higher plants.

Amino Acids by Hydrolysis

Extraction of pollen with water, alcohol and ether yields a residue with 2–6% nitrogen. This residual nitrogen, usually determined by the Kjeldahl method, is commonly used as a basis for computing the protein content, i.e. percent x 6.25 = protein. When the residue is hydrolyzed in strong acid, e.g. 25% HCl at 105° C for 24 hrs, free amino acids are produced from proteins and polypeptides. Precautions must be taken for interpreting certain amino acids. For example, cysteine may be destroyed or changed during acid hydrolysis; others may be incompletely separated during chromatography. But the detected patterns are usually quite reproducible.

The amino acids in pollen proteins vary with the species and generally differ from those freely extractable. Concentration of all amino acids is considerably higher in the bound than the free fraction; but the percent distribution tends to follow the same pattern, i.e. when an amino acid is abundant in the soluble fractions it is present in relatively high levels in the protein hydrolysate. However, in pine as in other species, several amino acids such as phenylalanine and leucine, usually not found free, may occur in a fairly high percentage in protein (Table 11–4). Proline, tyrosine and tryptophan were only found at trace levels in *Pinus sylvestris* and *P. nigra* (DJURBABIC et al., 1967), although, as indicated, they have been found in other pines (HATANO, 1955), and are commonly well represented in protein hydrolysates of other species.

Table 11-4. Comparison of *Pinus* pollen bound and free amino acids

Acid	P. sylvestris				P. nigra			
	bound		free		bound		free	
	mg/g	%	mg/g	%	mg/g	%	mg/g	%
α-alanine	5.20	5.9	0.45	3.7	11.10	9.2	0.70	6.3
arginine	16.80	19.0	4.10	33.2	24.80	19.3	3.25	30.0
asparagine	26.70	30.0	2.80	22.6	23.20	18.8	2.00	18.4
cysteine	2.00	2.2	0.80	6.6	8.00	6.6	0.60	5.4
phenylalanine	4.00	4.4	0.00		4.50	3.7	0.00	
glutamic acid	16.80	19.0	1.30	10.5	26.50	21.6	2.10	19.3
lysine	9.20	10.0	1.56	12.8	9.45	7.7	1.84	16.9
leucine	4.61	5.1	0.70	5.7	9.70	7.8	0.00	
methionine + valine	4.01	4.4	0.60	4.9	6.70	5.5	0.40	3.7

(Adapted from DJURBABIĆ et al., 1967).

Pollen proteins contain large amounts of basic amino acids, glutamate and arginine and the amide asparagine, resembling storage proteins in many seeds. When the levels of 12 amino acids in corn pollen hydrolysate were compared to those isolated from the purified corn proteins, zein and gluten, the percent of lysine was higher in pollen while leucine and glutamine were lower. The internal configuration and functional capacities of the protein may provide the rationale for the evolutionary predominance of these amino acids. Direct comparison of pollen to storage organs such as seeds, is of limited value. In seeds, storage protein is often concentrated in spezific cytoplasmic granules. Such organelles have not been found in pollen, where protein is primarily distributed among the many organelles including ribosomes, mitochondria and nuclei.

The pollen amino acid hydrolytic data is especially meaningful in bee and animal nutrition studies (Chapter 7). VIVINO and PALMER (1944) concluded that protein in bee-collected pollen was deficient in tryptophan and methionine for the bee requirements. However, insect nutritional studies afford no information about the role of amino acids for pollen growth. The pollen growth experiments usually correlate gross changes in free and bound amino acids with tube extension *in vitro*. Only occasionally are the hydrolytic data from pollen used to design an amino acid nutrient supplement for pollen growth studies *in vitro*.

Generally in most pollen 2 to 5 amino acids constitute over 50% of the protein weight. The other 50% of the proteins are generally built from the remaining 15 to 18 common amino acids. The different subcellular particles have different protein components relating to their function. Pollen of related species do not differ greatly in their amino acid composition; the spectral quality differences are broadest between genera and families.

Proteins

Several different system are used to classify proteins. One system compares simple to conjugate proteins; another compares enzymes to structural proteins. Some system, i.e. those based on extractive properties, have provided nomenclature used to group plant proteins.

Simple proteins include the globulins, albumins, prolamines and glutelins; they are differentially extracted and yield only amino acids on hydrolysis. Conjugated proteins are combined with a non-protein group; these include nucleoproteins, phosphoproteins, lipoproteins and glycoproteins.

One pollen grain contains several thousand different enzymes and isozymes, but probably only 100 or so non-enzymatic proteins. Some proteins in the plastids, membranes, ribosomes and organelles may also act as enzymes. In pollen, proteins and enzymes are also present in the mature grain wall (STANLEY and LINSKENS, 1965; KNOX and HESLOP-HARRISON, 1970). As the microspore approaches maturity, proteins different than those present in prior stages form. Gene expression during the advanced stages of microspore development in *Lilium henryi* generally superimposed new protein upon prior protein; however, some early developed proteins were completely absent from the mature pollen (LINSKENS, 1966). For reasons of simplicity, enzymes will be discussed subsequent to this general review of proteins (Chapter 14).

Protein values between 5.9 and 28.3% of pollen residue have been reported (Table 8–1). As already noted (Chapter 8), exine residue in gymnosperms may be high and thus reduce the percent protein. However, the variation in protein content per mg cytoplasm is probably similar in different species and might afford a more valid index. TODD and BRETHERICK (1942) reported pollen of date palm, *Phoenix dactylifera*, as containing 35.5% protein. This pollen is high in ash (Table 8–2) and low in carbohydrate (Table 9–1). But *Populus nigra* var. *Ital.* contains 36.5% protein and is not abnormal in content of ash or carbohydrates (STANDIFER, 1967). In *Cycas revoluta* pollen from different trees the crude protein varied from 32.9 to 33.8% dry weight, but in purified protein fractions it varied from 24.4 to 27.4% (NISHIDA 1935). Thus within a single species, even grown under similar conditions, variations of 5% are common in protein content of pollen on different plants. Qualitative differences in protein have also been observed and may be highly dependent on species variation. The albumin content of *Zea mays* pollen protein varied between 3.8 and 20.3% depending on the variety (ROSENTHAL, 1967).

Efforts have been made to correlate the protein content with other pollen characteristics. Differences in crude protein content do not correlate with whether it is from an entomophilous or anemophilous plant; the nutritive value for bees being also not directly correlated with the protein quantity, a qualitative factor is of greater importance (TODD and BRETHERICK, 1942).

Methods of characterizing proteins by disc gel acrylamide electrophoresis have been applied to pollen (BINGHAM et al., 1964; LINSKENS, 1966). Such analyses indicate 15 to 25 recognizable protein bands can be extracted (Fig. 11–1). Basic polyacrylamide gels separate about twice the number of pollen proteins as acid gels (PAYNE and FAIRBROTHERS, 1973). Location of protein bands on the gels was correlated with species and intrageneric differences; the method, in some cases, offers an additional tool for genetic analysis. Age and prior handling of the pollen can modify the protein pattern (BINGHAM et al., 1964). On storage, *Betula* pollen increased the positive charge of proteins extracted with salt solution (PAYNE and FAIRBROTHERS, 1973). As will be discussed, gel electrophoretic characterization is a particularly powerful tool for enzyme analysis and relating, by serological techniques, proteins to genetic differences (Chapter 14).

Fig. 11-1. Patterns of soluble proteins from pollen of *Lilium longiflorum* separated on acrylamide gels by electrophoresis (DESBOROUGH and PELOQUIN, 1968). Ace PG: pollen grains from cultivar Ace, Cro. PG: pollen grains from cultivar Croft

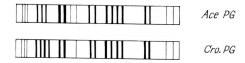

Ace PG

Cro. PG

Serological Reactions

Immunological assay techniques developed by bacteriologists have been applied to pollen. In this procedure, antibodies formed in animal blood serum in response to injected pollen antigens are tested for reactivity against other pollen. Unfortunately, the serological method of protein recognition is not absolutely specific. Although this method has not yet provided real chemical information about pollen proteins, it is a sensitive tool for comparing and separating related species.

Serological-type methods were applied in plants by PICADO (1921). He injected water and sucrose suspensions of maize pollen into developing cactus (*Opuntia*) fruit. After 8 days fruit juices were extracted, filtered, centrifuged and the clear extract tested against suspensions of maize and sorghum pollen. The agglutination reaction was observed after 2 hours. The injected pollen was related to antibody formation in the *Opuntia* serum. These pioneering immunological-type experiments in plant tissues were not extended to animal serums; also PICADO (1921) did not examine the normal plant tissues in which pollen grows. If this early study of PIGADO's can be reproduced, and plants are shown to be capable of synthesizing protein antibodies, in response to allergens, a whole new branch of serological studies may result. Some 30 years later KESZYTÜS (1950) and LEWIS (1952) applied animal serological techniques to characterize pollen, and pollen-style interactions; that work was based upon postulations by EAST (1929) relating to self-sterility in plants.

EAST (1929) stated: "If we assume that the secretions of a pollen tube bearing a given gene, say S_1; act as antigens against the stylar tissue bearing the genes S_1; if we further assume that the stylar tissue in which the gene S_1 is present forms antibodies against such a pollen-tube and thus inhibits its growth; then all requirements (for a reaction analogous to reactions found in immunology) are satisfied."

Serological reactions were applied by KESZYTÜS (1950; 1957) to show the degree of commonality within *Rosaceace* fruit cultivars. Antiserum was prepared for three cultivars of each genera: pears (*Pyrus*), apples (*Malus*) and plums (*Prunus*). Apple antiserum did not react with pear or plum pollen extracts, but did react with the different apple cultivars. Tests with the immune serum of each genus reacted only with related species and not with pollen of other genera in this family. KESZYTÜS (1950) observed no serological differences between selections of cultivars in their reacting precipitin titres.

LEWIS (1952) studied reactions between pollen from *Oenothera organensis* with recognized incompatible allelic genotypes. Antisera were prepared by injecting rabbits with a saline solution of lipid-free pollen. After about 40 days the serum

Test antigen pollen extracts

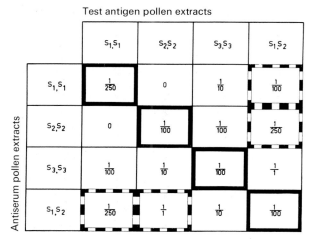

Fig. 11-2. Serological precipitation tests between pollen of *Petunia hybrida* (LINSKENS, 1960). Precipitate is induced following exposure of serum to test antigen at the concentration of the antiserum titre dilution given in the box. Reaction: Heterologous — no S alleles common; fully homologous — both S alleles common; half homologous — one S allele common

containing antibodies was isolated from the rabbits. Antisera from the rabbits, at various dilutions were tested with antigens from various pollen extracts. The cross-reactivity is considered to represent the number of homologous genes present. A less complex genetic system than *Oenothera* was analyzed in *Petunia hybrida* (LINSKENS, 1960). The test antigens reacted most strongly, i.e. required lower concentration of titre, when mixed with the homologous antisera (Fig. 11–2). This technique was used to test the potential of *Petunia* styles to form reacting „defense" bodies to specific gene-controlled diffusates from the pollen in *in vivo* growth. The protein antigens reacting with the female tissue are presumed to be associated with the pollen exine and intine (TSINGER and PETROVSKAYA-BARANO-VA, 1961; KNOX et al. 1972). Allegenic proteins, similar to antigen-E (Chapter 12) from *Ambrosia* intine, and other antigens and enzymes were observed to be released into mucilageous polysaccharides coating in *Cosmos* and other pollen grains shortly after contacting the stigma surfaces (KNOX, 1973). The human reacting antigens from *Cosmos* were apparently not factors involved in blocking tube growth in the incompatible reaction.

The pollen-protein antisera reaction can also be developed on agar plates, i.e. the Ouchterlony technique, or on acrylamide gels (HAGMAN, 1964; LEE and FAIR-BROTHERS, 1969). When serological reactions are developed on acrylamide gels, the reacting bands are not as precise nor as readily interpretable as those produced by immuno-diffusion (Fig. 11–3). Fluorescent dyes can also be introduced into specific serum antibodies to permit localization of antibody-antigen reactions *in vivo*, such as on the surface of germinating pollen tubes (HAGMAN, 1964; KNOX and HESLOP-HARRISON, 1971).

Because antibody-antigen reactions lack absolute specificity, several alternative interpretations of serological reactions are possible. Antibodies may be

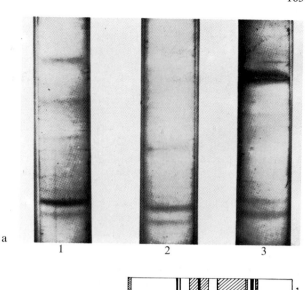

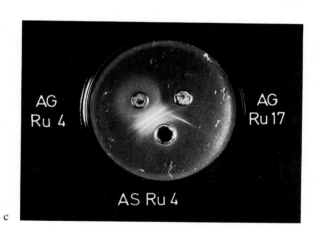

Fig. 11-3 a-c. Serological reactions pollen antiserum. (a) *Betulaceae* antiserum on acrylamide gel. The electrophoretic separation on acrylamide gel of the soluble protein of birch pollen. (1) Pollen of *B. verrucosa,* tree no. 2. (2) Pollen of *B. pubescens,* tree no. 13. (3) Pollen of *B. Mewediewi.* (b) *B.* antiserum reaction on agar slide by immunodiffusion. (1) Pollen of *B. verrucosa.* (2) Pollen of *B. pubescens.* (From MAX HAGMAN, 1964). (c) Serological reaction between two pollen species, precipitation in agar gel plates, carried out as double diffusion (method HAGMAN, 1971); AG Ru 4 = antigen *B. verrucosa,* AG Ru 17 = antigen *B. pubescens,* AS 4 = antiserum *B. verrucosa.* The birches were from Ruotsinklä Finnland). (From MAX HAGMAN, 1971)

formed to polysaccharides or glycoproteins as well as to proteins diffusing from the pollen. The reactivity, binding capacity, of an antibody also depends upon the animal species (rabbit, horse, rat, etc.) from which it was prepared as well as its condition of preparation and storage. However, while there are limitations in the application of serological techniques for understanding the genetics of pollen, they have been the basis for great progress in understanding and treating human allergy responses.

Chapter 12. Pollinosis

Human allergic reactions can be caused by many environmental substances. Common carriers of allergenic substances include pollen, mold spores, epidermoids (i.e. human dandruff, animal hair, feathers and dander), house dust, ticks, mites, insect and reptile venoms, drugs and certain foods. However, the allergic response to pollen, called hayfever or pollinosis, is the most widely recognized allergy. Pollen primarily affects the mucous menbrane of the upper respiratory tract. Clinically, hayfever is described as allergic rhinitis or conjunctivitis and is characterized by intense sneezing, watery eyes, nasal obstruction, itchy eyes and nose, and often coughing. Hayfever reactions usually occur minutes after exposure to the offending pollen (LINSKENS and v. BRONSWIJK, 1974).

Allergic reactions may occur in many different organs or tissues. If the lower respiratory tract, lungs or bronchi are affected, the clinical reaction will be that of asthma. Other reacting organs may be the skin (urticaria), the gastrointestinal tract, the central or peripheral nervous system, or the cardiovascular system. The organ involved in the allergic reaction is commonly referred to as the "shock organ".

Although the portal of entry of an allergen largely determines which shock organ will be involved in the allergic reaction, remote reactions do occur, e.g. allergic rhinitis from oral ingestion or parenteral injection of an antigen, or gastrointestinal symptoms from inhalation. Many different antigens are potentially responsible for such reactions. The most important allergic antigens are proteins or polypeptides, although polysaccharides, glycoproteins, and lipoproteins can also be effective antigens.

For a substance to be an important cause of a respiratory allergy such as hayfever, it must be contained in the inhaled air in relative abundance and release a chemical antigen (allergen) which fulfills certain chemical criteria (WITTIG et al., 1970). The antigen should:

1. be foreign to the species;

2. in general, have a molecular weight over 10,000;

3. the molecular structure should possess a certain rigidity as is usually conferred by aromatic groups, disulfide linkages or double bonds;

4. the molecular surface configuration must afford "polar groups" for attracting antibodies and conveying specificity;

5. be metabolized by the body in a specific period of time.

Proteins and certain other chemical moieties which can be eluted from pollen and mold spores fulfill most of the above requirements; their chemical structure makes them effective antigens.

Plant Sources

Pollinosis may be caused by any one or more of several hundred plant species. However, *Ambrosia* (ragweed) and *Phleum pratense* (timothy grass) pollen, widely distributed in norther temperate zone floras, are among the most disabilitating pollen and cause about half the hayfever cases in many large cities. Pollen allergy has been reported to occur in about 1% of the population of Sweden (HELANDER, 1960). In 1969 in the United States, the cost to industry for lost wages due to airborne pollen allergies was about $400 million (SANDERS, 1970). Species of *Ambrosia*, ragweed, are wind pollinated; their pollen, as well as pollen of two closely related, but entomophilic families, are all allergenically active (DURHAM, 1951). The most troublesome plant species in terms of pollinosis are listed in Table 12–1. Pollen causing allergy are generally grouped as derived from trees, grasses or weeds. Entomophilous pollen such as *Pyrus* and *Prunus* rarely cause allergenic problems, but they do contain antigenic proteins (KESZTYÜS, 1957).

Pollen from genera in the same family, or even related species, may differ in capacities for producing allergic reactions. *Avena fatua*, the wild oat, causes pollinosis, while pollen from the highly bred common oat, *Avena sativa*, produces few, if any, allergic responses. *Poa annua* is relatively non-allergenic, but *Poa pratensis* (Table 12–1) is one of the most troublesome plants, and was one of the original grass species for which the symptoms and name hayfever were described. Pollen calendars, seasonal index diagrams or charts of when troublesome plants in an area shed pollen, are commonly consulted by hayfever victims and physicians. Such charts represent good indices of potentially troublesome pollen. However, it should be recognized that pollen production from any one species usually varies independent of other species in the same environment (Fig. 12–1).

Hayfever. Our acquisition of knowledge about hayfever may conveniently be divided into five main historical periods (SCHADEWALDT, 1967). In the first, case reports on "this curious seasonal illness" focused more on human reactions to rose odors than to grasses. In the second period BOSTOCK (1819) gave an exact description of the clinical symptoms and later introduced the name "hayfever". Beginning in the mid-19th century, the third period, epidemiological research, gathered statistical data on prevalence of the illness. Monographs by PHOEBUS (1862) and BEARD (1876) were milestones of this period. A search, via exact scientific methods, for the sources causing hayfever characterized the fourth period. In 1831 ELLIOTSON suggested that grass pollen harvested in meadows caused hayfever. WYMAN (1872) and BLACKLEY (1873) experimentally substantiated this hypothesis and showed that the mucous membranes of hayfever patients react to trace amounts of pollen.

We are currently in the fifth stage, the therapeutic epoch. DUNBAR (1903) demonstrated that proteins are the allergenic factor and tried to treat hayfever patients with a special serum, Pollantin, obtained by immunizing horses with pollen solutions. WOLFF-EISNER (1907) contributed the concept that hayfever, the hypersensitive reaction to pollen, is in fact an allergic disease. The word allergic was introduced by PIRQUET (1906) from the Greek roots for different (allos) and action (ergon).

Table 12-1. Principal plants producing allergenic pollen. (Adapted from DUCHAINE, 1959)

Botanical name	English	French	German	Region where important[a]
Trees				
Alnus spp.	Alder	Aune	Erle	1, 2, 4, 5
Betula spp.	Birch	Bouleau	Birke	1, 2, 4, 5
Carpinus spp.	Hornbeam; Ironwood	Charme	Weissbuche	2, 5
Casuarina equisetifolia	Australian pine		Sumpfeiche	2, 3
Cupressus fragrans	Lawson cypress	Cyprès	Zypresse	2, 4, 6
Juniperus spp.	Juniper; Montain cedar	Genévrier	Wacholder	1, 2
Quercus ssp.	Oak	Chêne	Eiche	1–5
Carya pecan	Hickory	Noix Hickory	Hickorynuß	1, 2
Morus alba	White mulberry	Mûrier	Maulbeere	1, 2, 3, 5
Broussonetia papyrifera	Paper mulberry	Mûrier du Japon	Japanische Maulbeere	2, 5
Olea europaea	Olive tree	Olivier	Olivenbaum	2, 3, 5
Platanus spp.	Plane tree; Sycamore	Platane	Platane	2–5
Populus spp.	Poplar; Cottonwood	Peuolier	Pappel	1–5
Ulmus spp.	Elm	Orme	Ulme	1–5
Grasses				
Carex paniculata	Tussock sedge	Carex panuculé	Rispen-Segge	1, 2, 4, 5
Agrostis alba	Common bent; Red Top	Traine	Straußgras	1–5
Anthoxanthum odoratum	Sweet vernalgrass	Flouve odorante	Ruchgras	1, 2, 4, 5
Avena fatua	Wild oat	Folle avoine	Wildhafer	2–5
Bromus sterilis	Barren brome; Hungar. brome grass	Averon	Trespe	1–6
Cympognon spp.	Turpentine grass			6
Cynodon dactylon	Bermuda grass	Herbe des Bermudes	Bermudagras	2–6
Dactylis glomerata	Cocksfoot; Ochard grass	Pied de poule	Wiesen-Knäuelgras	1–5
Eragrostis spp.	Love grass	Amourette	Liebesgras	5
Holcus lanatus	Yorkshire fog; Tufted or velvet grass	Houque laineuse	Wolliges Honiggras	1–5
Hypyrrhenia spp.	Tambookie grass			6
Lolium perenne	Rye-grass; Ray-grass	Ivraie; Raygrass anglais	Lolch; Engl. Raygras	1–5
Phleum pratense	Timothy; Cattail	Fléole des pres; Timothée	Wiesenlischgras	1–5
Poa pratensis	English meadow grass	Paturin des pres	Wiesenrispengras	1–5
Secale cereale	Rye	Seigle	Roggen	4
Sporobolus pungens	Drop-seed grass	Sporobole	Samenwerfer	3, 5, 6

Weeds

				[a]
Salsola pestifer	Russian tistle	Salsola	Salzkraut	1, 2, 4
Ambrosia spp.	Ragweed	Herbe-à-poux Sariette	Traubenkraut	1–3
Artemisia spp.	Wormwood; Sagebrush	Armoise	Beifuß	1–5
Cyclachaena xanthifolia	Prairie ragweed; Burweed			1, 2
Iva spp.	Marshelder; Povertyweed			1
Taraxacum officinale	Dandelion	Pissenlit	Schlagkraut Löwenzahn	1, 3, 4, 5
Plantago lanceolata	English Plantain; Ribgrass	Plantain	Spitzwegerich	1–5
Rumex spp.	Dock; Sheepsorrel	Oseille; Surette	Ampfer	1–5
Parietaria officinalis	Pellitory	Pariétaire	Glaskraut	4, 5

[a] 1 – Canada
2 – U.S.A
3 – Central or South America
4 – Temperate Europe
5 – Mediterranean
6 – Africa

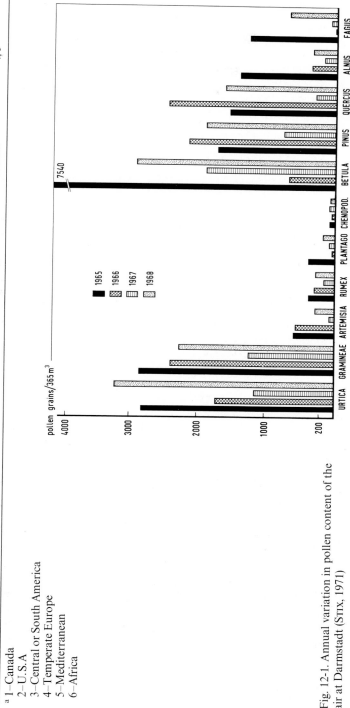

Fig. 12-1. Annual variation in pollen content of the air at Darmstadt (STIX, 1971)

Noon (1911) successfully treated hayfever by desensitizing the patients with pollen injections. Since then, many experiments have been aimed at isolation of pollen antigens. Methods devised by chemists to fractionate proteins have been promptly applied by allergists in an effort to purify pollen extracts. An impure protein was isolated by Kammann (1912) from rye *(Secale cereale)* pollen and shown to induce an allergic reaction. This work was disputed by some workers who claimed that deproteinized extracts induced allergic reactions (Grove and Coca, 1923; Moore et al., 1931). However, progress in extracting and fractionating pollen components, including treatment with the protein hydrolytic enzyme trypsin (Loeb, 1928) and heat denaturation, (Bouillenne and Bouillenne, 1931) strongly suggested at that time, that proteins are the principal allergens.

Allergens

Chemistry

Based on immunological responses to serum derived from rabbits inoculated with ragweed antigen, Johnson and Rappaport (1932) concluded that ragweed contains at least two antigens, one a globulin, the other albumin-like. Antigens from many pollen reacted with the antibodies formed by the two ragweed protein fractions. Pollen stimulating strong allergic responses were the most reactive to the antibodies. This suggests that many pollen contain protein allergens which have some chemical properties in common. Heyl and Hopkins (1920) proposed that the active protein fractions in ragweed were albumin and glutelin, the histidineless derivative of albumin. The two protein fractions were electrophoretically separated into high and low molecular weight fractions which caused precipitin-immunological reactions and hayfever-type reactions (Roth and Nelson, 1942; Abramson et al., 1942).

The fact that each successive extraction of ragweed and other pollen with water, sodium chloride and a basic solution removes an antigenic protein fraction, indicates that many protein fractions can cause allergic responses. Because of the persistence of carbohydrate-linked components in many of the extractives, the strictly protein nature of the active principle in ragweed and other species initially eluded investigations (Stone et al., 1947; Johnson and Thorne, 1958). The rapid diffusion of proteins from allergenic pollen (Stanley and Search, 1971) and the proximity of the proteins to exine and intine carbohydrates (Knox and Heslop-Harrison, 1970; Hubscher and Eisen, 1972) suggests that the proteins may be glucosidically linked *in vivo*. This association would facilitate their extraction as a glycoprotein complex. Berrens (1967) found that N-glycosidic-lysine-sugar moieties are attached to many atopic, non-pollen allergens. Introducing a lysine-sugar moiety into a non-allergenic protein molecule can convert it to a skin reactive protein (Berrens, 1971), suggesting that the added nonprotein moieties help prime or facilitate the allergic reaction. But in view of the nature of the immunological reactions and the effect of hydrolysis on the extracts from which all carbohydrates have been removed, the protein moiety appears to be the primary inducer of allergenic reactions. The sugar moiety of the pollen allergen may accelerate diffusion or binding to the reactive intracellular site.

Definitive experiments by AUGUSTIN (1959) showed that in all solvents tested, when the sugars, peptides and pigments are removed from the allergens the biological activity remains with the protein; also, most of the antigenic activity of grass pollen was lost after heating for one minute at 100° C, and after treatment with the hydrolytic enzymes pepsin, trypsin, and chrymotrypsin. This suggests that a protective moiety, such as a sugar molecule, may bind more strongly to the protein allergenic components of some pollens than others (COLLDAHL, 1959). Direct evidence of differential binding in protein extracts of *Parietaria* pollen was obtained by CRIFÒ and IANNETTI (1969). They reported that sugars were more tightly bound to the one allergenic globulin than to the three allergen albumin-type proteins separated by electrophoresis. MILNER et al. (1972) reported no significant differences in skin test activities between extracts of viable and non-viable pollen of the 4 grass species *Bromus mollis, Agrostis gigantea, Phleum pratense* and *Holcus lanatus*. Reactivity of allergenic proteins is probably not related as much to their binding in living pollen, as to the chemical components with which they are combined.

The very low level of allergenic protein in pollen relative to total protein can contribute to low sensitivity of allergen fractions to protein inactivation experiments. Allergens usually comprise between 0.5 and 1.0% of the total extractable pollen proteins (MARSH et al., 1966; BELIN and ROWLEY, 1971). Thus, destruction of a considerable quantity of protein can occur before the allergenic activity is drastically affected.

Carefully separated lipids, sterols and pigment fractions from the surface of the highly reactive pollen of *Parietaria officinalis* were completely free of allergenic activity (CRIFO et al., 1969). The rye grass, *Lolium perenne*, allergen did not lose activity when the extract was treated with β-galactosidases, cellulases and other carbohydrate metabolizing enzymes (MARSH et al., 1966). *Secale cereale* protein extracts released raffinose, mannose, fructose, glucose and inositol but still retained their allergenic activity (SCHEIBE and LOEWE, 1969). In soluble protein fractions from *Secale cereale* pollen, the principal allergen, free of lipids and sugars, has a molecular weight of about 20,000 (JORDE and LINSKENS, 1972). These facts further indicate that the carbohydrate component is not essential for an allergenic response.

Application of column and paper chromatography, along with electrophoresis, have provided pure protein fractions for chemical characterization. One problem in applying such isolation techniques is that chromatographic solvent residues, as well as pollen and its hydrolytic products, may induce allergic resoponses (FRANKEL et al., 1955). Of 10 different proteins separated from *Betula* extracts only one, possibly two, induced a strong allergic reaction (BELIN, 1972); four out of 12 proteins separated by a high pressure press from pre-extracted *Parietaria* were allergenically active (CRIFÒ and IANNETTI, 1969). Six chromatographically distinct fractions were separated from water soluble extracts of ragweed, but only two showed different allergenic activities in immunological tests. Boron was essential for maximum allergen activity of the isolated fractions (RICHTER et al., 1957), again suggesting that either the allergen may move across membranes as a complex, or its activity may be influenced by other molecules. Most pollen allergens are water soluble; this enhances their diffusion in mucoid tissues and increases the reactive charge groups available on dissociated molecules.

WODEHOUSE (1955) applied the gel diffusion method to study the antigenic relationships among six common hayfever grasses: timothy (*Phleum pratense* L.), red top (*Agrostis alba* L.) June grass (*Poa pratensis* L.), sweet vernal grass (*Anthoxanthum odoratum*), Orchard grass (*Dactylis glomerata* L.), Bermuda grass (*Cynodon dactylon* Pers.). All species yielded one major antigen and a variable number of minor antigens. Bermuda grass was completely different from the other grasses, antigenically.

The best known allergenic proteins are from ragweed, timothy and rye grass. The main allergen in ragweed is called antigen E; it produces 90% of the antigenic activity. With a molecular weight of about 37,800, antigen E constitutes 0.5% of the dry weight, and 6% of the pollen's protein (KING et al., 1964; ROBBINS et al., 1966). The other 10% of the antigenic response in ragweed is caused by at least 12 other proteins. Antigen K, the second most active allergen in ragweed, contains fewer amino acids than E and constitutes 3% of the pollen's protein. Ra.3 is the third most active component.

The principle allergens in rye grass are designated: Group I, antigen proteins having a molecular weight of about 27,000 (JOHNSON and MARSH, 1965). Group II, with a molecular weight of about 10,000 (MARSH et al., 1966), and Group III are weaker allergens. In timothy grass the strongest allergens, antigens A und B, are proteins with molecular weights of about 13,000 and 10,500 respectively (MALLEY et al., 1962); antigen D with a molecular weight of 5,000 is also a strong allergen in timothy. The allergenic protein from *Betula* is about 20,000 molecular weight (BELIN, 1972a). These compounds, strictly speaking, are below the 35,000 molecular weight usually specified by definition as the minimum molecular weight for proteins. However, these smaller polypeptide fragments are the active fractions; they are composed almost entirely of amino acids and for convenience we shall continue to refer to these smaller allergens as proteins.

Analysis of amino acids in two different antigen fractions from ragweed pollen (Table 12–2) indicates that the stronger antigen, E, is higher than antigen Ra.3 in aspartic acid, alanine and glycine and lower in proline and tryptophan (UNDERDOWN and GOODFRIEND, 1969). The reason one protein induces an allergic response, while another does not, or does so less strongly, is still not understood. Presumably, it is related to the amino acids and their sequence in the antigen. Antigen fractions, such as Ra.3, E and K, can cross-react in varying degrees in immunological tests with antisera from humans and rabbits. All three, therefore, contain some antigen determinants in common. However, the nature of the allergenic factor in pollen protein; i.e. what it is that induces a response in sensitive cells, awaits further chemical chracterization, comparisons and improved recognition of the nature of human responses in pollinosis.

Bacteria and fungi isolated from pollen surfaces can also supply allergenic materials provoking reactions in sensitive people. COLLDAHL and CARLSSON (1968) isolated bacteria from *Betula alba* and *Phleum pratense* pollen; tests on patients sensitive to pollen showed that they could not differentiate between the reactions provoked by the pollen extracts and those provoked by microorganisms prepared from the pollen. It is possible that in the technique used in these experiments for isolating the bacteria, some pollen protein was exuded from the pollen grains onto the agar on which the pollen was deposited, and subsequently may

Table 12-2. Comparison of amino acid components in ragweed
pollen antigens (UNDERDOWN and GOODFRIEND, 1969)

Amino acid	Antigen Ra. 3		Antigen E	
	No.	% distribution	No.	% distribution
Lysine	7	6.9	18	5.3
Histidine	3	3.0	6	1.8
Arginine	4	4.0	16	4.7
Aspartic	7	6.9	49	14.5
Threonine	8	7.9	17	5.0
Serine	5	5.0	26	7.6
Glutamic	8	7.9	25	7.3
Proline	10	9.9	15	4.4
Glycine	8	7.9	37	10.9
Alanine	6	5.9	31	9.2
Half-cystine	3	3.0	7	2.1
Valine	6	5.9	24	7.0
Methionine	1	1.0	7	2.1
Isileucine	4	4.0	20	5.9
Leucine	8	7.9	21	6.2
Tyrosine	2	2.0	4	1.2
Phenylanine	7	6.9	12	3.5
Tryptophan	4	4.0	6	1.8
Total number	101		341	
Molecular weight	15,000		37,000	
% Nitrogen	13.5		17.1	
Total hexose + pentose	12.4		0.5	

have been included in bacterial extracts. However, extracts from pure cultures of related bacteria induced similar responses in sensitive patients. But patients sensitive to *Betula* pollen did not respond to bacteria from the *Phleum* pollen and vice versa. People non-sensitive to pollen did not respond to the bacterial protein extracts. Results applying immunofluorescent antibody techniques to pollen *in situ*, after the surfaces have been washed, indicate that reactive allergenic proteins are produced by pollen and are not merely due to contaminating microorganisms (BELIN and ROWLEY, 1971). The specific antigen E is localized in the intine of *Ambrosia* pollen (KNOX and HESLOP-HARRISON, 1971) and thus is obviously not bacterial dependent.

NEWELL (1942) reviewed 156 publications on the chemistry of hayfever producing substances in pollen and concluded, "The sum of all this work is difficult to evaluate because of contradictory reports and lack of uniform methods for quantitative studies." In 1968, COLLDAHL and CARLSSON, in spite of the analytical chemical progress made, could still summarize the state of knowledge by saying, "The nature of the allergenic substances in pollen is not known." The main characteristic of pollen antigens is their ability to produce allergic reactions at very low concentrations.

Through the separation of Antigens E, K and Ra.3, and by studying their amino acid sequences we are beginning to unravel the mystery. However, efforts at separating and characterizing pollen antigens have failed to answer the question of whether there are several antigens in a pollen extract, or only one antigen which is fragmented or which assumes alternate physico-chemical properties during release from the pollen, where it is bound to different protein carriers. It may be possible that the antigen is carried with proteins of different molecular weight, thus on separation it could occur in several fractions in different concentrations. The carrier protein would also induce its own antibody response. The experimental reduction of 99% of the allergenicity of rye grass Group I allergen, without a marked diminution in its capacity to combine with rabbit antibody formed against the untreated antigen (MARSH et al., 1970a) suggests that the antigenic component is indeed carried on the immunologically reactive protein. Characterization of the proteins and reactive sites involved in pollinosis is an area requiring further research.

Human Reactions

In 1913 DALE postulated that the allergic response involves an immunological complexing reaction between an antigen, i.e. pollen allergen, and an antibody. The complex induces the detrimental allergic reactions recognized as pollinosis. PRAUSNITZ and KUESTNER (1921) found the blood of hayfever sufferers to contain a characteristic reagin (antibody) which can be transferred by injection to a normally non-reactive person.

Only three groups of antibodies were known to be in human blood serum and other tissues before 1960. These gamma globulin proteins were the immunoglobulins A, G and M (Ig A, Ig G and Ig M). Ig G was recognized as the dominant protective antibody in blood serum. But a new skin sensitizing antibody in the immunglobulin class, Ig E, was discovered in 1966 (ISHIZAKA and ISHIZAKA). Ig E is almost absent from serum of non-allergic people but is very high in the serum and "shock organs" of allergic patients. Ig E is formed in the "shock organs". However, while the level of Ig E may vary with exposure in sensitized patients, the concentration is still quite low, constituting only about 0.005% by weight of the total serum gamma globulins. Although the exact function of Ig E is not defined, it is recognized as the principal complexing antibody in pollinosis. Another class of immunoglobulins, IgND, discovered by JOHANSSON and BENNICH (1967) exists in serum at even lower concentrations than Ig E. But elevated levels of IgND in sera of hayfever patients were detected by the ^{125}I-labelled radioimmuno-absorbant technique (WIDE et al., 1967).

Steps leading to the allergic reaction in pollinosis are generally presumed as follows: 1. Pollen enters the nose or eyes and lands on mucous membrane tissue of the upper or lower respiratory tracts. 2. Mucous liquid solubilizes pollen allergens which penetrate the mucous tissues. 3. Mucous tissues of allergic persons contain mast cells which have a high concentration of antibodies. 4. The antigens, e.g. pollen allergens, quickly complex with the Ig E antibodies. 5. The complex activates enzymes which cause the release of "mediators" from special organelles in these cells. 6. The chemical mediators, e.g. histamine, induce the allergic symptoms, i.e. dilation of blood capillaries, contraction of nasal or bronchial muscles, endema of mucous membrane, hypersecretion of watery nasal fluids, and constriction of nasal or bronchial passages—the hayfever reaction.

Reactions to pollen depend upon the species, concentration, method of contact and the person's prior history of exposure. Other factors such as abundance of plants, time of the year and weather during anther dehiscence also affect the allergic response of a sensitive person. Five properties of pollen, the Thommen Postulates, are generally considered essential requirements for a species to cause a reaction in man (COCA et al., 1931).

1. The pollen must contain a hayfever allergen.
2. The mode of pollination must be by wind.
3. The plant must produce large quantities of pollen.
4. The pollen must be buoyant and transportable.
5. The plant should be widely distributed.

Some limitations and applications of these postulates are easily recognized. Pine pollen meets all these criteria except number one; it has only on rare occasions been found cause an allergic response (ROWE, 1939; NEWMARK and ITKIN, 1967). Gardeners and people who handle entomophilous flowers may still become sensitized and responsive to those non-wind borne pollens. The buoyancy need of pollen allergen sources eliminates such species as *Zea mays* where the pollen is usually too large and has sticky viscid exine strands. Other factors such as prior diet can also influence the response of a person to pollen (DUCHAINE, 1959).

Since a highly purified antigen fraction has been isolated only recently, a reliable *in vitro* assay for allergic capacity has not yet been developed. But methods of detecting the sensitivity of a person to pollen have been devised, i.e. the availability of a high concentration of Ig E antibodies to combine with pollen allergens can be determined. Tests for response to pollen allergens and procedures for de- or hyposensitizing therapy employ pollen extracts and mixtures of varying degrees of protein purity.

Preparation of Extracts

Whole pollen can be used in allergenic tests, but solutions are preferred in clinical application. The best allergen preparations are those in which the extract retains the characteristics it has in freshly shed pollen. Ideally, a pollen extract intended for introduction into a human body should be easy to prepare and retain a high and constant level of activity after several months storage. It must also be sterile, at a pH that does not irritate a person, and should be in a form which allows comparative standardization (STRAUSS, 1960).

Pollen collected from known sources is rapidly dehydrated to avoid loss of allergenic activity (DURHAM, 1951). Aliquots are then usually washed 3 times with ether to remove oils and some pigments, then dried. Defatted pollen provides a clearer solution when extracted. Questions have been raised relative to the desirability of the ether washing (FUCHS and STRAUSS, 1959; GOLDFARB, 1968) since it may denature some allergens. The lipid solvent wash may remove external spores or virus-contaminating impurities from the pollen surface, but it could also remove some lipoproteins. To guard against loss of allergens by defatting, good clinical practice includes concentrating the oily ether extracts to dryness and suspending the residue in olive oil for testing on patients (PHILLIPS, 1967). Other workers avoid the lipid solvent extraction and subject the pollen to a washing of one minute or less with distilled water before making the antigen extract (AUGUS-TIN, 1967).

Weak alkaline buffers, pH 8, are generally used to extract defatted pollen. Pollen acids (Chapter 10) lower the extracting solution pH to 6.0 or less; it is usually adjusted to neutrality with bicarbonate. One commonly used procedure (COCA, 1922) employs 100 ml of a protein extracting solution at pH 8.2 (NaCl 5.0 gms; NaHCO$_3$ 2.75 gms, C$_6$H$_5$OH 4 gms all mixed to 1 liter distilled water) to 5 to 10 gms of pollen. The higher quantities of pollen are usually used for tree and other pollen yielding less active or more difficult to extract allergen fractions.

Extraction is usually accomplished by shaking macerated pollen with extracting solution for 72 hrs at room temperature. Bacterial contamination is controlled during extraction by adding 1–2 ml toluene, which is evaporated before concentrating the extract. The mixture is filtered through coarse filter paper, brought to a given volume, concentrated in a dialyzing cellophane membrane to about 1/10th the original volume and sterilized by filtration. Alternatively, simple extracts for clinical use are prepared by shaking dried pollen with equal volumes of glycerin and buffered saline solution and allowing the mixture to stand at about 0° C. Fresh pollen is occasionally added to maintain the allergen potency in solution. Allergens are stable in glycerin for a longer period of time than in aqueous extracts. Organic solvents which dissolve but do not dissociate the proteins have also been employed (GOODMAN et al., 1968). Such compounds, e. g. ethylene carbonate and acetonitrile, extract more protein than COCA's solution and may afford a broader allergen spectrum. Mixtures of several species of grass, tree or weed pollen is sometimes combined in a single extraction to assure a broad spectrum of allergens for testing or treatment.

Saline extracts lose 50% or more of their allergen activity in one year's storage (STULL et al., 1933). Chemical methods of binding the protein have been introduced to stabilize the allergen activity in extracts. Protein can be precipitated in tannic acid and ZnCl$_2$ (NATERMAN, 1965). Alternatively, a buffered alum precipitate which can be ingested as tablets or injected in a saline suspension can be prepared (MAURER and STRAUSS, 1961). Alum precipitated extracts release the absorbed allergen slowly; many believe it is a more effective preparation and requires fewer repeat injections for desensitizing hayfever patients (NORMAN et al., 1970; 1972). An alternative method to binding the protein is to freeze-dry, lyophilize the extract and store the powder until it is needed. The freeze-dried extract retains full biological activity until it is reconstituted with an appropriate saline buffer solution.

Standardizing Tests

A uniform test for potency of pollen allergen extracts has still not been developed. Efforts to establish a standard unit for calibrating allergen activity include correlating chemical or physical measurements with biological activity. Direct measurements often employ a Pollen or Noon unit (STULL and COOKE, 1932). These units are based on the amount of water soluble antigen extracted from 1.0 µg of pollen. The ratio of pollen weight to volume of extracting solution, or the dilution factor, has often been used to describe allergen concentration. Total nitrogen determination (Kjeldahl assay), or protein per ml solution, or protein nitrogen units (PNU or Cooke Unit) are more widely used indices.

Assuming that a standard uniform technique is used in preparing the extracts, such methods are still beset with outside sources contributing to inaccuracy. For example, a µg of pollen produced on the same plant at the same site will vary in allergen yield from one year to the next; also, different pollen yields allergens of different reactivity. The nitrogen content of an extract may remain constant in storage, but the protein may denature and allergen potency diminish. Allergenic components are only a small part of the total nitrogen content in the extract. Last, but far from least among the problems besetting efforts to establish a standard potency assay, is the fact that hayfever victims respond at different thresholds of sensitivity to different pollen allergens.

Cooke Unit. The PNU assumes that the active component in the extract is always directly proportional to the protein present. One (1) PNU equals 0.01 µg phosphotungstic acid precipitable nitrogen, or about 0.06 µg of protein. One (1) PNU equals approximately two (2) Noon units, or 2.6 times the total nitrogen or Pollen units. Unfortunately, phosphotungstic acid precipitates compounds with molecular weights in the small polypeptide range as well as all the allergenic and non-allergenic proteins (SHELDON et al., 1967) and glycoproteins in the crude extract. As already discussed, other chemical moieties in pollen, the carbohydrates, boron, etc., may facilitate a variation in response to allergens which will not be detected by PNU assays. The PNU content failed to correlate with either the antigen E content or the degree of allergen reactivity in ragweed extracts (BAER et al., 1970). In fact, as much as a 100-fold difference in antigen E content, biological reactivity and PNU value were found between 6 different commercially supplied samples.

International Unit. A new method and unit for standardizing allergen extracts has been proposed by BERRENS (1970). The proposed international unit, I.U., of allergen (in weight units) is that amount which causes 50% histamine release from normal human leukocytes sensitized with a fixed amount (in weight units) of a standard reagenic serum. The standard reagenic serum will be freeze-dried and distributed from one international center. The test for histamine release from leukocytes (see below) may yet be developed as a good assay for human potential reactions to pollen. Coupled with further refinement in preparing antigen E, this test may yet serve to provide the basic standard unit which allergiste have long sought.

Biological tests using highly purified allergens can provide good measurements of allergen activity. Standardizing tests which assay allergen extracts by immuno-electrophoretic and gel diffusion antibody reactions have been developed (WODEHOUSE, 1954, 1955a; AUGUSTIN, 1955). The gel diffusion technique allowed for a clearer identification of the antigen fractions. Choosing the correct antibody for reacting in immuno-electrophoresis is quite important. Since immunoglobulin (Ig) E was discovered in 1966, at about the same time that the first allergen fractions of ragweed pollen were purified and characterized, development of immuno-electrophoretic and related biological assay procedures are still being perfected (YUNGINGER and GLEICH 1972). Limitations include the time necessary for purifying the allergen extract. For ragweed pollen standard preparations of anti-Ig E serum may be supplied to assay potency. Double diffusion reactions between antiserum and antigen E correlate well with the allergenic potency of an extract (BAER et al., 1970).

At present, the allergist in clinical practice recognizes the many variables in, and the non-specific qualities of, his extract. He therefore initially exposes his patients to very small doses of an extract. The patient's responses are compared in successive exposures to different levels and types of extracts. Avoidance of negative reactions, such as anaphylactic shock, primarily depends upon the practical knowledge of the treating physician (WERNER et al., 1970). Each patient's response is thus the ultimate and best available standardizing test.

In summary: An accurate chemical or biochemical assay for allergenic potency in different pollen extracts is a research goal yet to be attained. The major difficulty in standardizing pollen allergens is that a variable complex of allergen materials, different from antigen E, are extracted from the different pollens. Our knowledge of the genetic specificity of the exine and pollen protein can be of little use until we know in greater detail the chemical moieties involved. However, while the search for the chemical and biological source of specificity of the response to pollen allergens continues, physicians must seek to apply the knowledge at hand to help relieve symptoms of pollinosis. Procedures, using pollen extracts which can be applied with varying degrees of success to both testing and relief of pollinosis, have been developed.

Testing Human Reactions to Pollen

The most common methods for testing human sensitivity to pollen allergens are by a dermal (skin) scratch test or by intradermal, intracutaneous injection. These and several other assay methods, i.e. by inhalation, nasal, or bronchial mucosa— provocation tests (SHERMAN, 1968), or histamine release by leukocytes, are the principle methods used. After a person is found to be allergic to a particular pollen or several pollen species, control treatments can then be initiated to eliminate seasonal discomfort (WERNER and RUPPERT, 1968).

Skin Test. The dermal test involves local exposure to a pollen extract applied to the arm, forearm, or back. The skin is scarified and drops of extract applied to the rest sites. The patient's response is generally observed 10 to 20 minutes after treatment, or 24 hrs later, if a delayed response is suspected. Primary responses noted are degree of redness, size of wheal area, and itching or pain. Intracutaneous tests are more sensitive than scratch tests. Tests are usually done on the arm or thigh by introducing 10 to 20 µl of extract under the skin. Injections are usually made at dilutions of about 1/10th the strength tested by the dermal-scratch test. Control tests with diluents alone are run. The control site of the scratch or injection should not produce a wheal, increase in bleb size or itchiness (Fig. 12-2).

Provocation Test. Nasal inhalation tests expose a subject to a more natural pattern of elicitating mucous membrane reactions. Such tests, called provocation

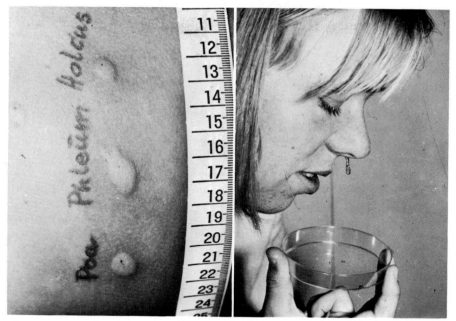

Fig. 12-2

Fig. 12-3

Fig. 12-2. Skin test with three different pollen species. (Photo W. GRONEMEYER)

Fig. 12-3. Provocation test after nasal inhalation. (Photo: W. GRONEMEYER)

tests, employ whole pollen or nebulized extracts (COLLDAHL, 1959). Respiratory sounds, air flow measurements, as well as visual symptoms are assessed. Some workers believe nasal reactions to pollen are a better diagnostic test than dermal applications. Nasal inhalation tests can replicate the pattern of a patient's seasonal pollen exposure, and test the spectrum of allergens found by intradermal tests (GRONEMEYER and FUCHS, 1959) (Fig. 12-3).

When ten species of dry, natural, non-defatted pollen were inhaled by an allergic patient, each pollen produced its own unique complex of hayfever symptoms localized in the nose, eye, or throat (MANDELL, 1967). This suggests that all pollen have their own immunologic identity, which also agrees with the biochemical evidence that pollen proteins confer a greater genetic specificity than is indicated by exine morphology. These results also suggest that hayfever symptoms must be treated with the specific pollen or extract to which a person is sensitized. Thus, *Quercus alba* pollen extract will afford little relief to a patient sensitized to *Quercus nigra* pollen.

The degree of relief by treatment with extracts of related species is a function, in part, of the cross reactivity of the protein in the two species, MARSH et al. (1970) used gel electrophoresis to separate proteins from seven grass pollen of six different botanical tribes. The separate proteins were tested on sensitized patients and by immunodiffusion reactions against antisera prepared from purified rye

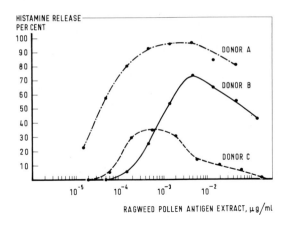

Fig. 12-4. Histamine release by leukocytes from ragweed-sensitive donors. Percent release of the total
available (acid extractable) histamine in leukocytes. (Adapted from
OSLER et al., 1968)

grass Group I, II and III proteins. Six of the grasses contained the three common
groups, in addition to other allergenic proteins. One species lacked all three
groups. The presence of common antigen groups elicited similar reactions in
individuals tested, and sensitized individuals reacted with a broadly similar response pattern to each of the pure similar antigenic proteins.

Histamine Release Test. The question of the nature and degree of common
cross-reacting antigens in related pollen is far from solved. Variations due to the
number of antigenic components in a pollen, and tendency of modifying moieties
to influence *in vivo* reactions, complicate the search for the key to the basis of
allergenic reactivity. The quest for a more sensitive, accurate test to relate clinical
pollinosis to the inducing chemical factor has moved to *in vitro* assays employing
leukocytes from allergic persons (BERGQUIST and NILSEN, 1968). Such leukocytes
release histamine on coming in contact with allergenic proteins.

A suspension of leukocytic cells from a ragweed-sensitive donor is used to
establish the limits of histamine release by ragweed pollen antigen. In this assay
the leukocyte suspension is incubated for 1 hr at 37° C with different levels of
antigens or other chemicals and the reaction is terminated by centrifugation in the
cold. The amount of histamine released into the buffer phase is measured fluro-
metrically and compared to the total histamine released by perchloric acid from
the original leukocyte suspension at zero time. The pathogenic antibody (reagin)
occurs in sensitized persons and renders their leukocytes reactive to the antigen.
Excessive amounts of antigen depress histamine release from leukocytes *in vitro*
(Fig. 12–4). Calculations with ragweed antigen E indicate a 50% release of available histamine is obtained from leukocytes of highly sensitized individuals at a
concentration of about 10^6 antigen molecules (OSLER et al., 1968).

Many limitations exist for the application of this laboratory *in vitro* assay.
Different donors each release different amounts of histamine at the same concentration of antigen E. About 20% of ragweed-sensitized individuals possess leukocytes which behave as those of Donor C, Fig. 12–4, yielding relatively little
histamine. Leukocytes release slightly more histamine when exposed to whole
pollen extracts than to just antigen E (MELAM et al., 1970). With such a wide

variation in histamine release from sensitized leukocytes, only comparative values are meaningful. But high release values do correlate with *in vivo* symptoms of pollinosis. The supply of test antigen or whole pollen extracts requires standardization. The variable nature of antigens released from different pollen results in induction of different antibodies sensitive individuals. Yet, in spite of these limitations in applications of the *in vitro* histamine assay, the technique does offer an opportunity to screen many patients rapidly and relatively painlessly for sensitivity to different pollen. Conceivably, if histamine release from leukocytes derived from a single genetic strain of test animal can be quantitized, we may ultimately gain a tool for standardizing the potency of pollen extracts. This may be the preferred solution for development of the I.U. proposed by BERRENS (1970), as discussed in the previous section.

The spectrum of protein antigens effective in pollinosis may be broader than formerly suspected. Common enzymes can have immunologic capacities. GREEN and KINMAN (1970) found that glucose oxidase from the fungus *Aspergillus niger* is a potent antigen. This enzyme, a glycoprotein with a molecular weight of about 150,000, is ubiquitous in pollen. A pronase-type enzyme is present in a chromatographically purified allergen from *Secale cereale* pollen (JORDE and LINSKENS, 1972). Whether or not allergen activity is associated with this enzyme component must still be determined. It is quite possible that enzymes essential for pollen germination and growth (Chapter 14) may also induce strong allergenic reactions in some individuals. If some pollen enzymes act as antigens, then discerning differences in isozyme and protein diffusion patterns from different pollen, may provide information about how pollen functions both in pollination and pollinosis.

Persorption

A situation exists in which even a sensitized person may come into direct contact with an allergenic pollen and not exhibit the usual symptoms of pollinosis. This occurs when pollen particles pass directly from the stomach into the blood stream. The passage of such large particles as pollen and spores directly from the intestinal lumen epithelial cells into the blood of dogs or man is called persorption (VOLKHEIMER and SCHULZ, 1968; VOLKHEIMER, 1972). In the subepithelial tissues such persorbed grains are transported in a basal direction via both mesentric veins and lymphatic vessels. Two hours after 18 dogs were fed pollen in milk cream, the intact pollen was observed in the blood of all and in the urine and cerebral spinal fluids of most of the dogs.

Pollen accumulation in man's gastrointestinal tract was demonstrated by WILSON et al. (1973), using pollen labeled with radioactive technetium, ^{99m}Tc. When labeled *Poa pratensis* is ingested orally, the round pollen of about 25 μ diameter is almost totally excluded from the lungs and respiratory tract, and instead accumulates in the lower part of the stomach. The delayed symptom response to inhaled whole pollen versus almost immediate response noted when sensitive patients are supplied concentrated allergenic extracts is presumably related to solubilization and circulation of the allergenic materials from the pollen.

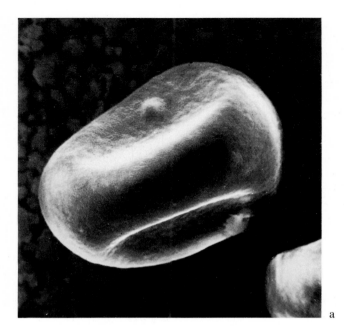

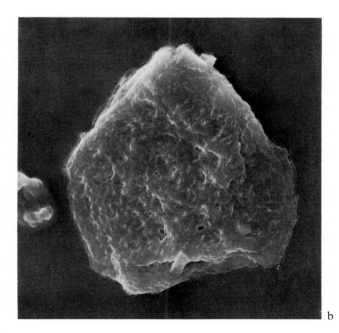

Fig. 12-5 a and b. Scanning electron micrographs of rye pollen before (a) and after (b) persorption. Pollen in (b) is recovered from blood 45 minutes after oral ingestion

LINSKENS and JORDE (1974) found that when 150 gms of pure rye *Secale cereale* pollen are orally ingested, at least 6,000 to 10,000 grains are persorbed into the blood stream. When the persorbed *Secale* pollen, or *Lycopodium* spores are removed from the blood stream after various periods of time, the exine appears progressively degraded (Fig. 12-5) (JORDE and LINSKENS, 1974; LINSKENS and JORDE, 1974).

The implications of persorptive distribution of pollen, on the allergic-response system of man must still be ascertained. Clinical procedures for building-up immunity to allergenic pollen by ingesting whole pollen as tablets before the season of pollen dehiscence may depend, in part, upon the persorption phenomenon. The effect of the circulatory, or nasal-mucoid fluids on solubilizing and metabolizing different pollen components is a subject that merits considerable study.

Chapter 13. Nucleic Acids

Information on pollen nucleic acid chemistry is very fragmentary. Since analyses are done by different methods, only limited comparisons can be made. Most of the standard assay methods were developed for animal and microbial material. Nucleic acid determinations on plant material in general, and pollen in particular, have special sources of errors because of contaminating pigment and polysaccharide compounds extracted with the nucleic acids (CHANG and KIVILAN, 1962; SÜSS, 1970). Since most nucleic acid determinations are based on direct spectrophotometry, as well as color reactions with ribose, the pigments which are not completely excluded distort such measurements.

Süss (1970, 1971) compared 13 methods, including enzymatic, acid and alkaline extraction, for isolating DNA and RNA from *Corylus avellana* pollen. Good yields were obtained by some methods, while others gave poor yields with as much as one-third of the DNA left in the RNA fraction. Of the methods Süss compared, alkaline hydrolysis gave the most misleading data, particularly when the quantitative assay was made by U.V. spectrophotometry. Attempts to remove all contaminating inorganic phosphorus from the extracted nucleic acids increased losses in the total nucleic acids recovered. Yields of DNA are dependent upon the time of homogenization, e. g. more is recovered after 4 and 5 than at 3 or 7 minutes, and salt concentrations (SÜSS, 1971a). The pH during extraction or precipitation procedures is also critical (SÜSS, 1972). Comparisons are not only difficult, but an ideal extractive procedure for separating RNA and DNA from pollen is still not available.

Nucleic Acids in Mature Pollen

Reports of the nucleic acid contents of mature pollen are summarized in Table 13-1. The computations are usually made on a dry weight basis. In most pollen, nucleic acids constitute about 2% of the total dry weight. In general, pre-1960 literature reported DNA content at levels of about 0.5% of the dry weight, and total RNA content varying from 0.6% to 10%. The high RNA value of 10%, Table 13-1, is questionable; an amount up to about 1.5% of the pollen dry weight is more probable. Earlier measurements must be accepted with considerable reservation because of the nonspecific assay methods used.

Pollen of two taxonomically different plant species may sometimes contain similar amounts of RNA; pollen from *Lilium henryi* (LINSKENS, 1958) and *Corylus avellana* (SOSA-BOURDOUIL and SOSA, 1954) both contain about 1% RNA. In contrast to levels of total DNA, RNA from tetraploid and diploid plants is proportional to cell volume (TURBIN et al., 1965). The DNA levels were found to be directly proportional to the degree of ploidy of the plants.

Table 13-1. Nucleic acid content of mature pollen

Species	DNA[a]	RNA[a]	Reference
Oenothera organensis	0.04%	0.6%	CHANG and KIVILAAN, 1964
Betula pubescens	0.5%	—	EULER et al., 1945
Betula pubescens	—	ca. 10%	EULER et al., 1948
Pinus montana	0.5%	ca. 9%	EULER et al., 1948
Pinus montana	0.2%	—	EULER et al., 1945
Pinus ponderosa	0.05%	0.5%	STANLEY and YEE, 1966
Corylus avellana	0.01%	0.56%	MACIEJEWSKA-POTAPCZYK et al., 1968
Corylus avellana	1.35%	—	SOSA-BOURDOUIL and SOSA, 1954
Corylus avellana	—	0.96%	VANYUSHIN and FAIS, 1961
Petunia hybrida	0.4%/anth.[b]	0.8%/anth.	LINSKENS, 1958
Agave americana	—	50 µg/anth.	MAHESHWARI and PRAKASH, 1965
Beta vulgaris 2n	13.4 pg/grain	31.84 pg/grain	TURBIN et al., 1965
Beta vulgaris 4n	22.89 pg/grain	56.38 pg/grain	TURBIN et al., 1965
Vitis spp.	—	0.6%	VANYUSHIN and FAIS, 1961
Magnolia precia	—	0.6%	VANYUSHIN and FAIS, 1961
Rosa canina	—	0.7%	VANYUSHIN and FAIS, 1961
Gleditsia triacanthos	—	0.8%	VANYUSHIN and FAIS, 1961
Typha shuttlewartii	—	0.93%	VANYUSHIN and FAIS, 1961
Sambucus nigra	—	1.1%	VANYUSHIN and FAIS, 1961

[a] Dry weight basis unless otherwise noted.
[b] One anther contains approximately 81,000 pollen grains.

Data (Table 13-1) indicate that the total DNA content of pollen does not vary much between different plant species. Among prominent exceptions reported are: corn pollen nuclei which contain one-tenth of the amount of DNA reported in *Tradescantia* nuclei (Goss, 1968). As expected, the amount of DNA in the sperm (male cell) nucleus corresponds to the chromosome number in a haploid cell (SWIFT, 1950). In *Beta vulgaris*, diploid pollen contains twice as much DNA as haploid (TURBIN et al., 1965). BORMOTOV (1966) reported that in *Beta vulgaris* a doubling of pollen DNA with polyploidy also resulted in a directly proportional increase in protein, plastids and other cell constituents; he related the changes to increased metabolism and phenotypic changes.

STANLEY and YEE (1966) found a lower percent of DNA and RNA in *Pinus ponderosa* than EULER et al. (1945, 1948) found in *P. montana* (Table 13-1). While this may possibly be attributed to a larger mass of enclosing wing and exine in *P. ponderosa*, it seems more likely to be the result of improved assay procedures; this is further borne out in studies by TOGASAWA et al. (1968) which reported that in *Pinus densiflora*, DNA and RNA occur at 0.016% and 0.14%, respectively. Recognizing that the more recently reported percent values of DNA and RNA, e.g. *Corylus avellana*, 0.017% and 0.56% (Table 13-1), also represent a considerable difference from the earlier reported values for this same species, may be further evidence suggesting that studies since 1960 are employing more accurate assay techniques or have less contamination than those methods used in earlier papers. A reassessment of the pollens used in pre-1960 studies is necessary to help resolve the wide differences listed in Table 13-1.

The relative DNA content of generative and vegetative nuclei in mature pollen of *Petunia* or other species, usually measured cytophotometrically, remains in dispute. According to HESEMANN (1971), DNA replication continues in the mature *Petunia* pollen grains and comes to an end only after the nuclei migrate into the pollen tubes. These findings disagree with the results of DNA measurements in *Lilium candidum* by JALOUZOT (1969) and in *Tradescantia* by WOODARD (1958) in which they reported that replication is terminated in the mature pollen grain prior to dehiscence.

Cytophotometric DNA measurements show that prior to dehiscence the DNA content exists at the 2C level in both the generative and vegetative nuclei of binucleate tobacco pollen, and in both sperm nuclei and the vegetative nucleus of trinucleate barley pollen. Vegetative nuclear DNA of *Tradescantia* as detected spectrophotometrically remained constant at a 1C level (SWIFT, 1950), or at the 1.3C level according to studies of RODKIEWICZ (1969). A 1C histone level occurs in the vegetative nucleus of *Hippeastrum belladonna* which has a 2C DNA level. This suggests that histone synthesis is independent of the DNA replication in the pollen vegetative nuclei (PIPKIN and LARSON, 1973). DNA duplication in the vegetative nucleus prior to dehiscence is assumed to play a role in the formation and growth control of the pollen tube (D'AMATO et al., 1965).

Differences occur in the intensity of DNA and RNA staining in the vegetative and generative nuclei. The DNA of the generative nucleus in *Pinus densiflora* stains intensely, while the tube cell nucleus stains lightly (KASAI et al., 1966). This difference, also observed in angiosperm pollens, may arise from several possible reasons. For example, the vegetative nucleus may lack a nucleolus and therefore contain quantitatively less detectable nuclear protein; or the generative cell nucleus may be more highly methylated and therefore more tightly condensed; or the DNA in one nucleus may be combined with different histones or amounts of protein than in the other. The nucleolus disappears from the vegetative nucleus, and the level of DNA and RNA detected therein by staining also decreases at various times after mitosis and germination.

Vegetative nuclei in many species show significant differences in the time of onset of the degeneration phase. Generally, the vegetative cell of pollen has a high metabolic activity prior to anthesis. Observations by JALOUZOT (1969) and WOODARD (1958) tracing the decrease in the amount of DNA and RNA in the vegetative nuclei of maturing *Lilium* and *Tradescantia* pollen do not agree with observations on other species. In many species the level of nucleic acids appears to remain stable in mature, dehiscing pollen. In *Lilium* the microspore nucleolus breaks down at metaphase, resulting in an increase in ribosomes in the cytoplasm and a decrease in further production of rRNA transcription (PARCHMAN and LIN, 1972). The decrease in nucleic acids detected by staining (JALOUZOT, 1969) could represent a change in the nucleolar chromosome volume or the nuclear organizing region which may occur when the RNA is released into the cytoplasm and the nucleolus distintegrates.

Some ultrastructural differences in these two nuclei may also be related to the nucleic acid staining patterns. A greater number of pores were observed by LAFOUNTAIN and LaFOUNTAIN (1973) in the nuclear membrane of the vegetative cell than in the generative nucleus of *Tradescantia*. This may not just indicate that the vegetative tube nucleus is more active in transporting messenger RNA (mRNA) during germination, but that a channel exists for the prompt outward diffusion of oligonucleotides from the disintegrating nucleolus in the vegetative nucleus. However, it is interesting to note that the same level of endonucleases, DNases, were isolated from both these nuclei as the microspore neared maturity (TAKATS and WEVER, 1971). Presumably these DNases could also affect the DNA in the generative nucleus as well as the vegetative nucleus.

Generative and vegetative nuclei analyzed by TAKATS and WEVER (1971) contained about 20% cross-contamination of one form in the other; the generative nuclei were purer than the vegetative nuclear fraction (WEVER and TAKATS, 1971). Yet they showed no differences in DNA-polymerase activity, although only the generative nucleus must transcribe, replicate and separate its new DNA into two cells. Both nuclei in *Tradescantia* still appeared to retain active DNA-polymerase just prior to dehiscence, even though the vegetative nucleus is in S-phase and unable to divide.

It is always possible that the measurements of DNA-polymerase activity, or the capacity to add terminal portions to the chromatid DNA chains via a type of "end DNA nucleotidyl transferase," in isolated nuclei do not represent conditions *in vivo*. However, assuming that such enzyme assays as those of TAKATS and WEVER (1971) are correct, the fact is that the two pollen nuclei after dehiscence function quite differently from those studied in microspores just reaching maturity.

As discussed above, the generative and vegetative nuclei in pollen stain differently and manifest certain differences in enzyme activity, as well as in the ratio of nucleotides from which their DNA and RNA are composed (Table 13-2). Another hypothesis that may explain these observed differences, and that could be tested on isolated nuclei, is that vegetative nuclei lack the specific nucleoside diphosphate kinase for synthesizing thymidine-5-phosphate, thus the metabolic pathway for production of thymidine from deoxyuridine monophosphate may be shut down at maturity of the microspores. To test alternative hypotheses, studies are needed employing standard techniques which would compare DNA, RNA and associated enzymes in gymnosperm to angiosperm pollen containing either two or three nuclei at maturity.

Nucleic Acids during Pollen Development

The pattern of DNA synthesis during microspore development seems to be essentially similar in the few species studied, i.e. in *Lilium* by TAYLOR and McMASTER (1954), MOSES and TAYLOR (1955), TAYLOR (1959), STERN (1960), STERN and HOTTA (1969), HOTTA and STERN (1971), LINSKENS and SCHRAUWEN (1968) and in *Triticum aestivum* by RILEY and BENNET (1971).

DNA synthesis takes place at three different stages:
a) in the early phase of the meiotic division (early prophase),
b) in connection with pollen mitosis,
c) before division of the generative nucleus in the case
of trinucleate pollen formation.

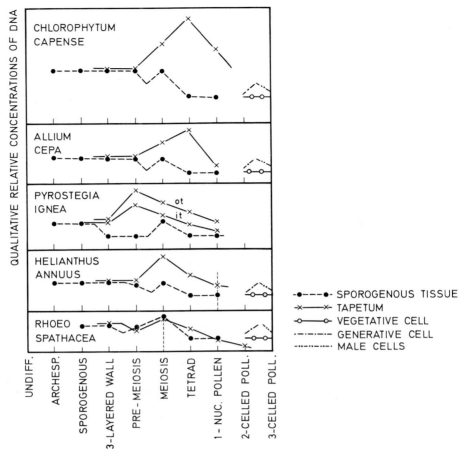

Fig. 13-1. Changes in DNA during microsporogenesis (NANDA, 1972). *ot*: outer tapetum, *it*: inner tapetum

DNA fluctuations in the microspore mother cells, MMCs (or PMCs) (Chapter 1) and in the vegetative and generative cells—including a decline in DNA content before meiosis (STERN, 1960) (Fig. 13–1)—were observed in *Lilium candidum* by MOSS and HESLOP-HARRISON (1967), in *Zea mays* by REZNIKOVA (1971) and in *Lilium candidum* (MAKHANETS, 1968; NANDA, 1972). Most DNA in the PMCs is duplicated or synthesized at the pre-meiotic S-phase, with a small fraction duplicated during zygotene and pachytene (STERN and HOTTA, 1969; NANDA, 1972). DNA content calculated per cell in *Lilium longiflorum* drops sharply at the end of meiosis with formation of the 4 microspores (OGUR et al., 1951). The DNA level gradually increases during interphase between meiosis and microspore mitosis. During microspore mitosis DNA content doubles rapidly (OGUR et al., 1951; TAYLOR and MCMASTER, 1954; MOSES and TAYLOR, 1955; TAYLOR, 1959; LINSKENS, 1958; MOSS and HESLOP-HARRISON, 1967; NANDA, 1972).

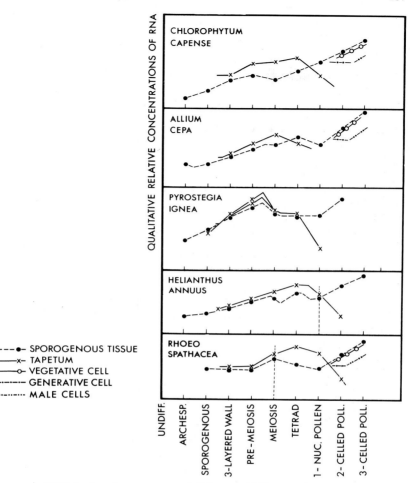

Fig. 13-2. Changes in RNA during nicrosporogenesis (NANDA, 1972)

RNA synthesis during pollen development was reviewed by MASCARENHAS (1971). Considerable differences occur in RNA synthesis in the various stages of pollen development (LINSKENS and SCHRAUWEN, 1968; NANDA, 1972), especially during and after meiosis (STERN, 1960). On the basis of autoradiographic studies in *Lilium longiflorum*, TAYLOR (1959) concluded that the bulk of RNA synthesis in the meiocytes takes place in the pre-meoitic and leptotene periods; this was later confirmed in other species and genera, i.e. *Lilium henryi* (HOTTA and STERN, 1963; LINSKENS and SCHRAUWEN, 1968), *Lilium candidum* (NANDA 1972), *Zea mays* (MOSS and HESLOP-HARRISON, 1967), and *Paeonia* (SAUTER, 1969). In young microspores, RNA synthesis rises again, followed by an intensive increase after pollen mitosis (OGUR et al., 1951; LINSKENS and SCHRAUWEN, 1968; SAUTER, 1969; NANDA, 1972). Interpretation of the results of different authors is somewhat controversial and difficult to compare since the experimental data are produced by different methods, e.g. autoradiographic, isotope incorporation studies and application of inhibitors (SAUTER, 1971) (Fig. 13-2).

DNA Transfer during Microsporogenesis

The question of the source of nucleic acids and the "nurse" function of the tapetum is still not completely resolved. Feulgen-positive-staining globules, indicating an appreciable content of DNA, were observed in the tapetum by Cooper (1952). Cytochemical evidence for transfer, confirmed by Linskens (1958) and Vasil (1959), was contradicted by others (Takats, 1959, 1962; Kosan, 1959; Heslop-Harrison and MacKenzie, 1967).

The evidence of nucleic acid synthesis in the tapetum, independent of developing pollen, can be explained in terms of the existence of a nontapetal pool of nucleic acid precursors. Tapetal DNA may break down and supply nucleotides for incorporation in the developing microspore nuclei. Experimental data suggests the latter interpretation. This evidence includes the following pathways: catabolism of tapetal DNA and uptake of the resulting precursors by the PMC, or/and the existence of a non-tapetal pool of DNA precursors delivered from other flower parts, or the vegetative organs of the plant. By following ^{3}H-labeled thymidine incorporation during microsporogenesis it may be possible to resolve some of these apparent differences in the tapetum DNA pool and help disclose the function it serves.

Nucleic Acids during Germination

Briefly summarized, in germinating pollen a general activation of metabolic processes takes place, which includes not only enzymes, but also RNA synthesis and assembly of polysomes (Mascarenhas, 1971). There is strong evidence that dormant, or masked forms of messenger RNA and transfer RNA are present prior to germination and are quickly activated upon germination (Linskens, 1971; Mascarenhas, 1971; Dexheimer, 1972). Also it is indicated, that mRNA is combined with a percentage of the ribosomes in some pollen. Ungerminated *Tradescantia* pollen contained about 38% of the rRNA as polysomes (Mascarenhas and Bell, 1969). The low oxidative enzyme activity (Chapter 14) found by Razmologov (1963) may reflect the need for formation of specific mRNA and proteins in the gymnosperm pollen before active growth can occur. The studies of the nucleoside pools in pollen may help explain the observed growth patterns.

The possibility exists that RNA and protein levels in germinating pollen are related to nuclear DNA activities. Gramineae and other 3-celled pollen all germinate very rapidly, minutes after dehiscence and landing in a receptive environment. Male cells in such pollen have already separated, with both cells containing DNA at the 2C level. According to this hypothesis, those intermediate in germination rate following pollination have already duplicated the DNA in the generative nucleus prior to dehiscence, but the generative cells have not divided. Other pollen, particularly among those which are slow to germinate may not have replicated their DNA in the generative nucleus prior to dehiscence. This hypothesis suggests that the rate of initial metabolic activity of pollen after it lands on a receptive female tissue may be related to the state of DNA in the generative cell. Such an hypothesis can probably be evaluated by comparing different pollen via both staining cytophotometric studies and radioisotopic labeling experiments of DNA and RNA during the initial phases of germination.

The function of the vegetative tube nucleus in pollen germination is still not definitely known. DNA of the vegetative nucleus has a different histone pattern from the DNA in the generative cell (SAUTER, 1969, 1971). The tube nucleus is involved in RNA-protein-synthesis. Proteins synthesized or activated in the pregermination stage of pollen growth are apparently required for the early stages of the tube development and germination. Molecular information for protein synthesis is primarily supplied through the mRNA-ribosomal complex present in the pollen which may be most actively released from the tube nucleus. The tube nucleus and its nucleic acid macromolecules generally disintegrate before the tube reaches the embryo sac and bursts.

Although the control mechanisms for initiating nuclear division or breakdown are presently not recognized, it is usually presumed that the information coded within the nucleotide sequences in the DNA and the mRNA facilitate the specific patterns of synthesis and enzyme activities underlying such changes. Efforts to understand the source of such changes require development of a holistic concept relating the substrates, enzymes and differences in the pollen DNA and RNA macromolecules.

Nucleotide Base Composition

Discerning the genetic relationships and mechanisms by which pollen DNA and cellular RNA control growth and transfer genetic information can be facilitated, in part, by recognizing the purine and pyrimidine base components of the pollen nucleic acids. Mole ratios of the nucleotide components in pollen are given in Table 13–2. Unlike yeast RNA, which has a guanine mole ratio of abouth 24%, and uridine near 27%, these base components are usually reported to occur in pollen at values of 28% and 23%, respectively (Table 13–2). However, the RNA nucleotide fractions in different subcellular moieties of pollen and in the different nuclei from the same pollen (Table 13–2), have different nucleotides proportions than those reported for whole pollen, or other plant tissues.

The purine:pyrimidine ratios of the RNA in microsomal fractions and in the generative nucleus are closest to unity, i.e. 1 (Table 13–2). This data suggests that cell fractions involved in information translation, those fractions which are base complimentary with nuclear DNA, are most likely to be composed of nucleotides in stable reproducible purine:pyrimidine ratios. A preponderance of tRNA or rRNA, or absorbed oligonucleotide fragments in a particular cell fraction or the whole pollen, could substantially alter the $A+U/C+G$ value from 1. The fact that RNA synthesized during initial germination of *Nicotiana* pollen has a different base composition than that from whole pollen extracts (TANO and TAKAHASHI, 1964) would be expected from the data on base composition in the RNA from different subcellular fractions (Table 13–2). Since the total RNA and the base mole fractions can vary so widely, then pollen nuclei and/or subcellular fractions should be carefully separated wherever possible before analyzing the RNA, if valid relative comparisons of evolutionary or metabolic significance are to be made between RNA from different pollen sources.

Table 13-2. Role base proportion of RNA components from pollen

Species	Fraction	Guanine	Adenine	Cytidine	Uridine	Purines Pyrimidines	Reference
Pinus sylvestris	Whole	27.6	23.4	23.4	23.7	1.12	VANYUSHIN and BELOZERSKII, 1959
Picea excelsa	Whole	28.8	22.5	22.5	23.3	1.18	VANYUSHIN and BELOZERSKII, 1959
Coryllus avellana	Whole	28.2	25.0	25.0	22.0	1.13	VANYUSHIN and BELOZERSKII, 1959
Sambucus nigra	Whole	28.7	23.8	23.8	21.2	1.11	VANYUSHIN and FAIS, 1961
Typha shuttlewartii	Whole	29.3	26.5	26.5	21.0	1.26	VANYUSHIN and FAIS, 1961
Pinus ponderosa	Microsomes	27.1	25.4	25.4	24.7	1.00	STANLEY and YEE, 1966
	Mitochondria	29.3	24.0	24.0	20.5	1.25	STANLEY and YEE, 1966
	Supernatant	28.9	22.8	22.8	26.6	1.04	STANLEY and YEE, 1966
	Cell wall	31.1	24.8	24.8	14.6	1.53	STANLEY and YEE, 1966
Tradescantia paludosa	Nucleus						
	Generative	16.2	34.3	23.7	25.6	1.02	LaFOUNTAIN and MASCARENHAS, 1972
	Vegetative	18.4	34.2	19.3	28.0	1.12	LaFOUNTAIN and MASCARENHAS, 1972

The presence of methylated bases, particularly in the tRNA and DNA may also distort the molar proportions of the various bases isolated from pollen RNA and DNA. The pyrimidine, 5-methylcytosine, has been found in pine pollen DNA (TOGASAWA et al., 1968), and the 6-methylaminopurine has been reported as a DNA component in *Salix caprea* and *Betula verrucose* (BUR'YANOV et al., 1972). Methylation of cytosine in *Lilium* DNA occurred in the three phases G_1, S and G_2 of the mitotic cycle of *Lilium*, which contained about 4.5% 5-methylcytosine in its DNA (HOTTA and HECHT, 1971).

None of the unique methylated purines or pyrimidine bases have been reported in pollen RNA fractions. Now that the different types of RNA, transfer, messenger and ribosomal are readily isolated from pollen (LINSKENS, 1967; STEFFENSEN, 1971) it should be possible to search for these distinct nucleic acid base components as possible clues to metabolic or genetic differences associated with pollen.

The rRNAs of pollen thus far isolated are generally associated with 40S ribosomes, separated by gel electrophoresis, and appear to dissociate into 23S or 25S subunits. The latter increase as pollen starts to grow. The presence of long-lived mRNA related to the initial capacity to germinate, has been suggested from experiments with inhibitors which act on RNA-polymerase. This pollen has no need for induction or formation of new mRNA and polyribosomes or tRNA before rapidly extending its tube (DEXHEIMER, 1968; 1972; MASCARENHAS, 1971; LINSKENS, 1971).

Free Nucleotides

In sugar beet *(Beta vulgaris)*, free adenine occurs at a level of 0.04% of the pollen dry weight (KIESEL and RUBIN, 1929). Such direct analyses of purine and pyrimidine bases in pollen have seldom been reported. Studies of NYGAARD (1972; 1973) on ungerminated *Pinus mugo*, and incorporation of ^{32}P into its nucleotides during the first 24 hours of germination showed that the adenosine-5'-triphosphate (ATP) initially present before germination, is exhausted as growth begins and then rapidly increases in the initial germination phases.

Conversion of RNA Precursors

NYGAARD (1973) proposed a tentative scheme (Fig. 13–3) showing the probable interrelation of the conversion of ribonucleotides and ribonucleosides in this pollen. Oxidative phosphoylation was suggested as the dominant source regenerating ATP in pollen rather than from source pools of 3-phosphoglycerate and polyphosphates often observed in other organisms.

Pollen in NYGAARD's studies expanded its nucleotide pool very rapidly during the initial phases of germination *in vitro*. This again suggests that nuclear and cytoplasm breakdown of some inherent or external RNA or DNA components may supply nucleotides for the build-up and resynthesis required for subsequent metabolic activity. Many of these nucleotides, in particular UDP and ATP, are involved in metabolism of the sugars incorporated into the pollen tube wall and membranes (Chapter 9). Yet, the potential pool as suggested from NYGAARD's

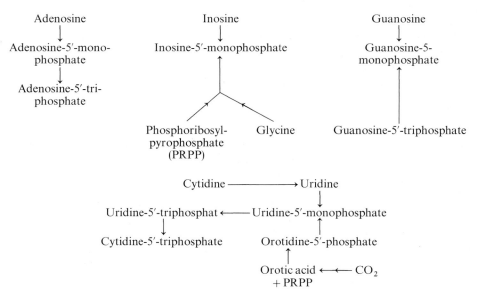

Fig. 13-3. Reactions probably involved in the conversion of ribonucleic acid precursors in pollen (NYGAARD, 1973)

studies on pine, which is a slow germinating pollen, shows that all the nucleotides and nucleosides required to form RNA are available at different levels in metabolic pools. Such substrates could facilitate translation of the DNA codons, and assemblage of ribosomal, or messenger-RNA moieties required for control of protein synthesis.

The actual patterns, level of synthesis and the possible role of RNA in pollen growth remains to be clearly discerned. It is also quite possible in pollen to trace the incorporation of $^{14}CO_2$ into metabolites which may also be involved in nucleotide and RNA synthesis (STANLEY et al., 1958). Such potential studies may shed additional information on the size and source of nucleotide and nucleoside pools in the different types of pollen as they prepare to extend a tube and convey the male cell to the egg. They can also help indicate which enzymes are functional in facilitating such metabolic interconversions in pollen.

Chapter 14. Enzymes and Cofactors

Pollen, a non-chlorophyllous tissue, contains enzymes to metabolize external and internal substrates essential for tube growth. In some species tube extension occurs minutes after a grain lands in the proper environment; in others, growth occurs several days or weeks after pollination. The tube may elongate and convey the male cells to the egg cell in minutes; in a few genera, e.g. *Pinus*, syngamy is delayed over 13 months. Such widely diverse growth patterns suggest several inherent characteristics for pollen enzymes.

The specificity of pollen enzymes is probably related to the tissues in which the tubes normally grow. Intra- and extracellular hydrolytic and anabolic enzymes would be expected at levels or activities sufficient to sustain the rates of wall formation and respiration observed in the rapid growing tubes. On the other hand, slow-growing pollen must retain, for varying periods of time, viable enzymes or the capacity to synthesize new enzymes. The narrow physiological conditions and chemical specificity of the tissues in which pollen grows suggests that pollen has a limited adaptive capacity which may be related to the highly differentiated and specialized nature of the tissue. Yet some haploid pollen can differentiate into tissue masses and whole plants (NITSCH and NITSCH, 1969; SUNDERLAND and WICKS, 1971). Thus, in some cases, the range of enzymes which pollen forms, or can be induced to form, corresponds to that of normal diploid cells. Enzyme comparisons have not been made between 1 n tissues differentiated from pollen and normal 2 n tissues. We can only conclude that pollen, while usually of limited growth expression, is endowed with a considerable range of enzymes or the capacity to form and/or activate many enzymes during growth.

The enzymes detected in pollen vary with the species and stage of development. During microsporogenesis, the most active enzymes are those associated with DNA replication, RNA synthesis and the developing intine-exine. Others relating to initial tube growth in the female tissue are active or easily activated in the mature microspore. Enzyme activities in developing microspores as well as those detected during germination and tube growth will be discussed. We will also attempt to relate a few of the enzymes detected in pollen to their physiological functions.

Methods of Detection

Starting about 1865, enzyme activities were detected in pollen by indirect experimental evidence, i.e. reactions in which pollen metabolized a substrate or formed a specific non-metabolic product. Only in the late 1960's were enzymes purified from pollen. Indirect enzyme assays are often done by reacting pollen, or a buffered pollen extract, with a specific substrate in the presence of a redox dye, e. g. triphenyl tetrazolium chloride (TTC), or 2,6-dichloroindophenol. These histochemical dye reactions detect the electron transfer which accompanies oxidative phosphorylation or hydrogen transfer such as the dehydrogenases mediate. The reactions are dependent primarily on pH, temperature, substrate and the presence of a compound with sufficient oxidation-reduction electromotive potential to act as an electron acceptor in the respiratory chain. The electron transfer induces a spectral light shift, a color change detected in the redox intermediate dye (Table 6–5).

Staining tests for enzymes depend upon the ability of the reacting chemicals to diffuse throughout the cell and its organelles. Some organelles and membranes impede passage of dye molecules. SAUTER and MARQUARDT (1970) reported that neither the callose layer in mature tetrads of *Paeonia* nor the exine, particularly in the pore regions, restrict inward diffusion of the tetrazolium solution. Thus, water soluble dyes readily penetrate this pollen. Yet in *Portulaca*, intense succinate dehydrogenase activity is detected only after initiation of germination and not in mature ungerminated pollen grains (SOOD et al., 1969). Visual limits of the light microscope may make it difficult to detect the initial low levels of enzyme-dye complex in ungerminated pollen. Dry pollen probably also contains only low levels of free and active enzymes. Histochemical detection is difficult until growth starts or the internal water level exceeds about 15%. Dyes are generally in a water solution; the time lag involved in their reaction may be related to enzyme activation as well as the proton interaction.

Another problem in using chemical agents to form reactive color complexes with enzymes or their products may arise when isozymes differ in their activity, pH and temperature characteristics. This can cause different responses to the dye. Application of fast garnet, to detect acid phosphatase in maize pollen, inhibited one of the three acid phosphatase isozymes. Other dyes reacted with all three phosphatases without any inhibition (EFRON 1969). Methods other than dyes, particularly those which activate or release enzymes by adding a substrate in buffer solution, may be more suitable for quiescent pollen.

Enzyme detection by placing pollen on a known metabolic substrate and determining the change in respiratory gas exchange is one such method. Enzyme cofactors are also often used with the substrate. Early experiments usually involved placing pollen in buffered media with different substrates and assaying CO_2 release or O_2 uptake in respiration (Table 14–1). Most enzymes essential for

Table 14-1. Pollen respiration on acid substrates (OKUNUKI, 1939)

Species (150 mg)	CO_2 release (mm^3)/30 min. pH 6.0, 25° C	
	Glutamic acid	Pyruvic acid
Camellia japonica	137	563
Paeonia albiflora	0	138
Pinus densiflora	185	140
Arisaema spp.	0	311
Thea sinensis	330	438

metabolism of Krebs Cycle acids were detected by this indirect method; the technique was also used to demonstrate capacity to metabolize specific sugars. Species differences, Table 14–1, were attributed to lack of either a specific enzyme or a cofactor for decarboxylation or reduction. Accuracy of this method can be improved by utilizing known radioactive substrates for metabolites and detecting the labeled products.

Radioactive sugar can be supplied to pollen and labelled products detected by isolation and chromatography. A whole range of metabolic intermediates and the required enzymes to synthesize them may be deduced by studies where very simple constituents such as $^{14}CO_2$ are incorporated by the growing pollen (STANLEY et al., 1958). However, the most accurate assays involve direct isolation of a functioning enzyme after ammonium sulfate fractionation or electrophoretic separation of the diffusates from whole pollen or pollen homogenate (Fig. 11–1).

Activities Detected

Early studies on pollen enzymes paralleled research on other plant tissues. VAN TIEGHEM (1869) reported invertase in pollen of *Narcissus*, *Hyacinthus* and *Viola*, while ERLENMEYER (1874) found diastase (amylase) in pine; STRASBURGER (1886) and GREEN (1894) focused on amylase and invertase, and the former also reported a cytase (hemicellulase). These studies showed that pollen allowed to stand with starch paste transformed starch to sugars (STRASBURGER, 1886). GREEN (1894) extracted *Lilium*, *Helianthus* and *Narcissus* pollen with a water-sodium chloride solution and showed, with careful elimination of microbial contaminants, that pollen was a rich source of diastase and invertase. To this early list of enzyme activities KAMMANN (1912) added protease, catalase and lipase in rye. These observations were correlated with the starch, sucrose and pectin content of epithelial tissue and other tissues of the style, stigmas or megasporophyll in which pollen normally grows.

Organization of enzymes on the basis of the chemical reactions they catalyze offers nomenclatural categories which facilitate discussion of the individual enzymes. The list of 34 enzymes which MÄKINEN and MACDONALD (1968) and 39 which BREWBAKER (1971) reported active in pollen can be justifiably expanded to at least 80 (Table 14–2). Undoubtedly more will be reported as new detection techniques are applied.

Enzymes not cited by reference in Table 14–2 are strongly indicated as being present on the basis of substrate metabolism experiments by various authors who have generally not searched for the specific enzyme, but rather only for products or growth of the pollen on a substrate. We have assumed from such experiments, or our own unpublished studies, that the essential enzyme to facilitate the metabolic conversion recognized in other higher plant tissues is present in pollen. Since these are only inferences, we have generally avoided stating that the enzyme was detected in a specific piece of published research. Hence, non-referenced enzymes are included on indirect, deductive evidence and must still be substantiated.

Table 14-2. Enzymes reported in pollen

Trivial name	EC number[a]	Reference
Class: Oxidoreductases		
Alcohol dehydrogenase	1.1.1.1	OKUNUKI, 1940
D-Arabinitol dehydrogenase	1.1.1.11	
Inositol dehydrogenase	1.1.1.18	
UDP-Glucose dehydrogenase	1.1.1.22	ROGGEN, 1967; DAVIES and DICKINSON, 1972
Lactate dehydrogenase	1.1.1.27	EULER and EULER, 1949
Malate dehydrogenase	1.1.1.37	THUNBURG, 1925
Isocitrate dehydrogenase (NADP)	1.1.1.42	DICKINSON and DAVIES, 1971
Phosphogluconate dehydrogenase	1.1.1.44	DESBOROUGH and PELOQUIN, 1968
Glucose dehydrogenase	1.1.1.47	
Glucose-6-phosphate dehydrogenase	1.1.1.49	DICKINSON and DAVIES, 1971
Triosephosphate dehydrogenase	1.2.1.9	OKUNUKI, 1940
Malonate semialdehyde dehydrogenase	1.2.1.15	LINSKENS, 1966
Succinate dehydrogenase	1.3.99.1	OKUNUKI, 1939a
Glutamate dehydrogenase (NADP)	1.4.1.3	OKUNUKI, 1939a
L-Amino-acid oxidase	1.4.3.2	EULER et al., 1948
Monoamine oxidase	1.4.3.4	SOOD et al., 1969a
Lipoamide dehydrogenase	1.6.4.3	OKUNUKI, 1940
Cytochrome oxidase	1.9.3.1	OKUNUKI, 1940
o-Diphenol oxidase, tyrosinase	1.10.3.1	ISTATKOV et al., 1964
Ascorbate oxidase	1.10.3.3	
Fatty acid peroxidase	1.11.1.3	
Catalase	1.11.1.6	LOPRIORE, 1928
Peroxidase	1.11.1.7	OSTAPENKO, 1960
meso-Inositol oxygenase	1.13.1.11	
Class: Transferases		
Aspartate carbamoyltransferase	2.1.3.2	ROGGEN, 1967
α-Glucan phosphorylase, P-enzyme	2.4.1.1	KESSLER et al., 1960
Maltose 4-glucosyltransferase, amylomaltase	2.4.1.3	KESSLER et al., 1960
UDP-Glucose-β-glucan glucosyltransferase	2.4.1.12	
Trehalosephosphate-UDP glucosyltransferase	2.4.1.15	GUSSIN and McCORMACK, 1970
α-Glucan-branching glycosyltransferase	2.4.1.18	KESSLER et al., 1960
UDP-Galactose-glucose galactosyltransferase	2.4.1.22	
Aspartate aminotransferase	2.6.1.1	BARTELS, 1960
Alanine aminotransferase	2.6.1.2	
Glycine aminotransferase	2.6.1.4	ROGGEN, 1970
Hexokinase	2.7.1.1	KATSUMATA and TOGASAWA, 1968
Glucokinase	2.7.1.2	
Xylulokinase	2.7.1.17	PUBOLS and AXELROD, 1959
Phosphoribulokinase	2.7.1.19	
Glucuronokinase	2.7.1.43	DICKINSON et al., 1973
Nucleosidediphosphate kinase	2.7.4.6	DICKINSON and DAVIES, 1971a
Phosphoglucomutase	2.7.5.1	DICKINSON and DAVIES, 1971
DNA Nucleotidyltransferase	2.7.7.7	WEVER and TAKATS, 1970
UDP-Glucose pyrophosphorylase	2.7.7.9	DICKINSON and DAVIES, 1971
ADP-Glucose pyrophosphorylase	2.7.7.x	DICKINSON and DAVIES, 1971
Ribonuclease (RNase)	2.7.7.16	LINSKENS and SCHRAUWEN, 1969
Class: Hydrolases		
Carboxylesterase (B-esterase)	3.1.1.1	MÄKINEN and BREWBAKER, 1967
Arylesterase (A-esterase)	3.1.1.2	MACDONALD, 1969
Lipase	3.1.1.3	KAMMANN, 1912

Table 14-2 (cont.)

Trivial name	EC number[a]	Reference
Cutinase		LINSKENS and HEINEN, 1962
Pectinesterase	3.1.1.11	
Alkaline phosphatase	3.1.3.1	PALUMBO, 1953
Acid phosphatase	3.1.3.2	GÓRSKA-BRYLASS, 1965
Phytase	3.1.3.8	BREDEMEIJER, 1971
Trehalosephosphatase	3.1.3.12	GUSSIN and McCORMACK, 1970
Phosphodiesterase	3.1.4.1	DICKINSON and DAVIES, 1971
Deoxyribonuclease (DNase)	3.1.4.5	GÓRSKA-BRYLASS, 1965
Arylsulphatase	3.1.6.1	GÓRSKA-BRYLASS, 1965
α-Amylase	3.2.1.1	HAECKEL, 1951
β-Amylase	3.2.1.2	ELSER and GANZMÜLLER, 1930
Cellulase	3.2.1.4	PATON, 1921
Laminaranase (Callase)	3.2.1.6	GORSKA-BRYLASS, 1967
Polygalacturonase (Pectinase)	3.2.1.15	VINSON, 1927
α-Glucosidase	3.2.1.20	DICKINSON, 1967
β-Glucosidase	3.2.1.21	GUSSIN et al., 1969
β-Glucosidase	3.2.1.23	GUSSIN et al., 1969
α-Mannosidase	3.2.1.24	LINSKENS et al., 1969
β-Fructofuranosidase, Invertase	3.2.1.26	ELSER and GANZMÜLLER, 1930
Trehalase	3.2.1.28	GUSSIN et al., 1969
β-N-Acetylglucosaminidase	3.2.1.29	LINSKENS et al., 1969
Oligo-1,3-glucosidase	3.2.1.39	
Leucine aminopeptiadase	3.4.1.1	MÄKINEN and BREWBAKER, 1967
Aminopeptidase	3.4.1.2	MÄKINEN and BREWBAKER, 1967
Pepsin, Protease	3.4.4.1	PATON, 1921
Trypsin	3.4.4.4	VENKATASUBRAMANIAN, 1953
Aminoacylase	3.5.1.14	UMEBAYASHI, 1968
Inorganic pyrophosphatase	3.6.1.1	HARA et al., 1970
ATPase	3.6.1.3	PALUMBO, 1953
ATPase	3.6.1.8	MALIK et al., 1970
Class: *Lyases*		
Pyruvic decarboxylase	4.1.1.1	OKUNUKI, 1939
Oxaloacetat decarboxylase	4.1.1.3	
Mesoxalic decarboxylase		GOAS, 1954
Glutamic decarboxylase	4.1.1.15	OKUNUKI, 1939
Phosphopyruvate carboxylase	4.1.1.31	
Phosphopyruvate carboxylase	4.1.1.38	
Ribulosediphosphate carboxylase, carboxydismutase	4.1.1.39	STANLEY et al., 1958
Ketose-1-phosphate aldolase	4.1.2.7	BARTELS, 1960
Fructosediphosphate aldolase	4.1.2.13	
Citrate synthase (synthetase)	4.1.3.7	ROGGEN, 1967
Phenylalanine ammonia-lyase	4.3.1.5	WIERMANN, 1972
Class: *Isomerases*		
UDP-Glucose epimerase	5.1.3.2	
Arabinose isomerase	5.3.1.3	
Xylose isomerase	5.3.1.5	PUBOLS and AXELOD, 1959
Ribosephosphate isomerase	5.3.1.6	
Glucosephosphate isomerase	5.3.1.9	
Class: *Ligases and Others*		
Carboxylases	6.4.1.(1,2)	OKUNUKI, 1939
Folic acid conjugase		NIELSEN and HOLMSTRÖM, 1957
D-glucose-6-P-cycloaldolase (NAD+)	4.1.2.x	LOEWUS and LOEWUS, 1971

[a] International Union Biochem. (Enzyme Commission). Enzyme Nomenclature. Amsterdam: Elsevier Publ. Co., 1965.

Oxidoreductases

The flavoprotein cytochrome c was one of the earliest natural redox components detected in pollen (OKUNUKI, 1939). SINKE et al. (1954) detected cytochrome a and b, as well as c, in 38 pollen species by spectrographic analysis of cytoplasm. Spectral bands, verified by sodium azide inhibition of the cytochrome oxidase, inhibiting the Nadi reaction, were prominent in angiosperm pollen. The cyto-chrome c band at 550 nm was particularly prominent. Gymnosperm pollen lacked nearly all the characteristic cytochrome absorption bands. In gymnosperm pollen the Nadi reaction indicated presence of cytochrome oxidase as follows: questionable in *Podocarpus*, weak in *Pinus*, and no enzyme could be detected in *Abies* or *Cryptomeria*. MURPHY (1973) detected several cytochrome oxidase iso-zymes in *Abies* pollen extracts. Presuming all grains permit an equal inward diffusion of reactants, then the level of cytochrome oxidase activity in viable non-germinated pollen is probably related to the capacity of pollen rapidly to initiate oxidative metabolic activities involved in preparation for tube growth. Gymno-sperm pollen is deficient in the cytochrome flavoproteins and they are also very slow to germinate. No one has correlated the presence or absence of such oxida-tive pathways with the capacity to initiate tube growth. However, ZINGER (TSIN-GER, 1961) attempted to correlate phylogenetic evolution of 28 angiosperm orders with the levels of cytochrome oxidase and peroxidase present in pollen grains and tubes as detected by PODDUBNAYA-ARNOLDI et al. (1959, 1961).

PODDUBNAYA-ARNOLDI and her group (1961) surveyed several oxidative en-zymes in pollen of 65 species distributed throughout 42 families. Polyphenoloxi-dase was almost totally absent from all pollen tested. Considerable variation occurred in enzyme levels of pollen from the same anthers. The greatest variation occurred between polyploids and the less active haploids; pollen of sterile hybrids is grossly deficient in enzyme activity. Earlier comparisons between *Salix* diploids and hexaploids showed that some enzymes increased while others decreased with increasing polyploidy (EULER and EULER, 1949). TSINGER (1961) found that in 14 of 28 orders he compared, the increasing peroxidase and cytochrome oxidase activities could be correlated to the relative systematic position. But as he ad-vanced to the higher orders, i.e. from Liliales to the Graminales, activity of these enzymes decreased. He presumed that this decrease represented degeneration of pollen with evolutionary progress. He reasoned that during evolution, polyphe-noloxidase, which is found in gymnosperm pollen, was eliminated first, followed by decreases in cytochrome oxidase while peroxidase activity was the most stable. Since evolutionary development in the Graminales and Caryophyllales includes an additional nuclear division before anthesis (BREWBAKER, 1959), the lack of a required energy-producing oxidative system to fulfill this complex reaction in these pollen species after germination might also suggest an evolutionary reason for a decrease in enzyme activity. However, activity was high in pollen from the Caryophyllales and variable to low in members of the Graminales; it was very low in the advanced Orchidales order with binucleate pollen. Within the more primitive order of Liliales, *Lilium* pollen was low or lacking cytochrome oxidase, *Gladiolus* was high; and all members of this order have good to high levels of peroxidase. Several members of the even less advanced order Ranales lacked peroxidase activities, in contrast to the presence of both these enzymes in many species of the more advanced Graminales. A trend toward simplification, particularly as evidenced by less cytochrome oxidase activity, occurred in some advanced orders, i.e. Asterales and Ericales, but there are many exceptions. The phylogenetic interpretation by TSINGER (1961) is hardly justified by the total data.

A less detailed intergeneric comparison based partly on oxidative enzymes within pollen of 9 different gymnosperm species was attempted by RAZMOLOGOV (1963). On the basis of lower peroxidase and cytochrome oxidase activity he classified *Zamia*, *Pinus* and *Picea* as more primitive than representatives of the genera *Taxus*, *Juniperus*, *Biota*, *Cupressus* and *Tetraclinis*. The lower oxidative activities were suggested as the reason for the slow pollen growth of the more primitive gymnosperms. He further suggested, without evidence, that the high level of oxidative activities in pollen of more advanced angiosperm plants accounted for their capacity to germinate rapidly.

Not all plant oxidases have been found in pollen. The rather common enzyme, tyrosinase, catechol oxidase, is relatively rare in pollen, or at least was not found in any of 83 species examined by PODDUBNAYA-ARNOLDI et al. (1959), although low enzyme activity was reported in corn pollen (Table 14–2).

Several reductases of the dehydrogenase type, when coupled to indicator dyes as well as oxidases, e.g. peroxidase, have been used to assay pollen viability (Chapter 6). At least 15 dehydrogenases have been detected in pollen (Table 14–2).

Cofactors or coupled reactors are required to detect some dehydrogenase activities. Succinic oxidase was detected in pollen on addition of the acid substrate and cytochrome c (OKUNUKI, 1939). Most dehydrogenases appear as two or more isozyme bands when separated by electrophoresis. Not all pollen release identical isozymes for the same enzyme. And while these differences may provide valuable genetic markers in breeding studies (BREWBAKER, 1971), they may also yield anomalies in enzyme analyses (see below "Sources of Variation").

Transferases

These enzymes have not been studied as thoroughly as the oxidoreductases. Many transferases are involved in the addition of glucose to polymers in formation of cellulose and pectins. The enzymes are inferred from radioisotopic studies in which glucose-^{14}C or inositol-^{14}C are incorporated into the tube wall polymers (STANLEY and LOEWUS, 1964; DICKINSON and DAVIES, 1971; ROGGEN and STANLEY, 1971). The maize pollen enzyme transferring glucose to quercetin (LARSON, 1971) is fairly specific and probably limited to related flavonols such as kaempferol (Chapter 15).

ATP-glucose pyrophosphorylase is associated primarily with starch synthesis. However, specificity of the GDP-glucose and UDP-glucose pyrophosphorylase, i.e. whether the reaction products are predominantly cellulose, via β-1,4- synthetase, or a β-1,3-polyglucan callose-like material (AXELOS and PEAUDLENOEL, 1969), appears to depend on the concentration of the substrates. When KESSLER et al. (1960) supplied germinating pollen of *Lapageria rosea* and *Impatiens oliveri* with glucose-^{14}C and fructose-^{14}C, they recovered not only radioactive sucrose but a series of 2 to 8 polymerized oligoglucosides of the β-1,3-D-glucose series, and a series of 2 to 4 units of α-1,4-D-oligoglucosides of the maltodextrin series. When they fed ^{14}C-sucrose to *Lapageria* pollen grown on D-mannitol they could isolate only labeled polymers of the β-1,3-D-glucan, callose-like precursor moieties, although tubes were not produced. Obviously, enzymes to synthesize both callose and starch can be active in pollen even prior to tube extension. But the products formed may vary not only with different species and available substrates, but also with the stage of development or growth.

Hydrolases

These include a wide range of enzymes that operate to supply many chemical building blocks for the pollen. Such pollen or stigma-style produced enzymes must be highly specific or they would autolyse the germinating pollen and inhibit growth. Cutinase, which might be classified as a fatty acid esterase, is secreted by pollen to hydrolyze long-chained hydroxymonocarboxylic acids, the major constituents of the cuticle covering the stigma of several species. Members of Cruciferaceae, in particular, have a well developed stigma cuticle. The original studies (HEINEN and LINSKENS, 1961) utilized the secreted enzymes of the fungus *Penicillium spinulosum*, which hydrolyze leaf cutins, to penetrate stigma cuticle and liberate free fatty acids. LINSKENS and HEINEN (1962) demonstrated that only pollen of plants with the waxy, cutinized stigma surface produced cutinase; pollen of plants which lacked a cutinized stigma such as *Petunia* and *Cannabis* do not possess enzymes to hydrolyze the wax. The latter pollen yield pectinase, polygalacturonase activity, suggesting that the wax hydrolyzing enzyme differs from the normal pectinase. Ultrastructural studies of the cutinized papilla cells at the stigma surface of Cruciferae (KANNO and HINATA, 1969) showed that even incompatible pollen secretes cutinase but lack enzymes, or is inhibited from growing through the cellulose-pectin wall below the cutin.

Although esterases and lipases are difficult to classify, they can be readily isolated from pollen by diffusion or several isozymes with one or more closely related activities (see below: Sources of Variation). Many pollen species listed as containing high esterase activity (MÄKINEN and MACDONALD, 1968) may, on closer study, manifest two or three related hydrolytic activities. MACDONALD (1969) found that *Zea mays* pollen esterases contained acetyl-acetic esterase, carboxylesterase (B-esterase) and arylesterase (A-esterase), Table 14–2. Most studies would generally report these activities as one hydrolase. Whether KAMMANN's (1912) reported lipase in *Ambrosia* is an esterase or lipase must still be confirmed.

Discerning different proteolytic enzyme activities in pollen is also quite complicated. Pepsin and trypsin activities are high in many pollen species, particularly those high in protein, i.e. *Nicotiana* and *Petunia* (BELLARTZ, 1956), and those growing rapidly, e.g. rice (VENKATASUBRAMANIAN, 1953). Enzyme activity is about twice as high in pollen as in the style and corresponds to protein and free amino acid levels in those tissues (BELLARTZ, 1956). However, the different proteinase activities in pollen have not been separated. Knowledge of such differences may help in understanding the release pattern of amino acids essential for synthesizing protoplasm, particularly in rapidly extending and long growing pollen tubes.

There have been no reports of cholinesterase (EC 3.1.1.8) in pollen. However, considering the relatively high level of choline in the phospholipid fractions of some pollen species (Chapter 12), it may be of value to search for this enzyme. This enzyme may assume importance if the proposed role of acetylcholine (JAFFE, 1971) in plant membrane permeability is correct.

Acid phosphatase has been found in practically all pollen in which it has been sought. It occurs in relatively high levels and is readily secreted by pollen (GÓRSKA-BRYLASS, 1965; KNOX and HESLOP-HARRISON, 1970). While acid phosphatase has only rarely not been detected in pollen (BELLARTZ, 1956), alkaline phosphatase is frequently reported as absent (MÄKINEN and MACDONALD, 1968; HAECKEL, 1951). Other investigators (PALUMBO, 1953; LEWIS et al., 1967; MARTIN, 1968) detected low levels of alkaline phosphatase rather well distributed throughout the pollen cytoplasm (GÓRSKA-BRYLASS, 1965).

Hydrolytic enzymes active on specific oligosaccharides, i.e. invertase on sucrose (VAN TIEGHEM, 1869), were among the first enzyme activities recognized in pollen. In dormant *Haemanthus albiflos* pollen, invertase occurs in a soluble and insoluble form, differentiated by slightly different pH optima (LENDZIAN and SCHÄFER, 1973). Invertase solubility differences may be related to where it is localized, as in *Lilium longiflorum* where DICKINSON (1967) found one form of this enzyme limited to outside the cell membrane.

Although pollen can grow well on lactose (HELLMERS and MACHLIS, 1956) hydrolysis occurs by β-glucosidase with incorporation of the resulting glucose and galactose moieties (STANLEY, 1958). The report of HITCHCOCK (1956) relating milk-lactose-digesting enzyme systems to bee-collected pollen was clarified when the fresh, aseptically collected pollen was found to be free of such lactose digesting activity; *Bacillus* in the bee gut was the source of the enzymes, not the pollen.

Lyases

Early studies (OKUNUKI, 1939) assayed O_2 uptake and CO_2 release by pollen growing on organic acids. With the availability of radioactive organic acids labeled in the carboxyl group, demonstration of a specific decarboxylase was easier and more accurate. The reverse reactions, incorporation of CO_2 in a molecule, have also been facilitated by use of $^{14}CO_2$ and separation of labeled organic acids and their derivatives formed by germinating pollen.

Addition of CO_2 to phosphoenolpyruvate can yield the dicarboxylic acid product, oxaloacetic acid, via either of two enzymes (1) phophoenolpyruvic (PEP) carboxykinase or (2) PEP carboxylase. When malic dehydrogenase-NAD is present oxaloacetic acid is readily converted to malic acid. Thus, studies of CO_2 fixation by germinating pollen of *Pinus ponderosa* (STANLEY et al., 1958) and *Ornithogalum caudatum* (GOSS and PANCHAL, 1965) showed a high level of activity fixed into oxaloacetate, malate and the derivative amino acids. Incorporation of $^{14}CO_2$ was stimulated by addition of PEP and ribulose diphosphate to the pollen. These results provide the basis for assuming the enzymes for these CO_2 fixing pathways, phosphopyruvate carboxylase, EC 4.1.1.31, and carboxydismutase, EC 4.1.1.39, are present (Table 14-2). Presumably, the pyruvic kinase, EC 2.7.1.40, a transferase, is also present in pollen to provide and recycle PEP. The condensing enzyme, citrate synthetase, from the Krebs Cycle, would be expected since all the organic acids of this pathway and simple sugar substrates seem to be oxidized primarily by this route. Mitochondria, as will be indicated below (Localization), are also well recognized in pollen and are the site of most of these enzymes.

Enzymes necessary to metabolize and interconvert nucleotides, purines and pyrimidines in pathways demonstrated in pine pollen (Fig. 13-3) by NYGAARD (1973), may also be assumed as present; however, these enzymes have not as yet been isolated from pollen. NYGAARD's data suggest that pollen contains at least 8 enzymes involved in metabolism of glycine to inosine monophosphate (IMP), and that an adenylosuccinate synthetase and adenylosuccinate lyase convert IMP to guanosine-5-monophosphate. Still other enzymes in pollen are involved in the conversion of orotic acid, as NYGAARD and others have done, to uridine- and cytosine-5'-monophosphate and RNA (Chapter 13). While these enzymes could be included in Table 14-2, we have delayed such a listing until further research isolates them, or directly demonstrates their presence in pollen.

Isomerases

Enzymes in this group, among the most active in pollen, operate as catalysts in the metabolism of carbohydrates and their derivatives. PUBOLS and AXELROD (1959) showed that corn pollen converted xylose-^{14}C to xylulose-^{14}C. This reaction, primary to the synthesis of pentosans in pollen tubes, yielded about 4% substrate conversion. They also inferred, from inhibitor experiments, that xylulokinase (EC 2.7.1.17) is active in pollen and provides xylulose-5-phosphate which is metabolized by enzymes of the pentose phosphate pathway.

Evidence for the existence of the pentose pathway, the aerobic hexosemonophosphate shunt, and the anaerobic glycolytic pathway of glucose metabolism was also obtained from studies on pine pollen (STANLEY, 1958). Differences in rates of metabolism of differentially labeled glucose-^{14}C showed a predominance of aerobic, glucose oxidase activity releasing ^{14}CO$_2$ from the carbon-1 of glucose; the activity increased with time (Table 14-3). This indirect evidence for the presence of the enzymes of these two pathways in pine pollen suffers from lack of a detailed analysis of the levels of recycling of labeled ^{14}CO$_2$, and an evaluation of ^{14}C-label incorporated into intermediary metabolites. Studies of OKUNUKI (1937) on respiration in limited O$_2$ concentration suggested that a Pasteur effect can occur in pollen. The observation of STANLEY (1958) noting a shift in glucose-^{14}C metabolism in pine pollen growing under nitrogen gas also suggests that the rate of glucose utilization in glycolysis can be temporarily increased in the absence of oxygen, supporting the earlier observations on the Pasteur effect.

Labeled glucose and inositol are incorporated in tube pectins, cellulose and callose (ROGGEN and STANLEY, 1971), supporting the inference that the glycolytic and pentose pathways and the Loewus inositol by-pass (LAMPORT, 1970) probably function in pollen. Not all the essential isomerases implied by these assumptions

Table 14-3. Oxidation of glucose-^{14}C by pollen of *Pinus ponderosa* (STANLEY, 1958)

Hours germinated 30° C	Substrate glucose (G)	% $^{14}CO_2$ recovered 60 min.	Ratios	
			C_6/C_1	C_6/C_2
4.5	G-6-^{14}C	4.2		
	G-1-^{14}C	3.2	1.3	
	G-2-^{14}C	1.3		3.7
15.0	G-6-^{14}C	17.3		
	G-1-^{14}C	28.0	0.6	
	G-2-^{14}C	9.4		1.8

are listed on Table 14-2, e.g. glyceraldehyde-3-phosphate isomerase and others are omitted. To do so would merely catalogue the obvious and expected, but without rigorous experimentation. WIERMANN (1972a) reported a chalcone-fla-vonone isomerase which catalyzes the conversion of 2',4,4',6'-tetrahydroxy-chalcone to 4',5,7-trihydroxyflavone in maturing *Lilium* pollen. This ring-closing enzyme was not isolated; but it was shown to be active on a substrate normally present in and around developing pollen. The enzyme probably accounts for the build-up of highly colored flavones in mature lily pollen (Chapter 15).

Ligases and Others

The ligases, often called "synthetase enzymes," catalyze the combining of two molecules with the release of pyrophosphate in ATP, GTP, or similar triphos-phates. Ligases include the RNA synthetases involved in polysome development (MASCARENHAS, 1971) or production of sRNA moieties present in pollen for which enzymes EC 6.1.1.1.-12 may or may not be active (STEFFENSEN, 1971) (Chapter 13).

The carbon-carbon bond-forming carboxylases were explored by testing sub-strate metabolism of pyruvate and other Krebs Cycle acids in 9 different pollen spe-cies. Activity suggesting rapid utilization was fairly high in the pollen tested. How-ever, the citrate synthetase step occurs at a low rate in *Petunia* pollen (ROGGEN, 1967), suggesting that pathways other than Krebs Cycle oxidation are active, or that the proper assay conditions for this enzyme have not been developed.

Several non-classified enzymes have been reported (Table 14-2). Folic acid conjugase probably represents a group of enzymes producing folic acid, pteroyl-glutamic acid, and its combinants. This enzyme, initially reported by NIELSEN and HOLMSTRÖM (1957) in such widely diverse pollen species as *Pinus*, *Zea mays* and *Secale*, probably produces an active folic acid coenzyme transfer group. Glucose-6-phosphate cycloaldolase, which converts D-glucose to myo-inositol, should probably be classified as a lyase (EC 4.1.2.x). As demonstrated in *Lilium longiflo-rum* by LOEWUS and LOEWUS (1971), it is a key enzyme in pathways to pollen tube wall synthesis. The pollen cell wall degrading enzyme, exinase, indirectly shown in the digestive tract of the *Collembola* by SCOTT and STOJANOVICH (1963) (Fig. 7-12), has been reported in *Pharbitis* pollen (GHERARDINI and HEALY, 1969) and possibly in secretions of certain fungi (Chapter 9). That pollen which lacks an aperture, pore or furrow, is probably split open by internal pressures and the action of other types of wall solubilizing enzymes.

Sources of Variation

Pollen enzyme activities can vary depending on the species, environment during pollen development, age of the plant when pollen developed, nutrient status of the plant, methods of pollen extraction (Chapter 4) and storage (Chapter 5). Moreover, whether a pollen germinates and protrudes a tube immediately, and its morphology, i.e. exine thickness and presence of one, two or four pores (Chapter 2), may influence the pattern of initially detected enzymes. PATON (1921) found slight amylase activity in fresh *Lilium longiflorum* pollen, but very high activity when it started to grow. Likewise, as already described, GREEN (1894a) related development of amylase activity and changes in the starch granules in germinating *Zamia*.

   Isoenzymes. Many enzymes, when separated by gel acrylamide and starch electrophoresis, appear as several different isozymes (Fig.14-1). Isozymes can arise as isolation artifacts resulting from partial proteolysis during separation from the pollen, and thus may not always be related to their functional condition in the pollen. Either genetic variation of physiological age, as indicated by different organelles varying in quantity and activity during development and growth, may be a source of different isozymes (LINSKENS, 1966). VEIDENBERG and SAFONOV (1968) found isozyme patterns from pollen of three species and a hybrid of *Malus* allowed a more precise species separation than the patterns of soluble proteins. HAMILL and BREWBAKER (1969) compared the isozymes of *Zea mays* pollen to anther and style patterns to detect different genetic alleles. When SCANDALIOS (1964) compared *Zea mays* pollen esterases to those from the leaves and other tissues of the same plants, the pollen yielded 3 isozymes of esterase with different mobilities than those from all other tissues. When PAYNE and FAIRBROTHERS (1973) compared esterase isozymes between and within 7 physiographic populations (ecotypes) of *Betula populifolia* they found an average of 87% of the gel-electrophoretic bands coincided between the different populations, but an

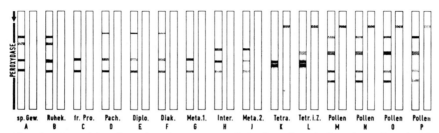

Fig. 14-1. Peroxidase isozymes from *Petunia hybrida* pollen during the pollen development. A = Sporogenic tissue; M/P = mature pollen grains. Left zymogram: tapetum fraction, right zymogram: pollen fraction

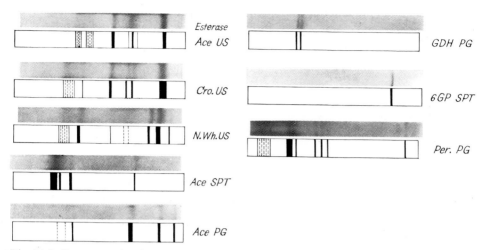

Fig. 14-2. Representative isoenzyme photographs and diagrams of esterase, glutamic de-hydrogenase (GDM), 6-gluconic phosphate dehydrogenase (6GP) and peroxydase (Per.) from pollen grains (PG), pollen tubes removed from styles following self pollination (SPT), in comparison with extracts from styles from unpollinated flowers (US) in various cultivars of *Lilium longiflorum*: Ace, Croft (Cro.), Nellie White (N.Wh.). (From DESBOROUGH and PELOQUIN, 1968)

average of only 66% of the isozyme bands coincided from pollen of trees within a single population. The isoenzyme pattern varies also within a species between different cultivars (Fig. 14-2). Pollen grain extracts from four cultivars of lily yielded a greater number of protein bands, up to 25, than pollen tubes or styles. Pollen grains from all cultivars appeared to have similar protein pattern (DESBOROUGH and PELOQUIN, 1968).

Extraction Procedure. Collecting and handling procedures influence the levels of enzymes, as well as the sugars, acids and enzyme cofactors that are detected. Unless pollen is extracted and collected aseptically, it generally has a surface bacterial flora which may supply enzymes and lead to spurious results in pollen enzyme assays. The example of HITCHCOCK (1956), previously discussed, which related the milk lactose-digesting enzyme systems in bee-collected pollen to the *Bacillus* in the bee gut is one such example.

Since hydrogen ions affect enzyme activity, variation in endogenous H^+ con-centration from pollen to pollen may, in part, account for the range of enzyme activities found when pollen is assayed *in vitro*. Using indicator dyes, OSTAPENKO (1960) found that a range of values from pH 3 to 7 occurred in different grains of *Prunus cerasus*. The pH of most *P. cerasus* pollen, from the same collection, varied between 3.0 and 5.0; about the same values occurred in 12 other angio-sperm pollens. The exine and intine layers were more acid than the protoplasm. Furthermore, OSTAPENKO (1960) showed by indicator dye assays that the redox potential of pollen varies widely ($E_o' = +372$ to -240 mv), as does the oxidizing potential of the individual grains. Isozymes of individual enzymes can have differ-ent pH optima. One phosphatase from ragweed had a pH optimum of 4.2, the

Table 14-4. Relative enzyme activity in germinating *Lilium* pollen
(DICKINSON and DAVIES, 1971)

Enzyme	Activity as: nmoles of product/min/mg pollen[b]
Malate dehydrogenase	1227.0
Isocitric dehydrogenase	10.2
Glucose-6-P dehydrogenase	1.8
Hexokinase	5.1
Nucleoside diphosphate kinase[a]	86.0
Phosphoglucomutase	5.5
UDP-glucose pyrophosphorylase[c]	131.0

[a] Assayed in non-centrifuged homogenate. All other enzyme assay were of clear supernatant after centrifugation at 38,000 X gravity, 15 min., 0° C.
[b] One mg pollen contained approximately 170 µg protein in 5,600 pollen grains.
[c] One mg pollen contained 131 µg protein.

other, 5.6 (NAKAMURA and BECKER, 1951). Such isozymes could be located at different cell sites in pollen and function at different times and different levels of activity. However, genetic variation rather than intracellular binding sites would probably explain why several of the 4 glutamic dehydrogenases from *Pinus echinata* and *P. taeda* pollen differ in R_F's (HARE, 1970).

Inherent variation in enzyme activities within a single species exists, but the significance or source of such variation is poorly understood. Differences in the endogenous pH values have not been correlated with growth capacity of the pollen or specific enzyme levels. Moreover, little application has been made of the electropotential levels detected and associated with respiratory enzyme activities, other than to apply redox dyes as indicators of pollen viability (Chapter 6). Such differences would be expected to reflect functional metabolic variation between grains. The fact that the maternal tissue in which pollen normally grows affords a strongly buffered environment, presumably optimizes and influences enzyme activities. Differences in rate of tube growth commonly observed *in vivo* compared to *in vitro* would, presumably, reflect inherent variations in pollen enzyme activities, although cell factors or metabolites still unknown may also strongly affect enzyme activities and growth.

Different enzymes assayed in pollen from one species generally vary greatly in their relative activities. In lily pollen extracts, Table 14-4, malate dehydrogenase is one of the most active enzymes. This enzyme, assayed in the reaction direction oxaloacetic to malic acid, may be either mitochondrialbound or soluble in the cytoplasm. It is a key enzyme in the tricarboxylic acid cycle, but it may also be involved in the interconversion of the product of dark CO_2 fixation which produces oxaloacetic acid from phosphoenol pyruvic acid (PEP). The fixation of CO_2 in pollen is well known (STANLEY et al., 1958; GOSS and PANCHAL, 1965). However, two different PEP-CO_2 incorporating enzymes occur in plants (Table 14-2). One, PEP-carboxykinase, yields ATP and is reversible; the other enzyme, PEP-carboxylase, does not directly produce a high energy phosphate compound and is practically irreversible. A highly active malic dehydrogenase, an enzyme common in some pollen, would conveniently utilize the oxaloacetic acid product of the PEP carboxylase reaction; its presence might thus explain the reason for the evolution of such a high level of this enzyme.

Destruction. Most workers take precaution to avoid enzyme destruction during their isolation. DICKINSON and his associates concluded that the seven enzymes (Table 14-4) and five others tested in lily pollen over a 3 hour germination period did not increase in activity. This suggested that no new synthesis occurs. Yet, in some species, inhibitors of protein synthesis also inhibit pollen germination and incorporation of leucine-^{14}C into macromolecules (TUPÝ, 1966). This implies that in some pollen, enzymes or structural proteins may be synthesized, while in others, including species with two male cells present at dehiscence, all enzymes required for initial tube growth may be present preformed or in a bound, inactive form at one stage of pollen growth, followed by activation at another stage. Phosphorylase and hexokinase activities increased in *Pinus densiflora* pollen when it was placed in a germination medium with or without sucrose. However, in *Zea mays* and *Cucurbita* pollen, activity of these enzymes increased only when sucrose was present (KATSUMATA and TOGASAWA, 1968a). The fact that pine germinates in plain water while the other two pollen species require a metabolizable sugar and a controlled osmoticum may be related to activation or synthesis of enzymes that must occur for growth initiation.

Release. Enzymes localized in exine and pore areas, which easily diffuse into the germination environment, probably represent primarily degradative enzymes, such as those acting on cutin (LINSKENS and HEINEN, 1962; KROH, 1964). Factors strongly influencing the solubilization and activation of such enzymes would be the surrounding osmotic pressure and the ionic concentration of hydrogen and salts. Increasing concentration from 0.33 M to 0.66 M by adding mannitol in the media of *Petunia* pollen, before a tube was present, decreased the capacity to metabolize sucrose-^{14}C by 25%. In using an electrolyte, salt solution, to extract enzymes from *Cycas revoluta* pollen, there was considerable variation between concentration of salt and the amounts of ribonuclease and alkaline phosphatase (phosphomonoesterase) released (HARA et al. 1970). No enzymes were solubilized by very weak or non-electrolyte solutions. Although the first elution with 0.5 M NaCl yielded over 50% of the enzyme activity, the residue after 4 elutions with new salt solutions, still held 20% of the alkaline phosphatase activity and 8% of the total ribonuclease activity. These results (HARA et al., 1970) indicate the differences in binding or accessibility to solvents of pollen enzymes. While salt extraction differs from simple solubilization of loosely bound exine-intine-associated enzymes, both the interacting osmotic and ionic factors would influence enzyme release. *Oenothera* pollen released many enzymes into an agar medium, but LEWIS et al. (1967) did not detect acid or alkaline phosphatase in the medium. These enzymes are present and were detected when the intact ungerminated grains were placed in liquid medium (MÄKINEN and BREWBAKER, 1967). As MARTIN (1968) showed, considerable variation in release of enzymes by different species exists. Leakage of substrates, and presumably enzymes, is enhanced in some pollen when calcium is lacking in the medium (DICKINSON, 1967).

Inhibitors. Pollen enzyme levels initially depend upon the quantity synthesized or activated and the rates of enzyme degradation and release from growing tubes.

While the above discussion focused on artifacts induced during formation and extraction of pollen, inherent factors can also lead to variations in enzymes detected. An abnormal inhibitor reaction may occur during preparation of the enzyme for assay. If phenolic compounds, compartmentalized in pollen before grinding, diffuse throughout the enzyme extract preparation they may combine and inactivate protein molecules.

Studies of soluble glutamic dehydrogenase (GDH) in *Petunia* pollen showed that the enzyme is inhibited by phenolic compounds commonly found in pollen (BREDEMEIJER, 1970). Only careful extraction of the enzyme with a phenolic binding agent, polyvinylpyrrolidone, facilitated detection of high activity. Further increase in GDH activity was obtained by incubating the solution extracted from the pollen with enzymes for up to 20 hours. Maltase, α-amylase, and galactose oxidase were particularly effective at releasing GDH activity, but not all enzymes stimulated activity. Trypsin, RNAse, β-glucosidase and glucose oxidase were among those enzymes that generally depressed the amination ability of GDH from *Petunia*. However, activity of purified GDH from the pollen was not affected by added enzymes (BREDEMEIJER, 1971). Merely storing the total enzyme extract increased GDH amination activity 66 fold. Other metabolic inhibitors or enzyme control factors may exist which, on tissue disruption or at different developmental stages, modify enzyme activity.

Enzymes operating in sequential metabolic pathways leading to tube wall intermediates, or other macromolecules, often draw upon the common metabolite glucose. The rates of activities and levels of various enzymes on the different pathways are regulated by control mechanisms which help insure that the final end products required for assemblage into wall polymers arrive at the right place at the time they are needed. Internal controls regulate the level of enzymes synthesized, destroyed or activated. Several mechanisms such as product feedback inhibition, allosteric regulation, and availability of an essential cofactor are examples of such regulatory controls. Any of these, functioning in pollen, could modify the level of certain enzymes. HOPPER and DICKINSON (1972) found evidence that sugar nucleotides act as control inhibitors for UDP-glucose pyrophosphorylase activity in *Lilium longiflorum* pollen. They isolated and purified the enzyme from ungerminated pollen and tested various uridine diphosphate sugars as inhibitors of the enzyme. The UDP-glucose product of the enzyme reaction was the strongest inhibitor. These sugar nucleotides, and related sugar acids, which are probably precursors for tube wall polysaccharide biosynthesis, were proposed as the *in vivo* feedback control mechanisms of the UDP-glucose pyrophosphorylase reaction, regulating the channelling of glucose and other carbon substrates to the tube wall. This hypothesis, if proven correct, helps explain the role of these metabolites in facilitating pollen growth, and might also clarify a source of variation observed in certain enzyme levels in pollen at different growth stages.

Detection of enzyme activities as discussed in this section suggest that assay of an isolated enzyme is subject to considerable error and may be influenced by available combining sites for interfering or activating compounds present in the extract. Such modifying factors, as well as inherent differences in cofactor levels with species, handling and developmental stage may all contribute to variation in enzyme activities detected.

Cofactors

Many enzymes require a component in addition to substrate for the reaction to proceed. These "cofactors" are usually unchanged after the reaction. There are two main types: 1. an organic molecule, a specific coenzyme; 2. an inorganic ion which acts as an activator. Cofactors can act as electron, hydrogen or phosphate donors or acceptors, and they can also influence the stereospecific configuration of the enzyme. Unless the required cofactor is present, the enzyme is nonfunctional or active at only a low level. Microspores developed on plants growing under adequate light and nutrient conditions will tend to accumulate or synthesize sufficient cofactors during development. But pollen lacking these cofactors must acquire them from the female tissue or synthesize them after growth is initiated. Certain cofactors, e.g. biotin, occur at higher levels in the stigma and ovary tissues than in the pollen which grows in those tissues (FILIPPOV, 1959).

Vitamins

Based on nutritional experiments with insects or animals, some cofactors in pollen are components of, or are themselves, classified as vitamins.

Ascorbate, Vitamin C, is one such cofactor. Coenzyme I and II, more correctly called nicotinamide-adenine dinucleotide (NAD) and NAD-phosphate (NADP), contain nicotinamide of the nicotinic acid B vitamins. Flavin, in the cofactor flavin adenine dinucleotide (FAD) is contained in vitamin B_2, riboflavin; cofactors phosphopyridoxal and phosphothiamine contain derivatives of vitamin B_6 (pyridoxin) and B_1, thiamine, respectively. The PEP carboxylase utilizes biotin, vitamin H, as a cofactor. In pollen these vitamins probably serve as a part of the coenzymes.

Levels of some pollen vitamins recognized as enzyme cofactors are listed in Table 14–5. Although cofactors such as riboflavin and cytochrome can contribute to pollen color, the common yellow color arising from carotene, the lipid-soluble pro-vitamin A source, will be discussed in the review of pollen pigments (Chapter 15). Pollen riboflavin almost always occurs in the ester form, and mostly as flavin mononucleotide (FMN). The levels of FMN and FAD decreased and break-down occurred as the viability decreased (TAKIGUCHI and HOTTA, 1960). Riboflavin was found in a very narrow concentration range around 6.4 µg/mg in *Phoenix dactylifera* pollen, while other vitamins often varied as much as 20% between samples of the same species (RIDI and ABOUL WAFA, 1950). A survey of bee-collected pollen (ROSENTHAL, 1967) also showed riboflavin occurred as the most constant in concentration of any of the vitamins although, as noted in Table 14–5, some rather large differences are found even between the levels of riboflavin in certain species. Some vitamin-like components, i.e. ascorbic acid and inositol (the latter discussed in Chapter 9 along with related cyclitols) occur at mg/gm dry weight levels instead of µg amounts. Inositol is particularly high in corn and grass pollens. Apparently it is not fuctioning primarily as a vitamin but as a substrate source for D-glucoronic acid which serves as cell wall substrate via UDP-glucuronic or UDP-galacturonic acid (LOEWUS and LOEWUS, 1971).

Table 14-5. Pollen vitamin content (NIELSEN et al., 1955; TOGASAWA et al., 1967)

Vitamin	μg/gm dry weight		
	Pinus montana	Alnus incana	Zea mays
B₂-Riboflavin	5.6	12.1	5.7
B₃-Nicotinic acid	79.8	82.3	40.7
B₅-Pantothenic acid	7.8	5.0	14.2
B₆-Pyridoxine	3.1	6.8	5.9
C-Ascorbic acid	73.1		58.5
H-Biotin	0.62	0.69	0.52

B-Vitamins. A high level of B vitamins were recognized as present in corn pollen by DUTCHER (1918). Extracts from *Zea mays* pollen rapidly relieved symptoms of polyneuritis in pigeons, an early biological assay for B vitamins. Ascorbic acid, which occurs in relatively high levels in pine and palm pollen probably functions in hydrogen transfer, although its role in cell metabolism is still poorly understood.

Vitamin E, tocopherol, was reported (GRUNT et al., 1948) to range between 21 (in *Carya*) and 170 μg/gm dry weight in pollen of *Echinops* (yellow thistle). With most of the 15 species they analyzed ranging between 40 and 90, e.g. *Malus domestica* 80 μg/gm dry weight, a pattern was not observable upon which to classify pollen into related families or genera on the basis of tocopherol content. The average value of vitamin E in bee-collected pollen (VIVINO and PALMER, 1944) was 320 μg/gm pollen fat. Fungal and yeast culture assays showed that *Nicotiana* pollen grains, but not the tubes, released fairly high levels of thiamine (vitamin B₁), biotin, pantothenic acid and nicotinic acid (KAKHIDZE and MEDVE-DEVA, 1956). The tube may function as a restricting membrane or possibly the vitamins which move off the grain before germination are localized exterior to the intine.

Some vitamins, i.e. FAD, consistently occur at a higher level in gymnosperm than angiosperm pollen (TOGASAWA et al., 1967). Also in pine, the FAD level exceeds that of FMN, unlike the pattern observed in angiosperm pollen by TAKI-GUCHI and HOTTA (1960). In general, feeding tests and chemical assays indicate that pine pollen is low in the water soluble B vitamins (SEKINE and LI, 1950; LUNDÉN, 1954). In many cases the variation correlates better with collecting and handling procedures than with species. Folic acid occurred in 10 different hand-collected pollen at levels of 0.42 to 2.20 μg/gm dry weight (NIELSEN and HOLM-STROM, 1957), and at the higher amounts of 3.4–6.8 μg/gm dry weight of defatted pollen collected by bees (WEYGAND and HOFMANN, 1950). It was more common in the combined form than free, which is to be expected if it functions in tetrahy-drofolate, a cofactor for transmethylating enzymes. These and other reactions involving this cofactor are essential to pectin synthesis and pollen wall development. The ratio however, in grasses of combined folate to uncombined was less. In these species either lower amounts of active coenzyme are initially present, or the cofactor is more sensitive to the handling to which the *Zea mays* and *Secale* pollen was subjected. In any case, breakdown and a decreased functional transmethyla-tion system could contribute to the ephemeral capacity for germination of grass pollens.

Analyses of vitamins in *Zea mays*, *Pinus* and *Alnus* pollen stored for one year showed some vitamins are more stable than others, e.g. biotin concentration changed the least, riboflavin and pantothenic acid the most; by comparison, the levels of most vitamins in *Zea mays* were relatively stable compared to those of *Pinus* and *Alnus* (SAGROMSKY, 1947). The vitamin levels at the time of harvest before drying were not ascertained, thus the *Zea mays* was already in a nonfunc-tional stage when initially assayed in the comparative study of NIELSEN (1956).

Non-vitamin Cofactors

Other non-vitamin-like cofactors also have stability patterns highly dependent upon species and handling. Glutathione, a tripeptide cofactor composed of cys-teine, glutamic acid and glycine, often appears in pollen extracts (Chapter 11). The haemoproteins, cytochromes a, b and c previously discussed, were recognized early as essential for cytochrome oxidase activity in pollen (OKUNUKI, 1939). The three or four distinct absorption bands in the ionized and reduced states of the cytochromes reveal which cytochrome cofactor is present and operative.

Nucleosides

The nucleoside-5'diphosphate, adenosine diphosphate (ADP), is the main cofac-tor involved as a phosphate carrier in pollen, as in practically all living systems. ADP is involved in the cell through the nucleoside diphosphate kinases; the nucleosides are particularly critical in pollen tube extension since they are active carriers of the sugars transferred to form di-, tri- or polysugar chains and in nucleic acid synthesis. The pool of ATP is very high in mature *Pinus mugo* pollen, rapidly decreasing on germination with an increase in the sugar nucleotide and ribonucleotide pools (NYGAARD, 1973).

The nucleoside derivatives are cofactors or substrates for transferases, e.g. nucleotides linked to sugars to form UDP-glucose, GDP-glucose, UDP-galactose, UDP-xylose, UDP-arabinose or GDP-mannose, which transfer the sugar group to build pectin, hemicellulose or cellulose. Maintenance of these pools for gluco-syl, galactosyl, arabinosyl, etc. transfer is essential to pollen tube growth. The ribonucleosides in *Pinus mugo* pollen had initial concentrations, n moles/mg of 1.0 ATP, 0.5 UTP, 0.25 GTP and 0.2 CTP. After 24 hours incubation with 10 mM of a sugar substrate, e.g. D-glucose or D-galactose, the free ATP and UTP were tempo-rarily depleted, followed by restoration of ATP; however, GTP and CTP levels were nearly unchanged (NYGAARD, 1969). No changes in any of the ribonucleo-side triphosphate pools occurred on incubation with non-metabolizable sugars. If similar results are found in other pollen it would suggest that the initially active sugar transfer system is primarily utilizing UDP not GDP.

Nucleosidediphosphate kinase from ungerminated lily pollen preferred the phosphate donor GTP compared to the reverse reaction usually with substrate

ADP, but which preferred CDP or UDP as the acceptor (DICKINSON and DAVIES, 1971a). The preferred substrates, or nucleoside levels, in germinating pollen would be expected to change with consumption of substrate as the tube extends and the proportion of the different components incorporated into the tube wall changes. It will be interesting to correlate the changes of the cofactor pool with components incorporated into the tube wall, and the potential cofactor level available in the female tissue. A limited energy or sugar transfer system at one stage of growth could very possibly account for cessation of, or limited tube growth.

Ions

Ions such as Fe^{++}, Co^{++}, Mn^{++} and Mg^{++} are absolute requirements for activity of certain enzymes; other ions, Hg^{++}, Ag^{++}, are toxic to most enzymes. Some ions, i.e. Na^+, may act as competitive inhibitors for enzymes activated by K^+. These ionic interactions are easier evaluated *in vitro* than *in vivo*. Initially, one assumes that the mature microspore has a sufficient pool of minerals to produce the ions essential to activate the enzymes necessary for germination. But the levels of inherent elements and organic molecules, as well as growth rates, vary throughout the year and with the site where the plants developed (Chapter 8). Levels of these ionic cofactors found in pollen place them in the micronutrient range; and the inherent levels of some such as boron and calcium, might be insufficient to meet all growth needs. However, several trace elements found in pollen (Table 8-4), i.e. chlorine, silicon, or titanium, have no known metabolic role and are probably passively accumulated by the microspores, just as they are passively absorbed by the roots. Other elements, i.e. iron and zinc, are quite high in pollen; these plus manganese and magnesium are well recognized as cofactors for specific enzymes. Some enzymes can be activated by two or three ions, interchangeably. Manganese and magnesium often act in this manner.

Cobalt was shown to activate aminoacylase in pollen (UMEBAYASHI, 1968). Optimum concentrations of Co^{++} occur in the lily style through which the tube grows (YAMADA, 1958). Availability of cobalt was suggested as a requirement for protein synthesis in the pollen. Obviously, any deficient cofactor or substrate supplied to pollen at the proper concentration would facilitate tube growth and protein synthesis or activation. The presence of boron in tissues in which pollen normally grows is well documented (GLENK and WAGNER, 1960), and was, in fact, among the clues which led to the recognition of the importance of boron in pollen growth.

Few studies, except for those of Co^{++} in the activation of aminoacylase from lily pollen (UMEBAYASHI, 1968) have focused on the effect of an ion on a specific pollen enzyme. Most studies have just added the element to the growth media and assumed it was stimulating or inhibiting a specific enzyme or metabolic pathway. Studies such as those of UMEBAYASHIL (1968) generally relate the activating effects observed as a result of specific ion concentration in the female tissue where pollen grows. Whether or not the tube comes in contact with a particular element, as found by total assay of a massive plant tissue, has not been determined. Also, enzyme activity and the effect of the cofactors depend, in part, upon where the enzymes are localized in the pollen. Spatial compartmentalization or membrane separation of a cofactor from an enzyme, or the enzyme from a substrate, can diminish or obviate the effectiveness of these molecules.

Localization and Function

Intracellular enzyme sites are usually determined by histochemical methods or direct enzyme assay of the particulate cell fractions isolated by differential centrifugation. Both methods have certain limitations. In histochemical assays, fixation methods may destroy or redistribute the enzymes. Phosphatases are among the enzymes most reliably determined histochemically; enzymes diffusing from pollen, such as amylase, are also easily assayed by the disappearance of starch and increase in sugar. In particulate separation techniques aqueous homogenization may damage the particulate structures and solubilize enzymes normally associated with specific subcellular organelles. Fractionation techniques have been applied to pollen and microspores to separate nuclei (WEVER and TAKATS, 1970), mitochondria (STANLEY, 1958), microsomes-ribosomes (STANLEY and YEE, 1966; LINSKENS, 1967), and vesicles derived from Golgi apparatus or endoplasmic reticulum (VANDERWOUDE et al., 1971; ENGELS, 1973, 1974). Electron micrographs (GULLVÅG, 1967) show that organelles in pollen tube cytoplasm are similar to those in most plant cells and presumably have similar enzyme activities and metabolic functions. While chloroplast mostly are lacking, plastids, i.e. amyloplasts (Chapter 9) are present. FRANKE et al. (1972) demonstrated the presence of microtubules in pollen tubes using an improved fixation procedure for EM, although previous work (SKVARLA and TURNER, 1966) with different fixation procedures did not reveal this organelle in pollen tubes. The generative cell cytoplasm has a complement of organelles similar to those in the tube cytoplasm although plastids are usually absent and microtubules are present (SASSEN 1964a; HOEFERT, 1969). Generative cell organelles have not been isolated and no knowledge or their enzyme activities are known except by topographical enzyme assays.

   Vesicles. The tip growth mechanism of pollen tubes, with the apical addition of secretory vesicular material, has been a rewarding area of ultrastructural study by cytologists and biochemists, particularly those concerned with cell wall formation (ROSEN, 1968). The highly vesiculated tube tip region was recognized from electron micrographs showing membrane-like invaginations and there is an increased concentration gradient and accumulation of vesicles, at the site where *wall* material is actively incorporated (MÜHLETHALER and LINSKENS, 1956; DASHEK and ROSEN, 1966). Active growth is restricted to the first 5 μ of the tube (Fig. 14–3A), and is accompanied by the fusion of vesicles with the plasmalemma, by a reverse pinocytotic mechanism, resulting in a double layer wall with a thickness of about 0.1 μ (ROSEN, 1968). Two types of vesicles are reported in pollen, rough and smooth (LARSON, 1965; VANDERWOUDE et al., 1971). In general the smooth vesicles and their enzyme and carbohydrate components accumulate at the tip and form new tube membrane. CRANG and HEIN (1970) believe that in *Lychnis alba* tubes the pectin forms an inner layer with an outer fibrous

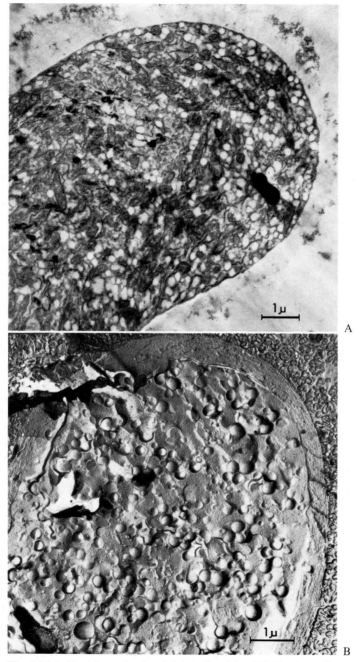

A

B

Fig. 14-3 A–E. Pollen tube tip addition of vesicles. (A) Electron micrograph of *Petunia hybrida*. (B) Freeze etching EM picture of *Petunia hybrida*. (C) Tip of pollen tube with vesicles, which are involved in the formation of the tube wall (top), enlargement (D) of the tip wall (w) showing loose outer structure, the vesicles are surrounded by a unit membrane, see double arrow (bottom). m = mitochondria, er = endoplasmic reticulum, g = Golgi apparatus, v = vesicles, pm = plasma membrane. (By courtesy of M.M.A. SASSEN, 1964). (E) Diagrammatic interpretation (After DASHEK and ROSEN, 1966)

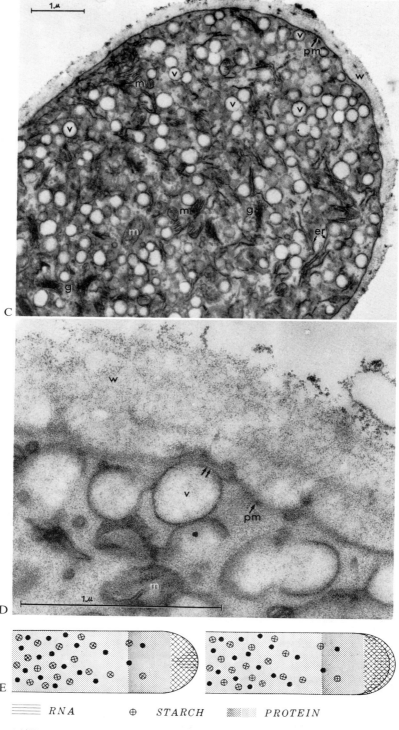

RNA

Fig. 14-3. C—E

RNA   ⊕   STARCH   PROTEIN

POLYSAC.   •   LIPID

cellulosic layer. This contrasts with the general view proposed in Fig. 14–3 E, where the tip is primarily pectinacious.

The smooth vesicles appear to incorporate or carry the precursor of pectin, myo-inositol (DASHEK and ROSEN, 1966). Tritium from ^{3}H-inositol is first visible in vesicles that accumulate at the tip, and then move out into the wall region. This suggests that the group of enzymes associated with the Loewus pathway (LAMPORT, 1970), preceeding pectin and hemicellulose formation from D-galacturonosyl and L-arabinosyl units, are in the vesicles or associated with the smooth membrane material. While not proven, radioautographic evidence supports the hypothesis that vesicle enzymes or carriers may be associated with the incorporation or conversion of proline to hydroxyproline in tube walls (DASHEK and ROSEN, 1966) and that vesicles are involved in the incorporation of other wall components. The common plant cell wall component, acetyl-D-glucosamine, for which active enzymes occur in pollen (LINSKENS et al., 1969) may also be transported or metabolized in these vesicles. Since soluble arabinose derived from sucrose-^{14}C decreases as insoluble tube wall material is synthesized, a transfer complex in the vesicles, possibly a UDP-derivative, may mediate this reaction (MASCARENHAS, 1970). Histochemical reactions have localized acid phosphatase in vesicles (CRANG and MILES, 1969), and a high level of ATPase and alkaline phosphatase in the pollen tube tips of *Portulaca* (MALIK et al., 1970). Indirect knowledge of other enzyme components associated with vesicles can gained by a chemical analysis of their membranes.

VANDERWOUDE et al. (1971) analyzed the components of isolated vesicles originating from the Golgi apparatus in pollen of *Lilium longiflorum*. These vesicles presumably contain enzymes internally or on their surfaces to form the chemical products they contain. Each gram of pollen yielded 1.1 mg dry weight of secretory vesicle material. In tubes which were elongating at the rate of 12 μ per minute, VANDERWOUDE and MÓRRE (1968) calculated that an increase in tube surface by plasma membrane addition at the rate of 600 μ²/min was matched by an equal area of vesicle membrane production from Golgi apparatus. Vesicles are probably the source of all tube membrane. Sugar components of the isolated vesicles were similar to those of the tube wall, although the proportions differed (Table 14–6). The L-fucose in vesicles and pollen walls may arise by enzymes acting upon-L-galactose, or from GDP-D-Mannose. Since little is known of enzymes involved in either pathway they were not listed in Table 14–2. The higher glucose level would be anticipated in the pollen wall (Table 14–6) where it is mostly alkali and acid soluble, not hot water extractable. Cellulose synthetase activity undoubtedly increased the microfibril and β-1,4-polyglucan in the walls compared to the vesicle composition. VANDERWOUDE et al. (1971) suggested that the vesicles might be containers of cellulose synthesizing enzymes. Conceivably, some vesicles may be rich in enzymes and components for pectin and hemicellulose, while others have a capacity for cellulose formation. ENGELS (1973a) has shown cellulose in pollen vesicles. A study of changes in tube vesicle composition parallel to changes in pollen walls has not been made.

Tip addition of vesicles originating from the Golgi apparatus or the endoplasmic reticulum located further back in the tube is experimentally well established. If these vesicles open when they join the plasmalemma, emptying

Table 14-6. Sugar distribution in pollen wall compared to that in the proposed originating vesicle source. From VanDerWoude et al., 1971)

Sugar	% of sugar recovered	
	Cell wall	Secretory vesicles
Rhamnose	2.1	6.2
Fucose	0.3	6.2
Arabinose	9.6	12.5
Xylose	1.4	6.2
Mannose	0.5	12.5
Galactose	6.3	25.0
Glucose	79.8	3.14

non-membrane bound enzymes and other contents into the external tip environment is not known. Such a mechanism could move large proteins out of the pollen tube, and may be supported by some interpretations of electron micrographs of the pollen tip region. This process, postulated in Fig. 14–3D, could explain the concentration of enzymes often observed in the tip membrane region (PODDUBNAYA-ARNOLDI et al., 1961).

Recently it could be shown, that the cell walls and the Golgi vesicles of pollen tubes of *Petunia hybrida* contain an alkali-resistant material largely composed of a glucose polymer, as revealed by thin-layer chromatography. X-ray patterns of the Golgi vesicles lead to the conclusion, that the vesicles contain the cellulose type of a polymer chain and are involved in its synthesis and transport. The cell walls of the pollen tubes contain native cellulose (ENGELS and KREGER, 1974) (Fig. 14–4).

Lily pollen tubes grown *in vitro* and prepared for viewing in the electron microscope by glutaraldehyde fixation usually show the tips covered with a loose cap. Since this cap is removed by glucuronidase and hyaluronidase it is probably composed of mucopolysaccharides (ROSEN, 1964; DASHEK and ROSEN, 1966), and possibly contains enzymes secreted from the vesicles. The role and chemical nature of this cap must still be determined. Different patterns of organelles and enzymes result when the fixation techniques are varied. Also, whether the pollen fixed is growing *in vivo* or *in vitro*; and if *in vivo*, whether it is growing in compatible or in incompatible female tissues (KROH, 1967; ROSEN, 1968), all influence the patterns seen. Enzyme distribution in the tube and diffusion from the pollen may be affected by different cytoplasmic organelle patterns. Crystalloid bodies in *Lychnis* pollen were suggested as a source of enzymes for metabolizing substrates during germination (CRANG and MILES, 1969).

Pollen Wall. Enzymes occur in the exine of pollen grains and may be present in high levels in the intine pore region. KNOX and HESLOP-HARRISON (1969, 1970) assayed, by histochemical techniques in the light and electron microscope, sections from 50 different angiosperm pollen and the gymnosperm *Pinus banksiana*. They found the mature microspore walls are high in esterase, amylase, ribonu-

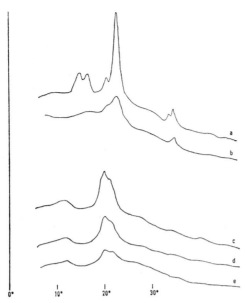

Fig. 14-4. Densitograms of X-ray powder patterns (Debye-Scherrer method) of cotton hairs
(a) and cell walls of *Petunia* pollen tubes (b) after extraction with N HCl, cotton hairs (c) and
pollen-tube walls (d) treated with N HCl and 20% NaOH and from Golgi vesicles after the
last treatment with 20% NaOH (e). The scale indicates the angle of diffraction. In
redrawing the curves the effect of film granularity has been smoothed out. (From ENGELS
and KREGER, 1974)

clease and acid phosphatase. TSINGER and PETROVSKAYA-BARANOVA (1961)
showed that the walls of *Peonia* and *Amaryllis* contain dehydrogenases, acid phosphatases and cytochrome oxidase. Since the intine is separated from the pollen
cytoplasm by the plasmalemma, the wall enzymes are extracellular. Pollen surface
enzymes, while generally highest in the intine, cellulose pore areas, may also be of
importance in nonaperturate pollen, i.e *Canna, Portulaca* and *Crocus* (MÜLLER-
STOLL, 1956). The diffusate from pollen contains enzymes associated with the
hydrolysis of wall polysaccharide constituents., i.e. pectinase, cellulose and β-1,3-
glucanase. Adding these enzymes to pollen *in vitro* during the initial phases of
germination shortens the time required for tube initiation and extension, and may
supplement the pollen enzymes, or enzymes occurring in the tissues where pollen
must grow (KONAR and STANLEY, 1969; ROGGEN and STANLEY, 1969). Callose
plugs, present in the pore region of many mature microspores, must be broken
down before the tube emerges (ROGGEN and STANLEY, 1971). These plugs contain
pectin and hemicellulose-like components in addition to callose, β-1,3-polyglucan
(PODDUBNAYA-ARNOLDI, 1960).
     Degradative enzymes secreted from pollen probably exert a limited effect on
the developing tube. The same enzymes which could, presumably, degrade the
tube tip are apparently secreted or activated during growth initiation. Since only
a limited number of vesicles join the tip per unit time (VANDERWOUDE and
MORRÉ, 1968), the quantity of enzymes added to the tip region may thus be

Table 14-7. Distribution of enzymes in *Cycas revoluta* pollen (HARA et al., 1972)

Enzyme	Activity detected (relative units)		% ratio A/B
	(A) at surface	(B) entire pollen	
Ribonuclease	14.65	35.15	41.7
Alkaline phosphatase (phosphomonestease)	0.66	5.03	13.1
Alkaline pyrophosphatase-Mg^{++}	0.16	48.92	0.3
Acid pyrophosphatase (acid phosphatase)	1.52	8.28	18.4
Invertase (β fructofuranosidase)	28.30	193.30	14.6

regulated; in addition, specificity associated with isozymes and the time of enzyme release could afford additional controls to avoid self-destruction, while allowing induction of sufficient wall plasticity (Fig. 14–3C) to sustain tube extension. Other degradative enzymes are associated with specific pollen organelles. The well documented decrease of RNA in the tube nucleus, with tube growth and aging (Chapter 13) is an example of action by a highly specific, localized enzyme. In one pollen nucleus, specificity of phosphatases has been found to occur in the chromosomal arms but not in the centromere region (PFEIFFER, 1956).

HARA et al. (1972) compared enzymes localized at surfaces to the total contained in *Cycas* pollen (Table 14–7). Surface-bound enzymes were determined by assay of the media after a short time incubation of cells with substrates for the enzyme reaction, activity of whole pollen was determined on cells homogenized after suspending and centrifuging 4 times, both washings and homogenate being assay for enzyme acticity. They noted, aside from differences in ratio of readily soluble enzymes (Table 14–7), that adsorption of ribonuclease was much more pH dependent than invertase; most RNase could only be released above pH 6.0, while the binding of invertase was not very pH dependent, over the test range of pH 4 to 9. Salt concentrations above 0.4 M NaCl released most of the invertase activity from the pollen walls and residue. Tween 80 at 5% (v/v) was used by DAVIES and DICKINSON (1972) to solubilize UDP-glucose dehydrogenase from particulates in pollen homogenates from *Lilium longiflorum*. Determination of the possible intracellular site of the Mg^{++} dependent alkaline (pyro)phosphatase reported by HARA et al. (1972) in *Cycas* pollen might be facilitated by a survey for levels of Mg^{++}, which can be readily detected by x-ray fluorescence (Fig. 8–2).

Mitochondria vary in size and number with location in the pollen. Those in the generative cell appear slightly smaller and contain less cristae than those in the tube cytoplasm (SANGER and JACKSON, 1971). There are also few mitochondria at the growing tip of the tube (Fig. 14-3 A, C). Presumably, the enzyme functions and energy-producing pathways in most mitochondria are similar. Several decarboxylation reactions are usually found associated with mitochondria. Some of these, or other enzymes may be solublized in growth or isolation. During tube growth, inherent changes in enzyme location may yield intramolecular modifications that result in isozymes. Thus, not all decarboxylases are accounted for in the mitochondria. GOAS (1954) discovered a decarboxylase for converting mesoxalic acid (ketomalonic) to glyoxylic in three species of Liliaceae pollen; however, its site has not been determined. It would be worthwhile to search for this site in association with mitochondria and different pollen organelles.

The decarboxylation of mesoxalic acid in the reaction:

$$
\begin{array}{ccc}
\text{COOH} & & \\
| & & \\
\text{C}{=}\text{O} & \longrightarrow & \text{CHO} \\
| & & | \quad\quad + \text{CO}_2 \\
\text{COOH} & & \text{COOH} \\
\text{Mesoxalic Acid} & & \text{Glyoxylic Acid}
\end{array}
$$

requires an -SH activated enzyme. *Tulipa* pollen was particularly high in this enzyme activity and in the amount of mesoxalic acid. Malic dehydrogenase isolated from bacteria has a modest level of activity toward this substrate (AKIRA, 1965). Malic dehydrogenase is one of the most active enzymes in pollen (Table 14-4) and probably in mitochondria; but how decarboxylation of mesoxalic acid is coupled to the dehydrogenase, if it is, is not known.

Plastids. The role of plastids, amyloplasts, is associated with their enzyme component for synthesizing and solubilizing starch granules or oil droplets in the tube cytoplasm. Thus, starch phosphorylase, probably pyrophosphorylase, and ADPG transglucosylase (α-glucan glucosyltransferase) are presumably localized in or on amyloplastids and form the α-1,4-linked glucosyl portion of amylose; presumably "Q" enzyme is also present to form the α-1,6-amylopectin portion. DICKINSON and DAVIES (1971) found that the *in vitro* rate of activity of ADP-glucose pyrophosphorylase was about one half the rate necessary to permit starch to accumulate in lily pollen *in vivo*; but, by adding the allosteric activator, 3-phosphoglycerate, the activity of this enzyme increased to four times the level necessary for starch formation *in vivo*. Such an allosteric control mechanism may conceivably operate *in vivo* and facilitate the accumulation of sugars into starch as a potential energy and carbon source for the growing pollen. The α- and β-amylases, and probably maltase (α-glucosidase), mobilize starch reserves and facilitate use of the glucose residues from starch in the amyloplasts. During starch accumulation the hydrolytic enzymes are presumably inactivated or isolated by compartmentalization. If they are external to the amyloplast they would have to pass through the surrounding double membrane, an unlikely translocation. Lipid pockets, potential enzyme sites, do occur in the stroma of amyloplasts and may relate this cytological observation to the required metabolic control mechanism.

IWANAMI (1959) suggested that among principle functions of the starch to sugar hydrolyzing enzymes in pollen is interconversion to maintain a proper internal osmoticum. He found that the osmotic value of the tube cytoplasm is slightly higher than the surrounding media, a condition essential for initation and continuation of tube extension. Such interconversions may be one of the main functions of amyloplasts in pollen metabolism and growth. Reversibility of starch phosphorylase is highly pH dependent (BADENHUIZEN, 1969). GREEN (1894) showed also that during tube extension in *Lilium* the metabolism of plastid starch occurred along a gradient. The closer to the tip, the more active the degradation of starch. Cytoplasmic conditions and acid concentrations may vary along the developing tube, although streaming would tend to mix and equalize such differences. In other plant cells proximity to the nucleus has been suggested as related to sequence of starch development and metabolism in amyloplasts (BADENHUIZEN, 1969). Whether the pollen starch crystalline x-ray spectrum pattern is of the A or B-diagram and its granule shape, has not been determined. A role in internal osmotic maintenance, i.e. withdrawing sugar from cell solution or adding it, is however, probably only part of the role of starch granules. The inward diffusion of sugar could be controlled without its accumulation in osmotically neutral starch. But extra sugars stored in the plastids as starch can function as a reserve source for energy and supply substrates for growth. The trehalose-trehalase system in cytoplasm surrounding amyloplasts may function as a glucose carrier system. This soluble enzyme system could possibly provide a mechanism for converting the fructosyl moiety of sucrose to glucose (GUSSIN and MCCORMACK, 1970) in the event of an enzyme system more active than the endogenous hexokinase and phosphoglucoisomerase being necessary.

Pollen tubes probably have a number of underdeveloped proplastids, which when given the opportunity to mature, as in an excess sugar environment, do so by activating or forming the enzymes necessary to function in the total osmotic system. Rapid development of starch granules occurs in germinating pine pollen when different sugars are supplied in excess of immediate growth needs (HELLMERS and MACHLIS, 1956). However, proamyloplasts are seldom identified in ultrastructural studies of pollen tubes. Internal and external concentration of water, the type of sugar and selective membrane permeability are other elements influencing this total pollen metabolic system. The mature plastids have a double membrane with a structureless, enzyme-containing stroma.

Irradiation experiments which modify enzymes in developing microspores (DE NETTENCOURT and ERIKSSON, 1968), may provide evidence about when the grains are initially forming plastids. They might also help evaluate the hypothesis of BADENHUIZEN (1969) that plastid DNA has a primary role in supplying specific enzymes to maintain stroma composition. Division of the plastids may occur by budding, or originate as vesicle-like organelles. By testing progeny plants in which the generative cell transmits the plastids, the enzyme complement and replication mechanism might be evaluated. No attempt has been made to study pollen plastid enzymes after various levels of irradiation, other than iodine detection of the amylose level in irradiated developing microspores (Chapter 9). The RNA-protein patterns, and the basis of genetic control and transmission in the plastids must still be determined. Species with the two different pollen plastid transmitting mechanisms, i.e. plastids localized in the generative cell versus those localized only in the tube cell would be a useful tool for further study.

Other plastids have not been reported in pollen. However, two which might be worth searching for are chromoplastids and elaioplastids. Chromoplasts contain the carotenoids and red pigments in many fruit and flower tissues. Oil-containing elaioplast-like organelles exist in pollen (Chapter 10) but vary with the species and stage of development (GÓRSKA-BRYLASS, 1962). The nomenclature and enzyme criterion of lipid-oil particles in plant cells must still be clarified before good comparisons can be made between different studies.

Lipid granules are scattered throughout the tube cytoplasm but are scarce or absent in the tip region. In a study of pollen tubes which contain predominantly

lipids and little or no starch, e.g. *Lathyrus, Hemerocallis,* TSINGER and PETROV-
SKAYA-BARANOVA (1965) showed that the lipid-containing particles, spherosomes,
have an enclosing membrane. This organelle, described only in plant cells, was for
a long time thought of as a simple oil droplet within the cytoplasm. However,
when tubes are treated with indophenol blue to induce swelling of the spherosomes, their surrounding membrance becomes very apparent. Spherosomes also
retain their particulate integrity when isolated free of the pollen protoplasm. The
cellular origin of spherosomes has been attributed to vesicles associated with the
endoplasmic reticulum (FREY-WYSSLING et al., 1963). But in developing microspores of *Lilium,* HESLOP-HARRISON and DICKINSON (1967) correlated large aggre-
gations of spherosomes with disappearance of the nuclear membrane. GÓRSKA-
BRYLASS (1962), in studies on 8 species of *Campanula* pollen observed the aggrega-
tion and fusion of oil droplets into a large elaioplastid, spherosomes cluster at the
callose plug and impart a dense fat reaction. TSINGER and PETROVSKAYA-BARA-
NOVA (1965) isolated spherosomes free of pollen and stained them for various
mitochondrial-like redox enzymes, and also fixed and viewed the lipid particles in
the electron microscope. They suggested that spherosome lipids serve as intermediate substrates for producing sugars utilized in callose plugs and tube wall
development. This is a reasonable hypothesis although experimental evidence is
still necessary, particularly in view of the well recognized rapid formation of
callose plugs under certain conditions. It is known that upon tube wall injury, or
at tube-bending sites where callose plugs are generally detected, among the first
cytoplasmic changes observed is an accumulation of spherosomes in those wall
areas where plugs form. While it is possible that spherosomes are involved in
formation of callose plugs, the metabolic pathways of this relationship must still
be discerned. Just as certain seeds primarily accumulate fats as their principal
carbon and energy source in germination, so the lipids in pollen such as *Lychnis
alba,* which lack starch and other storage carbohydrates, may function in a similar
manner (CRANG and MILES 1969).

Exine lipid deposits were correlated by PFEIFFER (1955) with the capacity of
lyophilized *Lilium* pollen to germinate. It would probably have been more valid
to correlate germination ability with the wall enzyme content, particularly since
reduction in enzyme activity with aging and during handling of some pollen has
been directly correlated with reduction of viability (HAECKEL, 1951).

The ultimate role of the different pollen enzymes is to facilitate tube growth
and fertilization. At a molecular level, enzymes represent the key mediators for
translating information from the pollen mother cell, and finally the generative and
tube cell nuclei, into growth. Our understanding of the molecular basis of pollen
growth is rapidly advancing from the classical type assays of strictly constitutive
chemicals. While much remains to be discerned, we are beginning to recognize the
regulating mechanisms in and around pollen which influence the activity of the
pollen enzymes. In this chapter we reviewed some of the types of enzymes present
and the factors influencing their activity. It is still not known if the rates of pollen
enzyme activities recognized *in vitro* can account for the rates of growth observed
in vivo, or which enzymes or cofactors must be synthesized or acquired by germinating pollen. But other chemical constituents, in particular pigments, are fairly
well described in many pollen species.

Chapter 15. Pollen Pigments

The two initial characteristics by which pollen species are distinguished are shape and color. Many people presume all pollen is yellow. But, in fact, not only are some pollen non-yellow, but they may not even yield yellow pigment on extraction. Pollen which is yellow may not contain any carotenoids or carotenes although such constituents may have been present during microsporogenesis (Chapters 2 and 9). Many pollen do contain carotenoids, derivatives of tetraterpenes; these include carotene or the oxygenated derivatives, xanthophylls. Others derive their color from phenolic based flavonoids. Pollen often contains both pigments, but while extractable carotenes may be absent, flavonoid derivatives are always present. A comparison of relative amounts of the two pigment groups was made for *Rosa damascena* by ZOLOTOVITCH and SEČENSKA (1962). When all the yellow extracted pigment in 100 mg was calculated as carotene, they computed 0.203%, yet the actual carotene content in the pigment extract was only 0.058%, suggesting that about 75% of the pigment in this pollen was flavonoid in nature.

Initially, patterns of inheritance of both flavonoids and carotenoids were studied as a key to understanding the mechanism of gene action. Pollen color, covering a range from white through blue and yellow-orange, is occasionally used as a genetic marker, along with the more commonly described color patterns of the flower petals and anthers. FERGUSON (1934), continuing the earlier *Petunia* studies of RASMUSON (1918) and NAUDIN (1858), demonstrated that genes controlling petal and pollen color are located on separate chromosomes. FERGUSON (1934) minimized variations in her comparisons of tint and shade by recognizing and controlling the effect of moisture content on the pollen color, and by using an index of color standards against a white background. More recently, chemical studies of pollen pigments have focused upon trying to understand their role and physiological function.

Carotenoids

Distribution

BERTRAND and POIRAULT (1892) using *Verbascum thapsiforme*, were the first to isolate carotene from pollen. Work now suggests that sporopollenin in the exine may be derived from carotenoid-like components or from a common isoprene-related precursor (Chapter 9). Transfer of lipid soluble carotenoids to the exine of maturing microspores, first recognized by VAN TIEGHEM (1884), can occur via polymerization of terpene precursors, or merely as a direct incorporation of preformed carotenoids from the tapetal tissues. Carotenoids in the anthers differ

Table 15-1. Carotenoids in anthers + pollen of Indian cress
(*Tropaeolum majus*) extracted from 2 gm samples (SYKUT, 1965)

Component	μg/g. dry wt.	% of total carotenoids
α-carotene	540.0	30.0
β-carotene	453.9	25.1
γ-carotene	311.9	17.3
δ-carotene	69.6	3.8
neo-β-carotene-U	136.0	7.5
neo-γ-carotene	74.6	4.1
hydroxy-α-carotene	19.2	1.1
kryptoxanthin	26.8	1.5
lutein (xanthophyll)	65.6	3.7
lutein-5,6-epoxide	19.6	1.1
antheraxanthin	52.5	2.9
unidentified	37.2	2.0

from those in the supporting filament tissues and other flower parts. Pollen and anthers are characteristically high in lutein (xanthophyll) esters, antheraxanthin and contain traces of carotenes (KARRER et al., 1950; GODWIN, 1965), with more free carotenes occurring in the anthers than in the pollen. JUNGALWALA and CAMA (1962) found that about 90% of the carotenoids in *Delonix regia* anthers was zeaxanthin, and only 0.12% was β-carotene; but, zeaxanthin is never found in such a high level in pollen. In *Lilium longiflorum* the lipid globuli carotenoid transfer bodies in the tapetum contain primarily α-carotene-5,6-epoxide (HESLOP-HARRISON, 1968c). This carotene derivative has not been reported in equally high or significant levels in the mature pollen. SYKUT (1965) correctly recognized that the sample he extracted contained both pollen and anther tissues, although the article title purported to disclose an analysis of "carotenoids in pollen of Indian Cress *Tropaeolum majus*". SYKUT (1965) identified most of the 12 carotenoids isolated, Table 15–1. Carotenes dominated in these tissues, comprising 72% of the total carotenoids. SYKUT (1965) compared the total μg/gm dry weight of carotenoids in anthers + pollen of Indian cress, 1804 μg, to that of green oak leaves 1065 μg, and to the 5700 μg reported in *Delonix regia* pollen by JUNGALWALA and CAMA (1962). The high levels of carotenes in Indian cress (Table 15–1) may be exceeded in some pollen where xanthophylls may occur as the dominant carotenoids. Separate analyses of anther and pollen are necessary for a valid comparison of the relative pigment levels in any particular pollen.

Within the pollen the cytoplasm may appear lightly colored, generally yellow. BECK and JOLY (1940) reported that the nucleus of *Hymenocallis tubiflora* was pigmented a reddish brown color. The nucleus color was different from that of the containing cytoplasm and exine. Unfortunately, no further studies have been reported relating to pollen nuclei color. Such studies of localized differences could possibly enhance our understanding of the role of pigments in pollen.

When pollen is extracted successively, first with a mild lipid solvent, then by a more intensive solvent refluxing, both solvents remove different quantities and types of carotenoids. This suggests that different carotenoids occur at different

Table 15-2. Carotinoids in pollen

Helianthus tuberosa	lutein esters α and β carotene cryptoxanthin xanthophyll flavoxanthin	CAMERONI, 1958
Acacia dealbata	α and β carotene epoxy-α-carotene lutein (trace) flavoxanthin (trace) epoxyxanthophyll	TAPPI, 1949/1950
Cyclamen persicum	lycopene β-carotene β-hydroxycarotene neurosporin	KARRER and LEUMANN, 1951
Lilium mandshuricum	β-carotene violaxanthin capsanthin capsorubin antheraxanthin	TAPPI and MENZIANA, 1956

loci in the grain. Obviously the sticky, thread-like strands on and in the exine—generally yellow in color—is an easily extracted pigment source which is more accessible than those incorporated in the exine or cytoplasm. MÖBIUS (1923), in studying pollen color in 120 different species, found that 80% were yellow colored, the pigment being mostly localized in the exine, but in a small percent the color was due to surface oils viscon strands. When the yellow surface oils of *Cyclamen* are volatilized the dry pollen lacks color and is white (TROLL, 1928). Carotenoids, anthoxanthin and anthocyanin were absent in extracts from 29 white, gray and green angiosperm pollen that were among some 60 species analyzed by LUBLINER-MIANOWSKA (1955).

While α- or β-carotene are generally the commonest carotenoids, many related pigments also occur. A few examples are listed in Table 15–2. About 0.06% of ragweed pollen dry weight are carotenoids. It varies from 0.05% to 0.08%, calculated as β-carotene (WITTGENSTEIN and SAWICKI, 1970). The β-carotene levels in Bermuda (*Cynodon dactylon*) and orchard (*Dactylis glomerata*) grasses vary from year to year, although the total percent of ether-extractable materials in each species was found to be relatively constant over two successive years. Grass pollen contains considerably less carotenoids and ether-extractable lipids than ragweed pollen, i.e. 1–6% lipid in grass vs. 12–15% in ragweed. The lipid fraction of ragweed contains mostly xanthophylls as the tri-hydroxy forms (0.02%), di-hydroxyxanthophyll (lutein) at the 0.3% level and only 0.005% β-carotene. A xanthophyll-related pigment was isolated from *Lilium tigrinum* pollen, associated with a yellow colchicine-like derivative (DUCLOUX, 1925). This interesting pigment-alkaloid complex has not been further studied; it is interesting to note that *Colchicum autumnale*, source plants for colchicine, are also in the family Liliaceae. In the genus *Rosa*, carotene content, as mg percent, was 0.76 in *R. damascena*, 1.67

in *R. gallica* and 2.08 in *R. alba* (ZOLOTOVITCH and SEČENSKA, 1963). These studies suggest that levels of extractable carotenoids vary with the species and the growing conditions. As initially demonstrated by STRAIN (1935), in exhaustive extractions of *Pinus ponderosa* and *P. radiata* pigment, pine pollen lacks free carotenes (TOGASAWA et al., 1967).

The extraction procedure can influence the pattern of pigments found. Photooxidation and bleaching of carotenoids may occur during extraction unless precautions are taken to protect the pollen and solvents from air and light, in particular from U. V. Poor handling may account for some observed differences in older studies. Source of pollen may also influence the pigments found. Bee-collected mixtures of pollen contain about two-thirds less carotenes than xanthophylls, i.e. about 47 μg carotenes vs. 141 μg xanthophylls per gram dry weight pollen (VIVINO and PALMER, 1944). Both bee-collected and wind-distributed pollen are subject to light and oxidative processes, although obviously for different periods of time. As will be discussed in considering the possible role of these pigments, efforts have been made to relate the presence of pigments, and in particular carotenes, to whether or not this pollen is collected by insects (LUBLINER-MIANOWSKA, 1955).

Animals and insects cannot directly use carotenes, but must first form the vitamin A derivative essential for their nutrition. One assay for pollen carotene is to assess the ability of pollen to supply this vitamin. Many studies have related carotene content to capacity of the pollen to serve as a vitamin A source in animal diets (MAMELI and CARRETTA, 1941). Feeding hens corn pollen markedly increases the egg yolk yellow color and vitamin A content. This research led to the suggestion that corn and other pollen is a useful source of carotenes, provitamin A, to improve the quality and appearance of commercially marketed eggs (TAMAS et al., 1970). In rat feeding experiments using *Pinus densiflora* and *P. thunbergii*, SEKINE and LI (1942) reconfirmed that carotenoids with vitamin A-like activity are absent in pine pollen.

Origin and Interrelations

Carotenoids are synthesized from acetyl-Co A through mevalonic acid and the common isoprenoid precursor, farnesyl pyrophosphate, to the C_{40} intermediate phytoene which is reduced to lycopene, followed by successive ring closures to form the ionone ring of γ-carotene or the two terminal rings of β-carotene. A precursor compound before lycopene may be metabolized by an alternative pathway to yield zeacarotene which can be converted to δ-caronte, which is readily converted to γ-carotene and possibly to α-carotene (BONNER, 1965; MACKINNEY, 1968). The enzyme to isomerize α-carotene to β-carotene, a reaction detected in very low levels in some plant tissues, has still not been isolated. But conversion of about 10% of the radioactivity in ^{14}C-lycopene to β-carotene was demonstrated in bean chloroplasts (HILL and ROGERS, 1969). Incorporation of ^{14}C-mevalonic acid into the exine of developing pollen was shown by SOUTHWORTH (1971). The pollen cement and tapetal residual deposits in *Lilium candidum* contain phytofluene (TAPPI and MONZANI, 1955), a colorless but essential intermediate between phytoene and the carotenes.

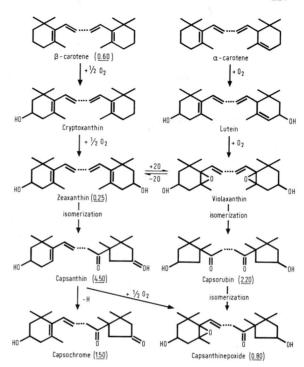

Fig. 15-1. Relationships between carotenoids in *Aesculus hippocastanum*. The quantity of identified component in the pollen, as µg/100 mg, is given (in parenthesis). (Adapted from NEAMTU et al., 1969)

The many different carotenoids in tapetum, developing anthers (Table 15–1) and pollen, afford a basis for speculating on the relationships between these constituents. NEAMTU et al. (1969) proposed a scheme (Fig. 15–1) for the biosynthesis of the carotenoid pigments of *Aesculus hippocastanum*. The α-carotene serves as the initial precursor for the xanthophylls (lutein) while β-carotene is oxidized to cryptoxanthin. Some of these components are isolated from anthers or the developing total floral bud extracts, as well as from the pollen. Out of 11 carotenoids extracted from the pollen, the 7 identified were present at the levels indicated in Fig. 15–1. The four unidentified pigments may represent other components in the scheme. Interestingly, no α-carotene was found in the pollen although it occurs throughout all stages of the developing flower bud and anthers. This lends further support to the hypothesis that considerable interconversion of precursor pigment components probably occurs during transfer of tapetal carotenoids or their precursor metabolites to the developing pollen wall (Chapter 9).

An obvious question still to be answered is what is the nature of the biochemical controls, or channelling mechanism, which diverts the precursor metabolites into these pigments and not into the many alternative steroids or other compounds derived from the common mevalonate-isopentenyl pyrophosphate precursor. Some examples of aborting, self-sterile pollen may be linked to carotenoid metabolism in the anthers. *Allium cepa*, onion plants, which produce sterile pollen show a decreased level of carotene, 1.7 mg/100 g, compared to the 2.9 mg/100 g observed in anthers producing fertile pollen (KRYLOVA, 1967; GENCHEV, 1970). In fungi, absence of the cofactor NAD^+ channels the metabolism of mevalonic acid-^{14}C to carotenoids instead of sterols (MacKINNEY, 1968). It will be interesting to learn if such control mechanisms exist in maturing anthers. Nothing is known of the sequence or origin of oxidized derivatives of the carotenoids, the xanthophylls, in developing pollen. Further analyses of carotenoid changes during development, such as those initiated on *Lilium* anthers by SAMORODOVA-BIANKI (1959), may help clarify the chemistry and role of these compounds in pollen. These initial studies were limited to changes in total carotenoid content in relation to meiosis.

Functions

The well-know role of carotene as the source of vitamin A in insect nutrition has been alluded to earlier in this chapter and discussed in Chapter 7. Pollen pigments also influence and control other aspects of the insect-plant relationship. The color of beeswax is due primarily to the β-carotene in the pollen source (TISCHER, 1941), although bee glue (propolis) from which beeswax is made also contains lutein and several other carotenoids from pollen. Pigments attract color-sensitive insects and may thus be the chemical source underlying the genetic selection and survival of a particular mutant or variant in a plant population. Anther and pollen colors are generally considered secondary in relative importance to the colors and odors of surrounding floral parts in attracting insects. A lutein ester isolated from mixed, bee-collected pollen was found to be an attractant to honey bees (LEPAGE and BOCH, 1968). This carotenoid ester was only one of several attractant chemicals isolated from the mixture. However, pollen color may be the primary attractant in the case of insects that directly grasp anthers and ingest or transport pollen. The passive role of attracting insect vectors that assure pollination and reproduction in many species has long been considered the principle evolutionary benefit of certain pigments in pollen walls (BERTRAND and POIRAULT, 1892).

Categorizing all entomophilous pollen as high in carotene and anemophilous pollen as low or lacking in carotenes, and therefore not attractive to insects (LUBLINER-MIANOWSKA, 1955) is an oversimplification. The general assay which tests the ability of pollen to serve as a vitamin A source in rat feeding experiments is less sensitive than direct chemical assay. EULER et al. (1945) detected carotenes in several pollen by chemical methods when they were not observed in feeding experiments. Many obvious exceptions exist, e.g. *Zea mays*, an anemophilous pollen, is high in β-carotene (TOGASAWA et al., 1967); *Begonia*, an entomophilous pollen, is low in carotene (LEBEDEV, 1948).

A function relating to pollen growth or survival might be a more reasonable explanation for the presence of carotenoids. Carotenes have been proposed as light screening, protecting substances in pollen (ASBECK, 1954); as growth stimulators for the male cells (EULER and RYDBOM, 1931); or as germination control compounds (LEBEDEV, 1948, 1955; SCHWARZENBACH, 1951, 1953).

The ability of yellow to screen against ultraviolet radiation suggests that evolutionarily pollen exposed to sunlight has evolved a yellow pigment in its exine. Regardless of the chemical source of the color, the primary importance is that the pollen walls be yellow. This requiememt is certainly met by most anemophilous species which are yellow, and yet may lack detectable carotenes. NEAMTU and BODEA (1970) correlated an increased level of epoxy-carotenoids and xantho-phylls in pollen with increasing altitude of plants growing on mountains. This further suggests that an evolutionary survival value may exist in certain forms of carotenoids in pollen walls. Instead of protecting nuclear components, the caro-tenes may be protecting against oxidation, saturation of unsaturated lipids and related compounds subject to auto- or photooxidation (NEAMTU, 1971). The ab-sorption maxima of carotenes shifts slightly toward the blue end of the spectrum with oxygenation. Also, the xanthophylls have an absorption peak at 270 nm. Plotting the action spectrum for decreased viability along with the absorption peaks of the different pigment systems might provide an additional means of evaluating this hypothesis.

The levels of carotenes in reproductive and other organs have been used to differentiate sexes in dioecious species. DZAPARIDZE (1965) indicated that the carotene content of female plant leaves is 66% higher than the content in leaves of male plants. Also, leaves of female plants accumulate carotene at a higher rate than those of male plants. These results were consistent in both herbaceous and arboreal dioecious plants. In hemp and other plants, both male and female reproductive organs accumulate different levels of carotene prior to dehiscence, after which the levels decrease (LEBEDEV, 1947; ZHUKOVSKII and MEDVEDEV, 1949). The pattern of high carotene content in female tissues, presumably where pollen grows, and lower levels in pollen cytoplasm and in the wall, suggested to some investigators that carotenes influence tube growth. LEBEDEV (1948) found maximum germination of hemp *(Cannibis sativa)* occurred *in vitro* at 0.005% β-carotene and inhibition above this concentration.

When SCHWARZENBACH (1951, 1953) compared the effect of 25 different caro-tenoids on *Cyclamen persicum*, he was able to group the compounds by their effect on tube growth *in vitro* (Table 15-3). Natural β-carotene and two other carotenoid fractions isolated from anthers with pollen, inhibited tube growth. Based on these observations, and the reversal of the β-carotene inhibition by the growth sub-stance, indoleactic acid (IAA) (Chapter 16), SCHWARZENBACH (1953) concluded that the role of carotenoids in pollen is to inhibit germination, and that once it lands on the stigma a reversal of the inhibition occurs by some naturally occur-ring growth substances.

GHAI and MODI (1970) showed in ripening mangoes that β-carotene was the regulator of carbohydrate metabolism by inhibiting the glucose-6-phosphate de-hydrogenase and 6-phosphogluconic dehydrogenase enzymes in the hexose mono-phosphate shunt. If such a mechanism exists in pollen, it may help explain the inhibition observations of SCHWARZENBACH and indicate one possible role for carotenes. The decrease in carotenoid-containing fat globules in orchid pollen tubes (TSINGER and PODDUBNAYA-ARNOL'DI, 1954) would thus be correlated with metabolic controls in metabolism of the tube, male cells, and developing embryo.

It is very probable that carotenoids in pollen have several functional roles depending upon the kind, location and other interacting constituents. Certain metabolic reactions parallel the sexual reproductive processes. The levels of caro-tene required for functioning of the male cell, originally postulated by EULER and RYDBOM (1931), may provide a clue to metabolic pathways not yet recognized.

Table 15-3. Influence of carotenoids and derivatives on *Cyclamen persicum* (SCHWARZENBACH, 1951, 1953)

Tube response *in vitro*	Compound
1. Greatly stimulates at 10^{-6} M	lycopene crocetin
2. Stimulates at $10^{-5} - 10^{-6}$ M	bixin capsanthin
3. Weak stimulation at 10^{-3} M	xanthophyll (lutein) lutein epoxide lutein dipalmitate cryptoxanthin zeaxanthin physaliene antheraxanthin eschscholtzxanthin dihydro-β-carotene
4. No effect	γ-carotene ε-carotene aurochrome vit. A alcohol deriv. vit. A aldehy. deriv.
5. Retarded growth at $10^{-5} - 10^{-6}$ M	mutatochrome luteochrome flavoxanthin α-carotene β-carotene

Flavonoids

Distribution

The water-soluble flavonoid pigments are related by a common C_{15} flavone molecule. Two phenolic rings, A and B, are joined by a three carbon unit as $C_6-C_3C_6$ (Fig. 15-2). Pollen species contain representatives of the colorless and yellow flavonoids, flavones and isoflavones; and of the red or purple anthocyanin pigments. LEWIS (1759) probably was first to report that the yellow color in alcohol or water extracts of yellow pollen intensified with alkali and turned red in acids. This reversible reaction indicates flavonoid compounds, similar to the anthocyanin red-blue transition in acid and alkali. Studies on the inheritance of anthocyanins in flowers provided much of the initial support for the one gene-one enzyme hypothesis proposed by George Beadle and his associates and now recognized as part of the foundation of molecular biology (HARBORNE, 1967). These pigments are not synthesized by animals. Thus, flavones in butterfly and moth wings are derived from ingestion of pollen and nectar (MORRIS and THOMPSON, 1963).

Most pollen flavonoids exist as the glycoside, i.e. a sugar is linked in a semi-acetal bond at one or more hydroxy groups. The sugar-free moiety is called the aglycone and arises *in vivo* by action of glucosidases. Sugar, when attached to

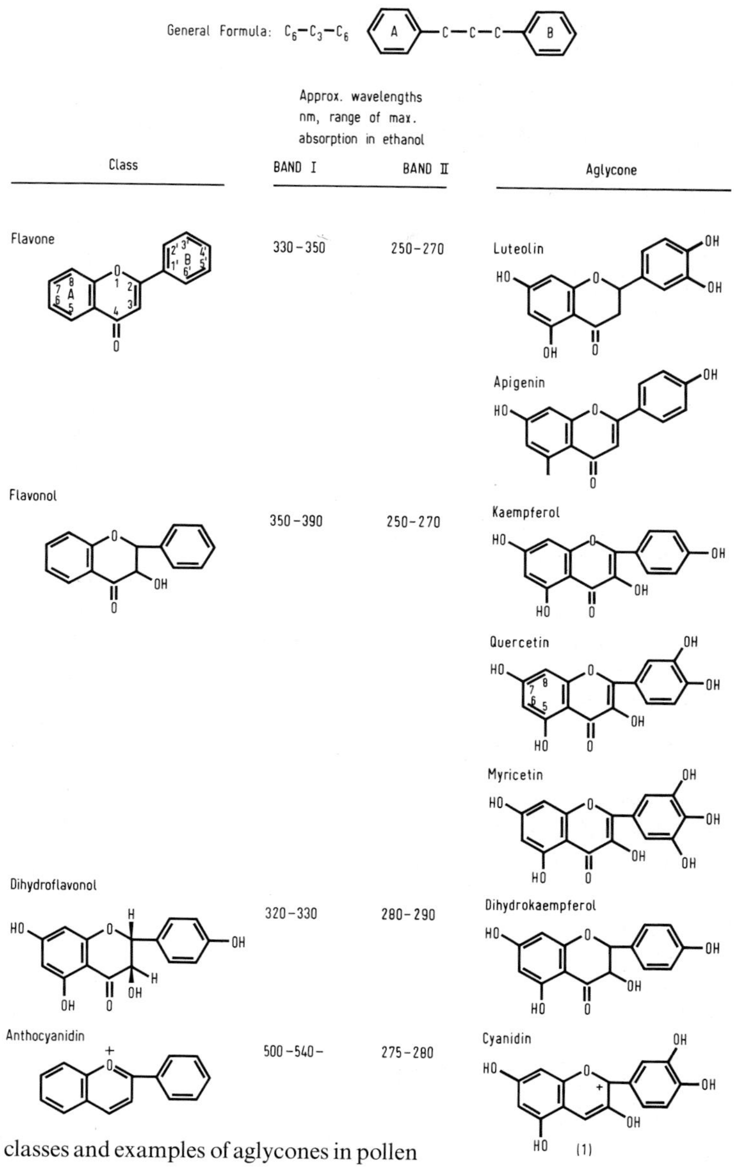

General Formula: $C_6-C_3-C_6$

Fig. 15-2. Flavonoid classes and examples of aglycones in pollen

most flavonoids, confers both cytoplasmic solubility and molecular stability to the resonating ring structure of the aglycone. D-glucose is the most common combined sugar, with D-galactose and L-rhamnose also frequently found; glucuronic acid, xylose and arabinose are rarely found. Some disaccharides and trisaccharides are occasionally attached to the flavonoids. One common disaccharide in pollen flavonoids is rutinose, L-rhamnosyl-D-glucose. In the early studies of pollen pigments precautions to avoid splitting off the glycoside were seldom taken.

A uniform nomenclature was not introduced until the mid-1960's (SWAIN, 1965). Therefore, the older aglycone derived namens found in most pollen literature will generally be used in this discussion. Flavonoids are characterized by spectral absorption bands in the 250 to 550 nm range, and by specific chemical reactions and derivatives. Representatives of all common classes of flavonoid glycosides occur in pollen. These include examples of the flavones, flavonols and anthocyanins. The chalcones and aurones, open ring precursors of flavones, have not been isolated from pollen, although they have been found in developing anthers (WIERMANN, 1970). The isomerase cyclizing tetrahydroxychalcone to trihydroxyflavone has been demonstrated in maturing *Lilium candidum* and *Tulipa* pollen (WIERMANN, 1972). The basic molecule of each flavonoid class in pollen, and a few related aglycones are listed in Fig. 15-2.

The amount of flavonoid, its location, and the presence of other pigments all contribute to the visible color of pollen, and the appearance of its extract. Small quantities of carotenoids can drastically modify the apparent color of pollen flavonoids. Identical flavonoids may be present in two species but because of variation in concentration and presence of co-pigments, the pollens may appear different in color. Complexing of flavonoids with Al^{+++} or Fe^{+++}, common metals in pollen (Table 8-4), will also modify the electron resonance and absorption spectrum of flavonoids, just as will the pH and chemical composition of the solvent extraction medium. Added to differences in color that may arise during microsporogenesis, SCHOCH-BODMER (1939) found that variations in *Lythrum* pollen color, the nature of flavonoids present and in particular anthocyanin content, depended upon the temperature and nutrient supply to the flowers during anthesis. The final pollen products are probably quite stable and are undoubtedly controlled primarily by genetic factors, with the environment during microspore development modifying the presence, absence, or mixture of different flavonoids.

Flavonoids usually occur in extractable concentrations of 0.04 to 0.06% of the pollen dry weight. TOGASAWA et al. (1966) isolated rutin and narcissin from *Lilium lancifolium* at yields of 0.056 and 0.036%, respectively. Higher levels of these pigments sometimes occur. Fresh *Crocus* pollen extracted by KUHN and LÖW (1944) yielded 0.56% flavonoid glucosides; but the yield from one-year-old pollen stored dry at room conditions was only 0.29%. WAFA (1951) showed that date palm, *Phoenix dactylifera*, pollen, long used as a dietary supplement in the Middle East, contained 2.4% quercetin. This compares very well to levels of one of the highest known natural sources, buckwheat seeds, which contains 2.0–2.5%. Quercetin or its derivatives is probably the commonest flavonoid in pollen; in palm it contains glucose and rhamnose. WIERMANN (1968) found, on analyzing over 140 different species, that it was present throughout most families and it was the principal flavonoid in genera in the family Betulaceae (Table 15-4), but not in other amentiferous species. He found also that catkins which produce the pollen contain a greater variety of flavonoids and not always in the same proportion as that in the mature microspore.

Pollen of dicotyledonous species usually are high in kaempferol (Table 15-5), although in the Hamamelidaceae kaempferol is replaced by isorhamnetin. In some Compositae both flavonoids are present at higher levels. To the broad survey of pollens from dicotyledonous species in Table 15-5, must be added the

Table 15-4. Flavonoids extracted from amentiferous species (WIERMANN, 1968)

Family Species	Pollen				Catkins							
	Qu[a]	Ka	Is	Na	Qu	Ka	Is	Na	My	Lu	Ap	Di
Myricaceae												
Myrica gale	3	1	3	T	3	1	3	T	1			
Juglandaceae												
Pterocarya fraxinifolia	3	2	1	T	3	2		T				
Juglans regia	3	2	1	T	3	2		T				
Carya alba	3	2		T	3	3		T				
Salicaceae												
Salix caprea	3	3		T	3	3	2	T		1		?
S. daphnoides		3		?	3	3	2	?		2	2	?
S. hastata		3		?	3	3	2	?			2	
S. repens	1	3		1	3	3	2	T		2	1	?
S. x smithiana		3		T	2	3	3	?		2		?
S. aurita ssp. angustifolia	1	2		T	2	2	3	T		2		?
S. triandra ssp. concolor	1	3		?	3	3	2	?				?
Populus tremula	1	3			2	3						
P. x euramericana	1	3		T	3	3	2	T	1			
Betulaceae												
Alnus cordata	3	2	1	1	3	2	1	T				
A. glutinosa	3	2		1	3	2	1	T				
A. glutinosa var. *laciniata*	3	3		1	3	2		T				
A. incana	3	2		1	3	2		T				
A. cf. hirsuta	3	2	1	1	3	2		T	2			
A. viridis	3	2		1	3	2		T	2			
Betula humilis	3	1		T	3	1		T				
B. populifolia	3	2	1	1	3	1		T				
B. pubescens	3	2	1	1	3	1		T				
B. japonica	3	2	1	1	3	1	1	T				
B. lenta		3		1	3	2		T				
B. verrucosa	3	2	1	1	3	2		T				
Carpinus betulus	3	1		1	3	1	1	T	1			
C. tschonoskii	3	2		1	3	2		T				
Corylus avellana	3	3		1	3	3		T	3			
C. acellana var. *atropurpurea*	3	3		1	3	3		T	2			
C. avellana var. *contorta*	3	2	1	1	3	3	2	T	2			
C. colurna	3	2	1	1	3	2		T				
C. colurnoides	3	2	1	1	3	2	3	T				
Ostrya virginiana	3	2	1	1	3	2		T	2			
Fagaceae												
Fagus silvatica		3		1	2	3		T				
Nothofagus antarctica	1	3		2	3	1		1	2			
Querus cerris	3	1	3	2	3	1	3	1				
Q. ilex	2		3	T	3	3	3	T				
Q. pubescens	2	3	3	?	3	3	2	?				
Q. robur	2	3	3	T	3	3	3	T				
Q. rubra	3	1	3	1	2	1	3	1				
Platanaceae												
Platanus acerifolia	3	2			3	2						

[a] Abbreviations used: Qu = quercetin; Ka = kaempferol; Is = Isohamnetin; Na = narin-genin; My = myricetin; Lu = luteolin; Ap = apigenin; Di = diosmetrin. Relative amounts: 3 = high; 2 = medium; 1 = low; T = trace; ? = possibly present.

thorough chemical analysis done at the University of Modena, Italy, on the *Acacia dealbata* in the Leguminosae (TAPPI et al., 1955; SPADA and CAMERONI, 1955, 1956). From methanol extracts of benzene-defatted pollen, they isolated morin(2'-hydroxy kaempferol) and several derivatives of myricetin and disaccharides of quercetin and naringenin. As mentioned previously, one must be carful, in discussing this work from Modena, to recognize that some of their studies of pollen pigment are actually extracted pigments of anthers containing pollen (TAPPI, 1947/1948).

In the monocotyledonous plants which WIERMANN (1968) examined, Table 15-6, kaempferol dominated only in the Amaryllidaceae, where in fact, except for a few traces of naringenin, it was the only flavonoid extracted from the pollen. REDEMANN et al. (1950) attributed the yellow pollen color in *Zea mays* to quercetin which they extracted and crystallized. However, the broader survey, Table 15-6, shows that the aglycone isorhamnetin occurs in the pollen at about the same level as quercetin, along with traces of kaempferol. Obviously, these other flavonoids also contribute to the pollen color along with quercetin.

Pigment is generally more difficult to extract from gymnosperm than angiosperm pollen, which is probably related to the lack of pigmented viscin-like coatings in gymnosperms, and also, the fact that the gymnosperm pollen is less sculptured, with fewer pores and thus is not easily penetrated by solvents. STROHL and SEIKEL (1965) found that the evolutionarily older gymnosperms usually yielded a greater variety of flavones and flavonols (Table 15-7) than occur in the angiosperm species. Gymnosperm pollen, it should be recalled, usually lacks free carotene.

Table 15-5. Flavonoids in pollen of dicotyledonous plants (WIERMANN, 1968)

| Family | Compound | | | |
Species	Qu[a]	Ka	Is	Na
Ulmaceae				
Ulmus glabra	2	3		
Zelkova serrata	T	3		T
Moraceae				
Cannabis sativa	3	3		T
Cactaceae				
Cleistocactus straussii	T	3		
Cereus speciosa		3		
Trichocereus macrogonus		3		T?
Selenicerus grandiflorus		3	T	
Polygonaceae				
Rumex acetosa	2	3		
R. acetosella	1	3		
Magnoliaceae				
Magnolia kobus		3	T	
M. soulangiana			3	
(M. denudata x liliflora)				
Ranunculaceae				
Anemone coronaria	3	T		
A. pulsatilla		3		
Caltha palustris	3	2		
Helleborus corsicus		3		T
Ranunculus repens	3	3	T	

Table 15-5. (cont.)

Family Species	Compound			
	Qu[a]	Ka	Is	Na
Paeoniaceae				
Paeonia obovata	T	2		
P. tenuifolia		1		
Papaveraceae				
Papaver orientale		3		
Cruciferae				
Brassica oleracea	2	3		T
Cheiranthus cheiri	T	3		T
Hamamelidaceae				
Corylopsis spicata	2		3	
Parrotia persica	2		3	
Rosaceae				
Sanguisorba minor	T	T		
Balsaminaceae				
Impatiens walleriana		3		
Malvaceae				
Pavonia multiflora	3	3		
Passifloraceae				
Passiflora coerulea		3		
Cistaceae				
Cistus purpurea	3	3	2	
Onagraceae				
Oenothera missouriensis		3		T
Haloragaceae				
Gunnera cf. manicata		3		T
Ericaceae				
Rhododendron fortunei		3		
Oleaceae				
Fraxinus ornus	2	3		T
Jasminum nudiflorum		3		T
Polemoniaceae				
Cobaea scandens	2	3		
Solanaceae				
Scopolia lurida		3		
Plantaginaceae				
Plantago lanceolata	2	3	2	T
Plantago media	2	3	2	T
Compositae				
Aster novae-angliae	2	3		
Bellis perennis	2	3	2	T?
Chrysanthemum leucanthemum	2	3	3	
Centaurea cyanus		3		
Cosmos bipinnatus	2	T	3	
Petasites hybridus		3		
Solidago canadensis		3		
Helianthus annuus	3	2	T	
Dahlia variabilis	2	3	T	

[a] Abbreviations used: Qu = quercetin; Ka = kaempferol; Is = isorhamnetin; Na = naringenin. Relative amounts: 3 = high; 2 = medium; 1 = low; T = trace; ? = possibly present.

Table 15-6. Flavonoids from pollen of monocot species (WIERMANN, 1968)

Family Species	Compound					
	Qu[a]	Ka	Is	Na	Lu	Ap
Potamogetonaceae						
Potamogeton lucens	3	1	2			
Liliaceae						
Lachenalia tricolor		2		T		
Lilium bulbiferum	3	3	3	T		
L. candidum	3	T	3	T		
L. regale	3	2	3	T		
Tulipa forsteriana	3	3	2	T?		
T. sp. cv. Apeldoorn	2	3	2			
T. silvestris	T	3	2			
Agavaceae						
Agave filifera		3				
Amaryllidaceae						
Clivia nobilis		3	T			
Crinum asiaticum		3	T			
Galanthus nivalis		3				
Hymenocallis speciosa		3	T			
Leucojum vernum		3				
Narcissus pseudonarcissus		3				
N. poeticus		3				
Iridaceae						
Crocus vernus	3	2	3			
Iris pseudacorus		T	2			
Juncaceae						
Juncus glaucus	3	T				
Luzula campestris	3	T	2			
L. nemorosa	3	3		T		
L. silvatica	3					
Bromeliaceae						
Abromeitiella brevifolia			3	T		
Hechtia argentea	2	T	3			
Gramineae						
Alopecurus geniculatus	3	2	3			
A. myosuroides	3	2	3			
A. pratensis	3	2	3			
Anthoxanthum odoratum	T	3	2			
Avena pubescens	T	2	3			
Briza media	1	1				
Dactylis glomerata	2	1	3			
Festuca pratensis	2	2	2			
Holcus lanatus	3	3	T			
Lolium perenne	2	3	2			
Phalaris arundinacea	3	2	3			
Secale cereale	T	2	2			
Zea mays	3	T	3			
Sparganiaceae						
Sparganium simplex	3	T	3	T		

Table 15-6. (cont.)

| Family | Compound | | | | | |
Species	Qu[a]	Ka	Is	Na	Lu	Ap
Typhaceae						
Typha angustifolia	T	3		T		
T. latifolia	2	3	3	T		
Cyperaceae						
Carex stricta		T		T		
C. acutiformis		T		T		
C. pendula		T		T		
Eriophorum angustifolium	3	3		T	2	2
E. vaginatum	3	2		T	2	3

[a] Abbreviations used: Qu = quercetin; Ka = kaempferol; Is = isorhamnetin; Na = naringenin; Lu = luteolin; Ap = apigenin. Relative amounts used: 3 = high; 2 = medium; 1 = low; T = trace; ? = possibly present.

Table 15-7. Flavonoids from gymnosperm pollen (WIERMANN, 1968; STROHL and SEIKEL, 1965)

| Family | Compound | | | | | | | |
Species	Qu[a]	Ka	Dihyqu	Dihyka	Lu	Ap	Na	My
Ginkgoaceae								
Ginko biloba		3					2	
Pinaceae								
Larix decidua		2					2	
Picea abies		T					3	
Pinus banksiana			T	1				
P. echinata			1	T				
P. elliottii			T	T				
P. cembra	T	2					3	
P. contorta	T	T	T				3	
P. montana	T	2					3	
P. nigra	T	T	T?	T?			3	
P. palustris			T	T				
P. ponderosa			T	T				
P. resinosa			1	T				
P. silvestris	T	2		T?			3	
P. strobus			T	T				
P. taeda			1	T		1		
P. thunbergii	T	T					3	
Taxodiaceae								
Cryptomeria japonica		2			2	3	1	
Cupressaceae								
Chamaecyparis lawsoniana	3	2			3	2		
Juniperus chinensis	3	2			2	3		
Taxaceae								
Torreya californica					?	3		
Taxus baccata	3	T			3	2		2

[a] Abbreviations used: Qu = quercetin; Ka = kaempferol; Dihyqu = dihydroxquercetin; Dihyka = dihydrokaempferol; Lu = luteolin; Ap = apigenin; Na = naringenin; My = myrecetin. Relative amounts used: 3 = high; 2 = medium; 1 = low; T = trace; ? = possibly present.

The methanol-extracted flavonoid glucoside in timothy grass is primarily isorhamnetin (INGLETT, 1956), which was first isolated from timothy and orchard grasses by MOORE and MOORE in 1931. These and other early studies (HEYL, 1919) were aimed at determining if non-nitrogen components carry the human allergenic-reacting material in hay fever pollen. The suggestion was made at that time (MOORE et al., 1931) that the glycoside extract contained the antigen material, a hypothesis which has not been verified. STEVENS et al. (1951), using highly purified crystalline isoquercetin from giant ragweed pollen, found it was inactive on cutaneous testing. While glycoproteins may possibly contain antigenic activity (Chapter 12), the crude procedures employed in the early studies, and the difficulty of removing pollen oils without removing surface localized proteins, make such extractive procedures too gross to establish any specific chemical cause-effect relationship. The improved analytical separation methods applied by WIERMANN (1968) detected several other flavonoids in these grasses (Table 15-6), none of which have ever been found to induce the hay fever reaction.

Identical, but uncommon sugar units are often found in different flavonoids. Cyanidin may occur coupled with either the disaccharide sophorose (2-0-β-D-glucopyranosyl-α-D-glucose) or rutinose (6-0-α-L-rhamnosyl-D-glucose), sugar moieties found in many flavonols. Rutinose is found in narcissin, the isorhamnetin-3-rutinoside of *Lilium auratum* pollen (KOTAKE and ARAKAWA, 1956) and occurs linked as quercetin-3-rutinoside, rutin, in other pollen (FUJITA et al., 1960). Sophorose is linked as quercetol-3-sophoroside in *Alnus cordata* pollen (SOSA and PERCHERON, 1965) and probably occurs in many other pollen with the quercetin aglycone (Table 15-4). The kaempferol-3-sophoroside is found in several pollen containing quercetol-3-sophoroside (PRATVIEL-SOSA and PERCHERON, 1972). However, such sugars are easily hydrolyzed from the aglycone and degraded during extraction, and thus frequently are not detected.

Isolating and characterizing flavonoids with three different attached sugars, heterosides, requires considerable care. A kaempferol-3-0-rhamnodiglucoside, first described in *Camellia sinensis* leaves, was finally isolated and the relationship of the sugars determined by SOSA and PERCHERON (1970) from the ethanol extract of *Populus yunnanensis* pollen. While the presence of the aglycone kaempferol has long been recognized in *Populus* pollen extracts (Table 15-4), the heterosidic sugars associated with this and other flavonols in pollen have not often been determined. The kaempferol-3-(p-coumaroylglucoside), tiliroside, occurs in very high concentrations in *Fagus* pollen (TISSUT and EGGER, 1969), which also has kaempferol-3-sophoroside but lacks quercetol-3-sophoroside (PRATVIEL-SOSA and PERCHERON, 1972). TAPPI and MENZIANA (1955), after extracting the carotenoids from *Lilium candidum*, were able to obtain a trisaccharide joined to isorhamnetin at the 3'-position. They showed that the sugars were rhamnose, glucose and galactose, although the linkages in the saccharides were not ascertained. Recognizing which cell moieties in pollen are binding sites for many of the simple sugars isolated by hydrolysis and extraction affords a challenge for future research.

The many different sugars isolated from *Typha* pollen by WATANABE et al. (1961), which included large amounts of rhamnose (Chapter 9) and many unusual di- and trisaccharides, e.g. isomaltose, kojibiose, leucrose, and nigrose, very possibly may have originated as glycosyl units in flavonoids. Tea (*Thea sinensis*) pollen

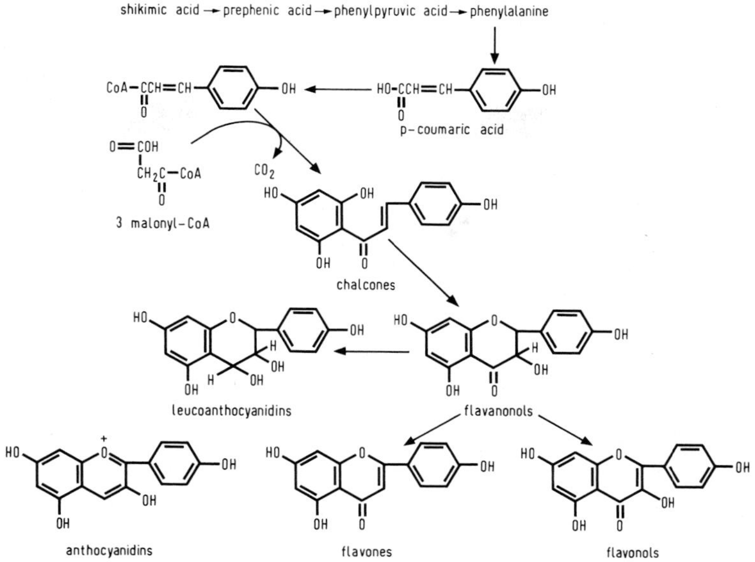

Fig. 15-3. Pathway for biosynthesis of principal flavonoids in pollen

contains a colored methanol-extractable complex of three flavonoids, pollenin *a*, *b* and *c*, all related by the common aglycone pollenitin, 3,5,8,4-tetrahydroxy-7-methoxyflavone (SAKAMOTO 1969). The pollenins differ by their associated sugars. Pollenin *a* is complexed with 3-rhamnoglucose, *b* with glucose and *c* with an unidentified sugar.

While most flavones and flavonols combine to yield yellow-colored pollen, some pollen is red or blue. Anthocyanins are generally the source of these colors. The blue pollen of *Anemone coronaria* in the Ranunculaceae is primarily produced by the monoarabinoside of delphinidin. TAPPI and MONZANI (1955a) also found lesser amounts of glycosides of the anthocyanins, pelargonidin, cyanidin and malvidin (Fig. 15-5). The pH of tissues containing these compounds influences resonant structures, association of the glycoside and the apparent color of the pollen.

Although LUBLINER-MIANOWSKA (1955) did not extract flavonoids from pine pollen, and KRUGMAN (1959) could not detect leucoanthocyanin in 13 species of pine pollen which he extracted; subsequent studies by HISAMICHI (1961) and STROHL and SEIKEL (1965) have shown flavonoids are abundant in pine and conifer pollen (Table 15-7). HISAMICHI (1961) detected isorhamnetin and quercetin in *Pinus thunbergii* and *P. densiflora*. STROHL and SEIKEL's data for 8 pine species, included in Table 15-7, was obtained by successive ether and methanol extractions, with separations in solutions of sodium acetate and carbonate followed by chromatography in different solvents. Absorption spectra were also run to confirm the identification of the flavonoids. The flavonoid fraction extracted in methanol showed the greatest intraspecific variation. STROHL and SEIKEL (1965) noted that poor handling of the pine pollen, and in particular high humidity and temperatures during collection and storing, hydrolyzes the flavonoids. This latter observation again suggests that negative or variable analytical results may be due to poor handling as well as to the extractive procedure used. Certainly pines, with proper treatment and vigorous extraction, yield a good range of flavonols and dihydroflavonols, their dominant color source.

Origin and Interrelations

The inheritance of flavonoids is the most thoroughly investigated aspect of plant chemicals. While genetic and biochemical studies usually focus upon flower pigment chemistry, the pattern of pigment formation has been studied during microsporogenesis. Intermediate and final pigment products in the tapetal transfer tissue with developing microspores have been analyzed. These results, in conjunction with evidence from metabolic studies in other plant tissues, provide an overview of how the pollen pigments are incorporated and how they differ in chemistry and color. Details of inheritance in flowers are well reviewed by HARBORNE (1967) and HESS (1968). This review will primarily be limited to aspects accounting for observed differences in pollen.

Coloring of pollen by flavonoids, in particular the anthocyanins, occurs in the final stages of microspore development. In *Lythrum*, SCHOCH-BODMER (1939) reported most pigment incorporation from the tapetum occurs one day before flower bud opening. WIERMANN (1969) found that *Narcissus* and *Tulipa* attain concentration maxima and the most rapid synthesis long before anthesis. In tulip, flavonoids are incorporated at a later stage of maturation than in *Narcissus*. The flavonols are incorporated into the exines before the anthocyanins, which are then followed by an increase in carotenoids after separation of the tetrads just before anthesis. WIERMANN and WEINERT (1969) followed color development in the final stages of microsporogenesis in *Tulipa*. They recognized two phases of synthesis in the tapetum, with deposition primarily occurring on the intine. Synthesis of flavonols occurs in the whole active tapetum, while anthocyanin synthesis increases during the degeneration of the tapetum, at about the same stage carotenoid biosynthesis reaches its maximum.

Precursors in flavonoid biosynthesis have been detected primarily by radioactive studies with leaves. HESS (1963) confirmed incorporation of the ^{14}C-amino acid precursor, phenylalanine, into anthocyanins of flowers of *Petunia hybrida*. Biochemical precursors for flavonoids, cinnamic acid and chalcone were found by WIERMANN (1970) in developing anthers of *Narcissus*, *Lilium* and *Tulipa*.

The biosynthetic pathway of the flavonoid $C_6-C_3-C_6$ carbon skeleton occurs in two components. The A-ring arises from 3-acetyl units through malonyl-CoA, the B-ring with 3 attached carbons arises from shikimic acid, through phenylalanine and p-coumaric acid. These intermediary steps are outlined in Fig. 15-3. WIERMANN (1970) isolated pentahydroxy-chalcone, p-coumaric acid, kaempferol, quercetin, isorhamnetin and the anthocyanidin, delphinidin, in *Tulipa* anthers during microsporogenesis. He proposed a scheme (WIERMANN, 1972a) relating the sequence of formation of the main pollen flavonoid pigments to the time of their appearance (Fig. 15-4), and demonstrated the enzyme in maturing pollen to cycli-

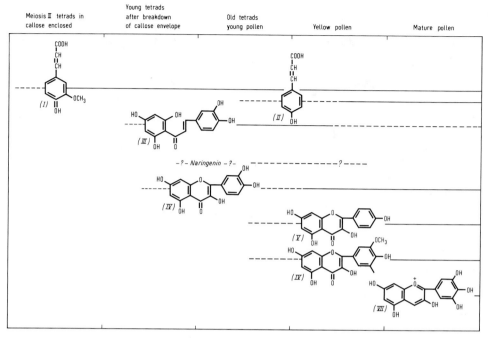

Fig. 15-4. Sequence of flavonoid pigment development in *Tulipa* var. Apeldoorn. (From Wiermann, 1970). I = ferulic acid; II = p-coumaric acid; III = pentahydroxy chalcone; IV = quercetin; V = kaempferol; VI = isorhamnetin; VII = delphinidin

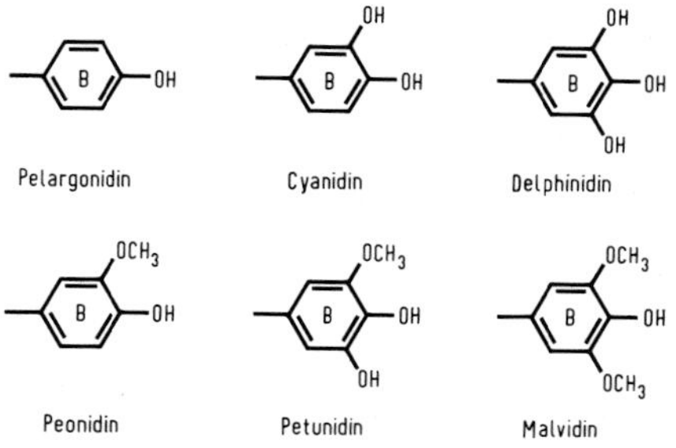

Fig. 15-5. Hydroxylation and methylation of anthocyanins found in pollen

cize the chalcones (Wiermann, 1972). Strohl and Seikel (1965) found o-coumaric and ferulic acid along with the flavonoids they isolated from pine pollen. The sequence of hydroxylation, methylation and glycosylation of the B-ring (Table 15-5) has not been determined, but glycosylation probably occurs last. Developing pollen affords an interesting tool for such studies.

A few pollens contain the complete sequence of hydroxylated anthocyanins and their methylated derivatives outlined in Fig. 15-5. Hess (1963) found that the ^{14}C-phenylalanine label was incorporated into all the anthocyanins in developing *Petunia* flowers, i.e. cyanidin, delphinidin, peonidin, petunidin and malvidin. The less hydroxylated and less methylated components were the first to be labeled, i.e. cyanidin labelling preceding delphinidin, then peonidin followed by ^{14}C-label incorporation into petunidin and malvidin. The S-adenoxylmethionine was shown to be the methyl donor to the anthocyanidins in *Petunia* (HESS, 1964). Again, sugars appear to be incorporated last. Enzymes for the transglucosylation, glucosyltransferases, are highly specific for specific positions, i.e. 3', 5' or 7' in the flavonoids. The specific enzyme for glucosylating quercetin, from UDP glucose, to form quercetin-3-glucoside (isoquercitrin) was identified in *Zea mays* pollen (LARSON, 1971). On the other hand, the hydrolytic enzymes, α- or β-glucosidase, are relatively nonspecific. Flavonoid-5-glucosides are relatively rare in plants and thus are frequently used as taxonomic markers (GLENNIE and HARBORNE, 1971). Except for preponderance of a specific aglycone (Tables 15-4 to 15-7) no such specificity for the glycosides has been found in pollen.

The triggering mechanism which initiates pollen pigmentation is not recognized. Light and nutrition influence developing anthers by increasing sugar substrates and possibly by inducing a conversion-type reaction in the red-far-red region of the spectrum through the phytochromes (HARBORNE, 1967). Understanding the possible role of light in pigment initiation may help to discern the role of the flavonoids in pollen and why little or no pigment is formed in the development of sterile pollen. STEIN and GABELMAN (1959) found a good correlation between the relative level of red and yellow pigment in mature *Beta vulgaris* anthers and the production of sterile pollen. When the anthers turned yellow, they dehisced a corresponding percentage of viable pollen; if they remained red the percent of viable pollen decreased. If stamens were allowed to take up sugars by incubation during microsporogenesis, the anthers formed red pigment and pollen viability decreased. In some still unrecognized way, genetic control of pigmentation via carbohydrate metabolism influences pollen development. Transition of phenylpropanes to the flavonoids, noted by WIERMANN (1970), Fig. 15-4, is essential to developing viable pollen.

Functions

Several theories have been proposed to explain the role of flavonoids in pollen. Some have suggested these pigments are involved in metabolism and growth, others have focused on a function in passive protection, or as an attractant of a pollinating vector.

Pigments obviously contribute to the viable color of the grains. But since only a few grains are deliberately ingested by insects or birds (Chapter 7), the color attractant hypothesis for the role of flavonoids probably has no basis. Of course, these pigments are important in providing color to petals or corolla, which attract many pollination vectors. But a biological and metabolic role in pollen is more difficult to establish.

Some flavonoids impart a bitter or sweet taste. Naringenin-5-glucoside is tasteless, while naringenin-7-glucoside is very bitter to man's taste. Whether a sufficient level of a bitter flavonoid exists in some pollen to discourage insect consumption and destruction has not been determined. On consuming flavonoids, animals readily hydrolyze the sugar from the flavonoid and produce CO_2 from the A-ring and phenylacetic acid from the B-ring (Fig. 15-2). The acid cannot be dehydroxylated in the mammalian liver or kidney, and is probably degraded by bacteria before excretion in the urine (HARBORNE, 1967). Animals have, of necessity, developed methods of coping with ingested flavonoids so that they are almost all non-toxic. In fact, rutin and some related flavonones were at one time called "vitamin P" and considered to have beneficial antioxidant activity in man toward ascorbic acid and adrenaline, and to be an inhibitor of certain enzymes and a relaxant of smooth muscles. The proposed vitamin role for these compounds has never been widely accepted.

The activity of flavonoids on growth is easily established by *in vitro* experiments. The simple expediency has been followed of adding pure flavonoids to pollen germinating media *in vitro* (MINAEVA and GORBALEVA, 1967). Growth stimulation was recorded at a low concentration (1.6×10^{-4} M) just as was observed with carotenoids (Table 15-3). Quercetin, isorhamnetin, rutin and isoquercetin all increased percent germination and usually tube length. The problem is to extend the results with pollen pigment extracts and isolated pigments to conditions where they act as growth regulators *in vivo*. *In vitro* data are difficult to extrapolate to *in vivo* growth conditions. Several hypotheses suggest a direct role for specific flavonoids in pollen growth in the pistil.

KUHN et al. (1942) found that a flavonoid glucoside from *Crocus* pollen, but not the aglycone, inhibited movement of gametes from the algae *Chlamydomonas eugametos*. At that time this assay was used to test for a hormonal or chemical activity on gamete mobility and attraction; it led to consideration of the *Crocus* flavonoid as a gynotermone. Two flavonoids with the aglycones quercetin and isorhamnetin were isolated from the *Crocus* stigmas and pollen (KUHN and LÖW, 1944). Only the 2-diglucoside of isorhamnetin inhibited mobility of the algae gametes. These investigators, in subsequent studies of pollen and styles of known incompatible plants of *Forsythia*, proposed an explanation for pollen-style incompatibility based on certain flavonoids (KUHN and LÖW, 1949; MOEWUS, 1950). Unfortunately, this work with *Forsythia* has never been supported by investigators who have attempted to repeat the original studies. Nevertheless, because of the novelty and implications of this for a specific group of chemical controls in plant reproduction—a long sought mechanism—the theory has been widely reproduced as originally proposed, often without reference to subsequent negative experimental evidence. To clarify the available facts with regard to this now disproven hypothesis, we will summarize the original theory, and evidence pro and con.

Basic to the proposal of KUHN and LÖW (1949) and MOEWUS (1950) are the facts that in distylic plants with heteromorphic incompatibility such as in *Forsythia*, *Primula*, *Oenothera* and other angiosperms, the short style-long stamen "thrum" form and the long style-short stamen "pin" form are self-sterile but cross-fertile. Pollen from "thrum" flowers will grow down a "pin" style and form seed but will not fertilize and form seed when deposited on its own stigma, and vice versa. The other previously observed fact which they extended to this incompatible system was that when inhibiting flavonol glycosides were hydrolyzed, the aglycone was inactive in inhibiting gamete mobility.

KUHN and LÖW (1949) reported that different flavonol glycosides occur in the two heterostylic pollens. Quercetrin (quercetin-3-rhamnoside), they indicated, occurred in the pollen from short style "thrum" anthers, rutin (quercetin-3-rhamnoglucoside) in pollen from long-styled "pin" anthers. In extracting rutin they also reported a small amount of lactose. MOEWUS (1950) proposed that self-incompatibility was due to the inhibitor-like nature of these pollen flavonol glycosides which could be overcome either by addition of boron to the pollen-germinating environment, or glycoside hydrolyzing enzymes in the stigma. His observation was that pollen from maturing "thrum" buds was self-fertile and germinated well *in vivo*. But pollen from open "thrum" flowers was self-sterile and only self-fertile if applied with a suspension of 100 ppm boric acid. A 10^{-3} M H_3BO_3 solution was shown, *in vitro*, to overcome a 10^{-6} growth inhibiting concentration of flavonoid. Quercetrin was suggested as the flavonoid in the pollen from buds, while rutin was present in mature dehisced pollen. He also reported that the *Chlamydomonas* assay, showed quercetin present in water extracts of mixtures of "pin" and "thrum" pollen or stigmas, but this flavonol was absent in water extracts of "pin" mixed with "pin" or "thrum" with "thrum" pollen and stigmas. According to MOEWUS, pollen of short stamen, long style "pin" *Forsythia* before and after shedding contained only quercetrin. He proposed that "thrum" stigma contained a quercetrinase which inactivated the inhibitory affect of the mature "pin" pollen, released quercetin and allowed cross pollination to occur.

The MOEWUS hypothesis was first openly questioned by ESSER (1953), ESSER and STRAUB (1954) and LEWIS (1954). Evidence against the proposal was summarized by ROTHSCHILD (1956), REZNIK (1957), RENNER (1958) and FUJITA et al. (1960). REZNIK re-examined the flavonoids in pollen from several heterostylic species and he could find neither qualitative nor quantitative differences between pollen of *Forsythia intermedia*, *Primula intermedia*, or *Lythrum salicaria*. REZNIK and RENNER concluded that flavonoids are not concerned with self-sterility. VISSER (1956) and RENNER (1958) criticized several procedural errors in MOEWUS' earlier work, and showed that so called inhibited pollen germinates in stigmas of the same flower, while immature, non-dehisced pollen will not. The block to fertilization in *Forsythia* must occur after pollen growth in the stigma. RENNER concluded that the basic report in which specific esters of the flavonoid crocin differentially inhibited flagella action in *Chlamydomonas* was in error. FUJITA et al. (1960) re-examined flavonoids of flowers and pollen of four species of *Forsythia*. Unlike KUHN and LÖW (1949), but in agreement with other investigators, they found no correlation between quercetin or rutin and self-sterility, and they could only isolate glucose and fructose, never lactose as the sugar moiety. Rutin was practically the only flavonoid found in all *Forsythia* pollen they examined. FRITZ (1960) also failed to detect lactose among disaccharides from flavonoids of *Petunia* pollen.

As recently as 1969 investigators have failed to confirm the biochemical basis of the hypothesis and evidence offered by MOEWUS (1950). SAKAMOTO (1969) tried to relate self sterility in tea to flavonol glucosides. No flavonoid hydrolyzing enzyme activity could be isolated from the stigmas or pistils. He concluded in his study, as practically all others have concluded, that in heterostylic species the mechanism of self sterility is not related to the presence of specific flavonoids.

The suggestion that flavonoids may interact with boron has received some indirect support. TAUBÖCK (1942) shifted the absorption spectrum of flavonols in solution with boric acid, separate from the shift induced by hydrogen ions. He also correlated a high level of flavonols in reproductive tissues with a high boron level. However, *in vitro* experiments which suggest or demonstrate the chelation of the –OH groups of flavonol, such as occurs with the borate anion and many sugars at high pH's, cannot be extrapolated to *in vivo* conditions. The precise localization of such components in pollen are still not known; although many pigments are known to occur in wall and membranes, they also occur in the cytoplasm where free boron may or may not be available. KUHN (1943) suggested that isorhamnetin, released by hydrolysis of a flavonol glucoside from *Crocus* pollen, combined with boron to stimulate tomato pollen growth. Experimental evidence to support this hypothesis is lacking.

Shifts in flavonoid synthesis in relation to boron were reported in sunflower plants (WATANABE et al., 1964). Flavonoids in corn plants decrease long before boron deficiency symptoms are apparent (KRUPNIKOVA, 1970). Boron, either directly or indirectly, affects sugar synthesis or translocation in plants, and moieties such as nucleic acids and flavonoids in which sugars are involved. The available evidence precludes reaching a definite decision of the possible role of flavonoids and boron in cell metabolism. Certainly, pollen affords a most amenable tool for investigating questions relating to the role of these cell constituents in plant metabolism.

Changes in flavonoids could be induced if boron were acting at a different point in the control or synthesis of flavonoids and not just chelating with the free molecule. For example, preliminary *in vitro* studies show some enzyme activities are influenced by fluorescence emitted from flavonoids. Thus, the well recognized influence of flavonoids on oxidation of ascorbic acid, by inhibiting oxidase activity, and the affect on other enzymes such as peroxidase or tyrosinase (STENLID and SAMORODOVA-BIANKI, 1969), may be an indirect control mechanism where flavonoids modify metabolism and growth. MALYUTIN (1969) suggested that flavonoids, by absorbing in different regions of the solar spectrum, raise the temperature and selectively catalyze enzyme activity in local areas of stamen and pistils, promoting pollen growth. Such a theory may find support if either the fluorescent emissions, or the thermal effect on enzyme turnover, influence specific reactions.

Screening against U.V. radiation, previously discussed as a possible role for carotenoids in pollen, merits consideration for the evolutionary role or selective potential which such protection may provide for pollen of one species over another. Pollen seems to lack many of the metabolic DNA repair mechanisms found in diploid cells, and the DNA in pollen is a readily accessible target for modifying genetic information (Chapter 12). Supplying these cells with a screen for harmful U.V. would be an ideal role for flavonoids and other pigments. WERFFT (1951, 1951a) showed that pine pollen is readily harmed by 3 to 4 hours exposure to U.V. and sunlight, and proposed a role for these pigments in protecting pollen exposed to atmospheric conditions for limited, brief periods. JOHNSON and CRITCHFIELD (1974) compared the germination of a white mutant from *Pinus aristata* to a normal yellow after both pollens were exposed to U.V. for 20 minutes to 1 hour. They found no difference in germinating ability or response of the pollen. Of course, white pine pollen could contain just as high a level of flavonoids as the yellow non-mutant pollen. Looking at flavonoid distribution inside pollen organelles, KRUPKO (1959) detected a high concentration of flavonoids in the generative cell. This is certainly a convenient site in which to localize flavonoids to screen DNA from harmful radiation.

CAPPELLETTI and TAPPI (1948) suggested that flavonoids in the ether extracts of *Lilium* stigmas are the factors which stimulate pollen germination, by acting as a growth hormone. The role of certain pollen pigments as enzyme cofactors, e.g. riboflavins, is well recognized (Chapter 14). This type of enzyme-substrate related growth stimulation must be differentiated from growth hormones and inhibitors which also exist in pollen. GALSTON (1969) and others believe kaempferol derivatives in plant shoots promote indoleacetic acid oxidase (IAA oxidase) activity and quercetin compounds inhibit growth hormone destruction. Based on flavonoid changes in intact shoots after phytochrome activation, a hypothesis has evolved that these flavonoids control growth by affecting functional levels of IAA. Since endogenous IAA is present in pollen (Chapter 16) and generally influences tube growth it may be valuable to test this hypothesis in Umbelliferae pollen tubes, and others shown by MINAEVA and GORBALEVA (1967) to respond to flavonoids added *in vitro*.

Complicating the discernment of a functional role for pollen pigments is the fact that many other chromatic compounds occur in pollen besides flavonoids and carotenoids. A few such as the flavoproteins, pyridine nucleotides and cytochromes are present in pollen, as in other plant tissues, i.e. tightly bound as a prosthetic or cofactor groups in a protein complex coupled to enzyme oxidation-reduction reactions (Chapter 14). But most of these components, such as the polyphenolic methoxy and hydroxy derivatives of benzoic and cinnamic acids, have been only slightly studied in pollen, generally via chemical extraction (MATHUR, 1969) or as tests on germination *in vitro*. Obviously, more information is needed about the color-imparting components of pollen before we can hope correctly to recognize their functions in development and growth.

Chapter 16. Growth Regulators

Six classes of plant growth regulators—auxins, gibberellins, kinins, brassins, ethylene and inhibitors—occur in pollen or can potentially influence their growth. Representatives of all six groups have not been reported in any one pollen, but many are probably present and active in each pollen, or tissues in which pollen develops or grows. Biosynthetic pathways leading to their formation are fairly well known. Yet little is known of the mechanism by which they function to control tube growth, embryo development and seed formation. This review will be primarily limited to naturally occurring growth regulators found associated with pollen.

Many of the conceptual foundations underlying the search for plant hormones can be attributed to Charles Darwin's (DARWIN and DARWIN, 1880) studies of directional growth in grasses. However, pollen provided one of the first sources of material for the isolation and characterization of a phytohormone. The replacement of natural pollen-produced hormones by synthetic chemicals to stimulate fruit production without pollination and fertilization, i.e. parthenocarpic fruit set, affords indirect confirmation of the role of growth regulators normally associated with pollen. Vitamins will not be considered in this category. They are present in fairly high levels in most pollen (Table 14-5) and often stimulate growth similar to the activity of hormones. However, pollen vitamins are generally recognized as enzyme cofactors (Chapter 14).

Historical Background

When C. DARWIN (1889) revised his book on insect pollination of orchids he referred to the report by the two Müller brothers, Fritz working in Brazil and Herman in Germany, that the orchid perianth wilts and dies within 24 hrs after pollination. DARWIN (1904) continued to reflect the concept originally proposed by F. Müller, that pollen from the same flower is impotent and in some cases "poisonous" to the stigma (ARDITTI, 1971). MASSART (1902) placed incompatbile and crushed pollen on stigmas of members of the Cucurbitaceae family, and cut open the ovary after the normal time for syngamy. He showed that the ovary was stimulated to grow although fertilization did not occur, and the fruit did not fully mature. He attributed the growth stimulation to a chemical produced by the pollen. H. FITTING (1909; 1910), after investigating the pollination reaction in orchids, concluded that the "poison" compounds of orchid pollinia were hormones. In using this term for the source of the growth reaction, FITTING was the first to apply the word hormone to plants. MORITA (1918) showed that hot- or cold-water-, alcohol-, or ether-extracts of pollen from the *Cymbidium* orchid all caused a closing of the stigma, initiation of withering, and a slight enlargement of the column. The column (gynostemium) in orchid flowers is an organ of fused stamens and pistils which normally enlarges after pollination.

Investigations in the 1920's and 1930's suggested that the "Pollenhormon" from orchid pollinia which induces column growth is auxin. LAIBACH (1932; 1933) applied orchid pollinia extracts to *Avena* (oat) coleoptiles following concepts and techniques developed by WENT (1928). LAIBACH showed that the hot-water extract of orchid pollinia and pollen of *Hibiscus* both contained a substance which, when applied to the stigma, caused extension of the orchid column. The extracts also stimulated growth of the oat coleoptile. He concluded that the hormone of orchid and other pollen was probably identical to the growth substance from the oat tip which stimulated coleoptile growth. LAIBACH (1933a), using pollinia extracts, was the first to apply a lanolin paste as the growth substance carrier, in place of the agar block used with all extracts prior to his studies. LAIBACH and MASCHMANN (1933) compared all reactions then recognized in *Avena* coleoptiles to those produced by extracts of orchid pollinia, pollen and growing pollen tubes of 9 different angiosperm species. All pollen tested yielded activity that stimulated a growth reaction in decapitated oat coleoptiles.

Other studies verified the parallel nature of the reactions produced by the growth substances from orchid pollinia and auxin. MAI (1934) applied orchid pollinia in distilled water directly to cut leaf petioles and observed induced swelling and inhibition of abscission of attached leaves. This observation was confirmed by LA RUE (1936) in debladed *Coleus* petioles. SEREISKII (1936) showed that orchid pollinia inhibited root growth of *Zea mays* and stimulated bean cotyledons in a manner similar to treatment with auxins. Typical examples, Table 16–1, of these growth results are those reported by MITCHELL and WHITEHEAD (1941) showing first internode elongation of young kidney beans following application of a lanolin paste containing the ether extract of corn pollen. Orchid pollinia extracts were used by DOLLFUS (1936) to induce roots on cut stem sections, and to stimulate rhizoid formation in *Marchantia* (MOEWUS and SCHADER, 1952).

Although many workers continued these studies correlating growth and development changes in the orchid column following pollination with auxinlike response (HSIANG, 1951; LEON, 1963), it was not until 1953 that MÜLLER (1953) showed that the growth-regulating substance extracted from 900 mg of orchid pollinia, which effected column expansion, was identical to β-indole-3-acetic acid (IAA). She made these comparative chemical separations about 20 yrs after KÖGL et al. (1934) isolated the first auxins. Indole-acetic acid, the principal natu-

Table 16-1. Comparison of the first internode, mm elongation of kidney bean seedlings (MITCHELL and WHITEHEAD, 1941)

Time after treatment (hours)	Treatment			
	None (control)	Corn pollen extract	IAA	NAA
52	12.9	26.2	15.6	13.2
100	25.9	39.8	20.6	17.6

rally-occurring auxin in plants (AUDUS, 1953), is widely reported in pollen. Early work with pollen extracts from plants other than orchids, e.g. *Sequoia* (THIMANN, 1934); *Petunia* (YASUDA, 1934), *Pinus* and *Zea mays* (GUSTAFSON, 1937), provided further foundations for the concept that pollen hormones such as IAA (HAGE-MANN, 1937) stimulate ovary growth after pollination. LUND (1956) further extended the earlier chemical isolation work by chromatographically demonstrating that IAA was the hormone secreted or activated by germinating *Nicotiana* pollen, and is also released by the pistil tissue. Different species and varieties of *Nicotiana* produce maximum pollen tube growth *in vitro* at different concentrations of IAA (CHERVONENKO, 1959).

The early observations relative to IAA were intimately associated with pollen. Other growth regulators were discovered much more recently than the auxins. But except for the brassins, the historical backgrounds of these other regulators did not involve pollen. For this reason, their historical backgrounds are omitted from this brief review.

Auxins

Angiosperms

Since the earliest work with orchid pollinia, many other angiosperm pollen extracts have been assayed for auxin activity. The assay results have not always been positive. LAIBACH and MASCHMANN (1933), while finding auxin in orchids, failed to extract auxin from pollen of *Abutilon, Acacia, Anona, Lilium, Hippeastrum* or *Strelitzia*; auxin was only recovered in the alkali extracts of *Antirrhinum* and *Nicotiana*. This suggests that auxin in some pollen may exist in a bound form. In the latter case auxin could be activated and released by specific enzymes or metabolites; or in some pollen may be synthesized either in the tube or the surrounding stylar tissues after germination begins. It is also probable that levels of other growth substances or inhibitors may influence the assay for auxin by growth tests. Reduced auxin content could result from improper tissue extract preparation, or the presence of inactivating enzymes or inhibitors.

Using paper chromatographic methods to separate the ethyl ether (neutral) and water (acid) soluble extracts from *Corylus avellana* pollen, MICHALSKI (1958) assayed the different R_f segments by the oat coleoptile curvature test. Both auxin-like stimulators and inhibitory substances are present in the growth regulator fractions of *Corylus* pollen extracts (Fig. 16–1). The R_f's of the inhibiting substances depend upon whether the neutral or acid growth regulator fractions are partitioned from the total ether extract. Neutral fraction inhibitors (Fig. 16–1a) appear masked when combined with the initial ether extract. The chloroform extract of this same pollen species, when originally studied by YAKUSCHKINA (1947), gave a similar growth stimulation-inhibitor response histogram pattern as that in Fig. 16–1b. This data suggests that response in a biological assay test such as oat coleoptile curvature or pea stem elongation represents a combination of extractable growth stimulating and inhibiting factors.

Growth regulators detected vary quantitatively and qualitatively, depending on the stage of maturity or germination. In addition, as KOTS (1971) discovered in

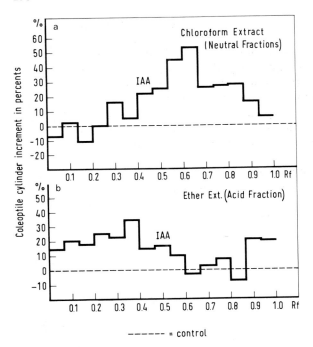

Fig. 16-1 a and b. Histogram of growth substances in the neutral fraction (a) and acid fraction (b) from the ether extract of *Corylus avellana* pollen. (After MICHALSKI, 1958)

comparing 10 species of *Populus*, the amount detected in pollen varies with age, storage, and species, even within the same genera. Auxin decreases with increasing time in storage. MICHALSKI and CHROMINSKI (1960) found that in initial growth stages, *Corylus* pollen contained many inhibitors and few auxin-like components. As growth progressed, the two main inhibitors decreased, while activity of stimulators and a fraction co-chromatographing with IAA increased.

Gymnosperms

Growth substances have not been as thoroughly investigated in the gymnosperm as in the angiosperm pollen. However, some progress has been made in studies with pine. YAKUSCHKINA (1947) compared the ether extracts of *Pinus sylvestris* with those from *Corylus* pollen and concluded that pine contained IAA and other growth-promoting substances. The ether extract of *Pinus montana* stimulated both *Ceratozamia* and *Pinus* pollen tube growth to about the same extent as optimum concentrations of IAA (ANHAEUSSER, 1953). However, TANAKA (1958) observed only inhibition of *Tradescantia* pollen germination by ether extracts from *Pinus densiflora*. Not only was germination reduced but considerable tube bursting resulted. Ether- as well as water extracts of *Abies* pollen stimulated *Abies* pollen tube growth (KÜHLWEIN, 1948). This latter study, primarily a test of the flavonoid hypothesis (Chapter 15), unfortunately did not differentiate between the stimulating factor as a growth regulator or a substrate metabolite.

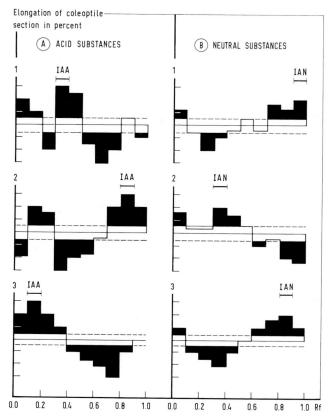

Fig. 16-2 A and B. Histogram of growth substances from *Pinus sylvestris* pollen in the acid fractions (A) and neutral fractions (B) (MICHALSKI, 1967). Solvent systems: 1 = isopropynol-ammonia-water (10:1:1); 2 = H_2O; 3 = n-butanol-ammonia-water (100:3:18)

Studies of auxin and other regulators require, as a minimum, separation and co-chromatographing of the extract, followed by a biological assay of the separated fractions. MICHALSKI (1967) prepared biohistograms of auxin-like components in extracts from 25 g samples of *Pinus sylvestris* pollen separated on paper chromatograms. Elongation of *Avena* coleoptile sections was used as an index of activity (Fig. 16–2). The place where IAA and indoleacetonitrile (IAN) ran on the chromatograms is indicated for each solvent system. *Avena* elongation responses to different concentrations of IAA, chromatographed in similar solvents, are presented on control histograms represented in the center of Fig. 16-2. However, the color reactions, of Salkowski's and Prochazka's reagent sprays for IAA and IAN, on the chromatograms were not the same as those typical of pure substances. This suggests that the extracted growth regulators are not in the same free state as the chemically pure IAA and IAN, or that these growth regulators are mixed with other components which chromatograph at the same R_f's and partially mask or modify the color reaction of known auxins.

Brassins

A new family group of plant growth regulators, called brassins, have been isolated from the lipid ether extracts of rape pollen, *Brassica napus*, (MITCHELL et al., 1970; MANDAVA et al., 1973). Separation of extracts from *Alnus glutinosa* and 15 other pollen provides further evidence for the possible general distribution of these compounds. These compounds, while still not fully chracterized, were suggested as unsaturated glycerides or glucose esters of fatty acids (MANDAVA and MITCHELL, 1972). Initially, 5 different chemicals were separated with growth-stimulating activity. The quantity isolated, about .002% by dry weight, was at a much higher percent level than other growth substances extracted from pollen. This may reflect the impure nature of the initial active extracts of brassin, which possibly include an inactive lipid carrier. MILBORROW and PRYCE (1973) indicate several possible sources of impurities and experimental error underlying the conclusions suggesting that brassins are a new class of plant growth hormone.

The initially reported brassin fractions from rape differ in certain physiological activities from fatty hormone extracted from bean pollen, *Phaseolus vulgaris* (MITCHEL et al., 1971). Brassins cause both elongation and cell division (WORLEY and MITCHELL, 1971) of tested bean stem internodes, while bean pollen extracts, like gibberellins, only induce elongation. Light and magnetic resonant spectral patterns showed that the fatty hormones extracted were free of indole or gibberellin-like ring components in their long chain; components of the active fractions had molecular weights of 250 to 580. In summary, there appear to be several related, lipid-like growth-stimulating components in angiosperm pollen. Assuming this is correct, then the determination of their structure and function is an interesting challenge for the future.

Gibberellins

Biossays, represented by the histogram in Figs. 16–1 and 2, show that several other growth stimulators are present. While non-IAA growth-stimulating regions of the chromatograms from pollen extracts have not been evaluated for brassin activity, they have been found to provide good evidence of gibberellin (GA) activity. Although MICHALSKI (1967) did not have standards of pure gibberellic acid (GA_3) available to him, by following chromatographic methods reported in the literature, coupled with a lettuce hypocotyl biossay, he concluded that the extracts of *Pinus sylvestris* pollen were high in GA_1 and GA_3. IVONIS (1969) found 5 different GA fractions in *P. sylvestris* cones and pollen, two of the 5 fractions present only in pollen. A new GA occurred in the pollinated cones when they began to droop in response to curved stem growth of the cone peduncle. IVONIS suggested that the new GA indicated a metabolic shift after pollination. While the levels of GA change and new GA is synthesized at different times in a tissue's life cycle, it is hard to attribute discontinuous pollen tube growth or cone elongation to the new GA factors alone. Cause and effect are probably not that simple, but rather represent balances between IAA and other growth substances and substrates in the female tissues as well as the pollen.

Levels of GA vary at different stages of pollen development and in the microsporophyll or anther tissues in which they develop. Pine microsporangia reached a level equivalent to 1.65 µg GA_3/100 gms dry weight just at pollen maturity but decreased to an equivalent of 0.77 µg GA_3/100 gms at anthesis (KOPCEWICZ, 1969). In three species of germinating pine pollen, KAMIENSKA and PHARIS (1971) found that the GA_3 increased from a mean of 0.013 µg GA_3/g pollen to 0.250 µg GA_3/g at 15 hrs, a time corresponding to maximum tube emergence. Pollen GA_3 decreased from 15 to 72 hrs growth when they concluded the study. The GA_3 level in the growth media remained constant, implying that the loss of pollen GA_3 was probably not due to simple outward diffusion. The three species differed in GA levels, with *Pinus attenuata* significantly higher than *P. coulteri*, and *P. ponderosa* intermediate. All three species yielded 5 GA peaks; 2 occurring at lower concentrations than the other 3.

Extractions by COOMBE (1960) of GA from anthers of seedless and seeded grape *(Vitis vinifera)* varieties at the time of anthesis showed that anthers of seedless varieties were as much as 10 times richer in GA than seeded grape varieties. A more detailed study of changes in the soluble GA was reported with *Lilium henryi* pollen where BARENDSE et al. (1970) showed that as pollen matured the equivalent GA_3 level per gm of fresh weight decreased. The GA_3 level continued to decrease during germination and a certain amount of activity leached out of the pollen. Following germination *in vitro*, GA_3 seems to break down (Table 16–2), suggesting that a lack of GA_3 may be among the factors inducing or accompanying a cessation of tube elongation *in vitro*. Also, it is possible that this apparent decrease represents an increase in a GA inhibitor, antagonist or complexing compound in the maturing pollen. The imporant observation is that pollen may not have its highest level of soluble GA at anthesis.

Addition of GA at low concentrations to some species of germinating pollen stimulates tube growth and increases percent germination *in vitro* (CHANDLER, 1958). *Digitaria* grass pollen with added GA_3 showed over a 100% increase in percent of grains germinating (CARMICHAEL, 1970). This suggests that maintenance of a critical level of GA is particularly essential for pollen germination; also, that GA may be required at higher levels in some species than others. Or, as BARENDSE et al. (1970) suggested, the great loss of GA from pollen that occurs in the first hour of germination may influence the capacity of the pollen to continue to germinate at a high rate. Possibly GA may influence the metabolism of the female tissue where pollen normally germinates. In some species adding GA may overcome this natural decline in GA and thus influence germination.

Table 16-2. Gibberellins from *Lilium henryi* pollen (BARENDSE et al., 1970)

Pollen condition	µg GA_3 eq./g fresh weight		% germination
	in pollen	in medium	
Immature (buds 1/2 max length)	1.79		
Immature (buds 3/4 max length)	2.19		
Mature-anthesis (buds 65 mm)	1.08		
Hrs germinated 1	0.22	0.02	48
2.5	0.24	0.03	65
3.5	0.17	0.07	70
6.0	0.05	0.07	—

Kinins

Very little is known of the cytokinins in pollen. The 6-furfuryl-adenine, kinetin has not been identified in pollen extracts. SWEET and LEWIS (1971) detected 4 different GAs in *Pinus radiata* and one compound with properties of both GA and cytokinin. However, using bioassays they were unable to detect a fraction with purely cytokinin-like activity in the pine pollen extracts. Since many nucleotides and their derivatives are found in pollen (Chapter 13) it will be of interest to learn if the adenine related zeatin riboside or isopentenyl adenosine, natural cell consti-tuents which manifest cytokinin activity, also occur in pollen.

Ethylene

Until the development of gas chromatographic analyzers the growth stimulant ethylene, C_2H_4, was difficult to detect *in vivo* due to the low physiological levels at which it is generally active. Indirect influences of C_2H_4 on fruit development were recognized before the gas was associated with flowers and pollination. In 1967 HALL and FORSYTH reported that *Vaccinium* and *Fragaria* flowers produced ethy-lene after pollination. Earlier observations by DENNEY and MILLER (1935) sug-gested that wounded flower tissues produce ethylene, and the initial amounts of gas produced in fading orchid flowers accelerates further C_2H_4 production (BURG and DIJKMAN, 1967). BUCHANAN and BIGGS (1969) showed that ethylene-producing compounds can stimulate pollen tube growth after pollination.

The first direct evidence that pollen produces ethylene was reported by COOPER (1972) in studies with Valencia oranges. The pollen produced about 18 nl of C_2H_4 per hr per gm fresh weight; which was about 50% of the volume pro-duced per gm of stigma tissue. HALL and FORSYTH (1967) also reported that *Vaccinium* stigmata and styles gave off over 100 times the volume of C_2H_4 than the anthers plus corollas.

Growth stimulation occurred during the first 30 minutes when *Pyrus commu-nis* pollen was exposed *in vitro* to relatively high concentrations of C_2H_4; tube growth was accelerated 22% and germination rate 46% (SEARCH and STANLEY, 1970). Peach pollen growth was also stimulated *in vitro* by ethrel, a source of C_2H_4 (SAULS and BIGGS, 1970). This C_2H_4 stimulating effect on pollen *in vitro* could not be detected after several hours' growth (SEARCH and STANLEY, 1970; SFAKIOTAKIS et al., 1972). Conceivably the stigma C_2H_4 level may influence initial pollen germination. However, it is still too early to make such a generalization. Assays of the endogenous C_2H_4 produced in different pollens and flowers might detect a critical threshold time for an initial surge in C_2H_4 level. This interval, if found, could provide a tool for helping to discern the metabolic function of this compound.

Inhibitors

At least two inhibitors can be detected by chromatography depending upon which solvent system is used (Fig. 16-2). Inhibitors from pine pollen have been extensively studied by TANAKA (1958, 1960, 1964, 1964a). Various solvent extrac-tions from germinated and ungerminated *P. densiflora* pollen were separated on paper chromatograms and the effect was tested on growth of the angiosperm pollens *Tradescantia*, *Impatiens* and *Lilium*. These studies showed that an active, soluble inhibitor in pine pollen decreased with germination (TANAKA, 1964a); moreover, the molecule was small enough to diffuse through a cellulose membrane and agar. The inhibitor extract reduced growth of pine pollen. Inhibition was not due merely to a lowering of the pH or shifting of the osmotic value in the germination media (TANAKA, 1958). Evidence indicated its nature was phenolic, but probably not flavonol-like. TANAKA (1964) reported that the pine pollen derived inhibitor also inhibited *Brassica* seed germination, just as a general cell metabolic inhibitor would be expected to act. Several different inhibitors were separated in the acid and neutral fractions from *P. densiflora* with R_f's corresponding to those observed in the *P. sylvestris* extracts shown in the histograms in Fig. 16-2. However, differences in growth response to pollen inhibitors occur since angiosperm pollen are not inhibited as strongly by both acidic inhibitor fractions as is pine (TANAKA, 1960). In all growth tests with *Avena* coleoptile and pollen the component(s) chromatographing at R_f 0.6–0.7 is the most inhibitory of all fractions separated.

Bioassay of chromatographed extracts from *Pinus radiata* pollen (SWEET and LEWIS, 1971) showed that of 13 active substances on the chromatogram, 5 gave auxin-like responses in *Avena* coleoptile tests, while 3 were inhibitors. One of the 3 inhibitors in the acid-ether fraction co-chromatographed with only a slightly lower R_f than pure abscisic acid (ABA). This inhibitor decreased significantly during germination. The other two inhibitors in the neutral-ether fraction either remained at the same level or increased during germination. Although no one has yet characterized the pollen inhibitors, the fluorescence pattern and R_f region of the chromatogram with the strongest inhibitors is the R_f region where ABA normally moves on a chromatogram. TANAKA (1958, 1964) suggested that the slowly decreasing pine pollen inhibitor was what slowed initial germination, e.g. the inhibitor had to disappear before tube growth could occur; pine required about 15 hrs in contrast to the rapid growth observed in one to 60 minutes in many pollenns. The repressor, growth-inhibiting action of ABA is overcome by other growth substances, in particular gibberellic acid. Total inflorescences of *Olea europaea* were high in ABA (BADR et al., 1971); while this study did not separate anther and pollen tissues, certainly microspores develop in tissues high in ABA, and with changing levels of GA. It will be interesting to learn if pollen is endowed with such a growth control, repressor-derepressor mechanism. In no case ABA has a major role in the regulation of pollen germination and tube extension (SONDHEIMER and LINSKENS, 1974).

Sources of Variation

Variations in quantity as well as quality of growth substances extracted from different pollen, and changes during microspore development and germination, are species specific. However, the levels of IAA and related metabolic products also depend upon the intermediary pathways leading to hormone synthesis. It is thus rather interesting to note that in *Brassica* a different metabolic pathway produces IAA than the paths followed in most plants. Generally, IAA is derived from tryptophan via typtamine or indole pyruvic acid intermediates. However, in the Brassicaceae IAA utilizes indole acetonitrile (IAN) as a precursor intermediate. The acetonitrile is converted to IAA by nitrilase (THIMANN and MAHADEVAN, 1958); but, tryptophan is still a precursor for the IAN (MAHADEVAN, 1963; KINDL, 1968), or IAN can arise from glucobrassicin (GMELIN and VIRTANEN, 1961). FODA (1966), using a barely coleoptile growth assay on chromatographed extracts of *Gossypium, Sorghum* and *Phoenix dactylifera,* found both IAA and IAN in all three pollen. He suggested that the two auxins function at different times in the cycle of pollen growth.

It is only speculation that differences in growth regulators in pollen extracts are related to differences in the precursors and metabolic pathways. But a comparative analysis of levels of substrate intermediates and enzymes in pollen may help reveal the nature and source of the observed differences. In the case of pollen yielding brassins and other lipid-related growth substances, it will be of particular interest to learn if these substances are confined to pollen with primarily high lipid storage products compared to those high in starch (Chapter 9). In these latter pollen, gibberellins might be a more prevalent growth regulator. Presumably, growth regulator levels will be influenced by metabolic patterns in the developing microspore and pollen, and the internal and external substrates all of which influence growth reactions during germination. Thus, the inherent pools of metabolites may be correlated not just with different substrates for incorporating into cell wall and protoplasm during growth, but with activation of specific pollen growth regulators.

The wide areas on chromatograms which yield positive stimulation indicate that growth regulators and stimulators other than those already discussed may occur in pollen (Fig. 16-1). The strongest growth stimulators were found in the neutral fraction. MITCHELL and WHITEHEAD's (1941) observation (Table 16-1) that corn pollen extracts stimulated internodal growth greater than any IAA concentration tested, suggests the presence of growth substances other than the auxin type, or of a synergystic interaction of compounds with IAA. Isolation and identification of these other growth factors has, in general, not been actively pursued. One simple compound reported to have a high level of growth stimulating activity in ether extracts of *Zea mays* pollen is a hydroxy pyridin (REDEMANN, 1949). No further work has been done on this compound and it has not been reported from extracts of other pollen.

Active, non-nitrogen containing compounds have also been reported in pollen extracts. Methyl esters of fatty acids with growth stimulating activity can be separated from *Zea mays* pollen (FUKUI et al., 1958; FATHIPOUR et al., 1967). These growth stimulators in corn clearly occur in addition to IAA and related indole compounds (FUKUI et al., 1958; TANAKA, 1958). Steroids, which function as hormones in man and fungi (BARKSDALE and MCMORRIS, 1967; TAKAO et al., 1970) also occur in pollen (Table 10–5), often in association with lipids (HOEBERICHTS and LINSKENS, 1968).

SCHOELLER and GOEBEL (1932) showed that follicle hormones may behave as plant growth substances. They suggested that a steroid hormone which they isolated from palm seed might serve as a plant oestrogen and affect flower and fruit formation (BUTENANDT, 1933). Such speculation and analogy to animal systems is still unsupported by experimental evidence. Androgen activity was absent from corn pollen (BUTZ and FRAPS, 1945). Thus, except for studies on the effect of sterol compounds on pollen germination (MATSUBARA, 1969, 1971; STANLEY, 1971), insufficient knowledge exists concerning the possible growth-regulating role of these compounds to consider them in this discussion of recognized plant growth regulators. Pollen provides an ideal tool for further work on the physiological role of the ubiquitous plant steroids. Hopefully, the role of steroids in plant growth and pollen (Chapter 10) will soon be established.

Roles

This discussion will not attempt to analyze the biochemical basis of growth regulator activity, as the reactions for any one compound have not yet been completely defined (CARR, 1972). Rather we will review a few of the physiological reactions associated with pollen growth *in vivo*, and possible approaches to the problems associated with growth regulators in pollen.

Evidence suggests that pollen tube growth and fertilization are controlled or regulated by growth substances (GIERTYCH, 1964). Merely adding IAA or other growth substances to pollen germinating *in vitro* can be misleading unless all other growth factors are optimal and levels of endogenous regulators are known. Loo and HWANG (1944) reported that, when compared to boron and manganese, auxin had only slight effect on pollen growth. As previously discussed, different pollen species have different growth requirements (Table 6-1) and contain different amounts of growth regulators. Pollen and stylar metabolites of a non-regulatory nature can inhibit or stimulate pollen growth. Phenolics can inhibit tube extension by binding to membrane protein; SCHWARZENBACH (1952) suggested from studies of *in vitro* germination of *Cyclamen* pollen that β-carotene can repress growth by combining with auxin, an observation not yet verified *in vivo*. However, IAA does combine with inositol (UEDA et al., 1970) which is a very common chemical (Chapter 9) associated with tube biosynthesis (STANLEY and LOEWUS, 1963).

When pollen lands on a compatible tissue, activation of growth substances is probably among the reactions preceding tube extension. Both bound and free IAA occur in most pollen and probably are released as the tube grows into the style. Ovary and style are generally low in auxin until after the tube begins to penetrate the style (MUIR, 1947). LUND (1956) reported that IAA is highest in style tissue just in front of the extending tube; also that enzymes released from the tube, primarily at the tip, can convert tryptophan to IAA. These and other observations permit us to ascribe a few specific roles to pollen growth regulators.

1. Inhibitors can repress growth until the grain is in the correct environment; then the inhibitors are inactivated or leached from the pollen.

2. Auxins may be involved in directional growth of the pollen to the egg (MASCARENHAS and MACHLIS, 1962).

3. Gibberellins can activate amylase and thus, as the pollen moves forward, starch can be hydrolyzed to provide sugars to sustain growth (COOMBE, 1960).

Starch granules form and disappear from stylar cells as pollen tubes grow down the styles (SESHAGIRIAH, 1941). Other polysaccharides are available in cells where *Lilium* pollen initially grows (YAMADA, 1965). Enzymes directly from the pollen, or activated by regulators such as GA, may solubilize such substrates.

4. Growth substances diffusing from the pollen may stimulate maturation or receptivity of the egg cell preceding fusion with the male cell.

5. Growth regulators can control tube extension in many ways, one of which is facilitating wall growth (POHL, 1951; LAMPORT, 1970).

Dead pollen stimulates certain growth phenomena, corresponding to exogenous applications of IAA. However, in many flowers the growth stimulus ceases after a few days and parthenocarpic fruit is not developed (GUSTAFSON, 1961; DOLCHER, 1967). In fact, IAA, GA, cytokinins, and ethylene have all been reported in one species or another to induce or increase parthenocarpic fruit development (CRANE, 1969). Different periods in fruit development are stimulated by different growth substances. This suggests that the changing endogenous levels of the different compounds following pollination is essential, since different inhibitors and hormones selectively affect some, but not all, post-pollination phenomena (ARDITTI and KNAUFT, 1969; ARDITTI et al., 1971). It is important to differentiate growth regulators supplied by pollen from those produced by the female tissue.

Pollen growth regulators can be grouped into activities involved in a) tube growth and delivery of the male cell and b) those involved in attraction or contact of the male cell to the egg cell. These activities mediate growth processes within the pollen and external to the tube. It may soon be possible to separate the two different types of chemically regulated activities more accurately. Techniques for the isolations of the two separate pollen nuclei in aqueous and non-aqueous solutions (SHERIDAN, 1972; LAFOUNTAIN and MASCARENHAS, 1972) and the cell walls, cytoplasm and vesicles (VANDERWOUDE et al., 1971; ENGELS, 1973) may provide improved methods to separate the sites of activity and characterize these growth regulators.

Literature

The index also serves as an author index. Italic numbers between brackets refer to pages on which the paper is cited.

ABRAMSON, H. A. (1947): Ann. Allergy **5**, 19 (*132*).

ABRAMSON, H. A., MOORE, D. H., GETTNER, H. H. (1942): J. Phys. Chem. **46**, 1129 (*168*).

ADAM, D. P., FERGUSON, C. W., LAMARCH, JR., V. C. (1967): Science **157**, 1067 (*45*).

ADAMS, J. D., MACKAY, E. (1953): Stain Technol. **28**, 295 (*77*).

AFZELIUS, B. M. (1955): Botan. Notiser **108**, 128 (*18*).

AIZENSTAT, J. S. (1954): Izv. Akad. Nauk USSR **4**, 42 (*64*).

AKIRA, Y. (1965): J. Biol. Chem. **240**, 1118 (*220*).

ALDRICH, W. W., CRAWFORD, C. L. (1941): Rep. Date Growers' Inst. Coachella Valley Calif. Rpt. **18**, 5 (*58*).

ALFONSUS, E. C. (1933): Arch. Bienenk. **14**, 220 (*103*).

ALLAN, G. S., SZIKLAI, O. (1962): Forest Sci. **8**, 64 (*42*).

ALLEN, M. D., JEFFREE, E. P. (1956): Ann. Appl. Biol. **44**, 649 (*103*).

AMELUNG, B. (1893): Flora **77**, 176 (*27*).

AMIN, E. S., PALEOLOGOU, A. M. (1973): Phytochemistry **12**, 899 (*151*).

ANDERSEN, S. T. (1946): Danmark Geol. Undersoegelse IV **4(1)**, 1 (*26, 27*).

ANDERSEN, S. T. (1970): Danmark Geol. Undersoegelse II **96**, 7 (*33, 34*).

ANDERSON, L. D., ATKINS, JR., E. L. (1968): Ann. Rev. Entomol. **13**, 213 (*93*).

ANDERSON, R. J. (1923): J. Biol. Chem. **55**, 611 (*150*).

ANDERSON, R. J., KULP, W. L. (1922): J. Biol. Chem. **50**, 433 (*119, 120, 125, 132, 154*).

ANDERSSON, E. (1947): Hereditas **33**, 301 (*9*).

ANDRONESCU, D. I. (1915): The physiology of the pollen of *Zea mays* with special regard to vitality. Ph. D. Thesis. Urbana: Univ. Illinois (*57, 120*).

ANELLI, G., LOTTI, G. (1971): Agrochimica **15**, 539 (*120, 125*).

ANHAEUSSER, H. (1953): Beitr. Biol. Pflanz. **29**, 297 (*250*).

ANIKIEV, V. V., GOROSCENKO, E. N. (1950): Dokl. Akad. Nauk. SSSR **74**, 272 (*29*).

ANTHONY, S., HARLAN, H. V. (1920): J. Agric. Res. **18**, 525 (*57, 59*).

ANTLES, L. C. (1920): Ann. Rep. Vermont State Hort. Soc. **55**, 18 (*59*).

ANTLES, L. C. (1965): A man's footprints, a guide to Orchard pollination. New York: Carlton Press (*42*).

ARABADZHI, V. I. (1973): Fiziol. Rast. **20**, 206 (*36*).

ARDITTI, J. (1971): Am. J. Botany **58**, 480 (*247*).

ARDITTI, J., FLICK, B., JEFFREY, D. (1971): New Phytol. **70**, 333 (*258*).

ARDITTI, J., KNAUFT, R. (1969): Am. J. Botany **56**, 620 (*258*).

ARIENZO, R., BRESCIAN, E., NERI, M. (1969): B. Ital. Biol. **45**, 1335 (*148*).

ARNETT, R. H., JR. (1963): Proc. 16th Internatl. Cong. Zool. **2**, 8 (*110*).

ASBECK, F. (1954): Strahlentherapie **93**, 602 (*49, 228*).

ASK-UPMARK, E. (1963): Z. Urol. **56**, 3 (*114*).

ASK-UPMARK, E. (1967): Acta Med. Scand. **181**, 355 (*114*).

ASLAM, M., BROWN, M. S., KOHEL, R. J. (1964): Crop Sci. **4**, 508 (*84*).

ASPINALL, G., KESSLER, G. (1957): Chem. Ind. **39**, 1296 (*15*).

ATALIAH, A. M., NICHOLAS, H. J. (1971): Steroids **17**, 611 (*153*).

AUDUS, L. J. (1953): Plant Growth Substances, 1st Ed. New York: Interscience (*249*).

AUGUSTIN, R. (1955): Quart. Rev. Allergy **9**, 504 (*175*).

AUGUSTIN, R. (1959): Immunology **2**, 1 (*169*).

AUGUSTIN, R. (1967): In: WIER, D. M. (Ed.): Handbook of experimental immunology, p. 1076. Philadelphia: F. A. Davis (*174*).

AUGUSTIN, R., NIXON, D. A. (1957): Nature **179**, 530 (*132*).

AXELOS, M., PEAUD-LENOEL, C. (1969): Chem. Zvesti **23**, 770 (*199*).

BADENHUIZEN, N. P. (1969): The biogenesis of starch granules in higher plants. New York: Appleton-Century-Crofts (*220, 221*).

BADR, S. A., MARTIN, G. C., HARTMANN, H. T. (1971): Physiol. Plantarum **24**, 191 (*255*).

BAER, H., GODFREY, H., MALONEY, C. J., NORMAN, P. S., LICHTENSTEIN, L. M. (1970): J. Allergy **45**, 347 (*175, 176*).

BAILEY, I. W. (1960): J. Arnold Arbor. **41**, 141 (*137*).

BAILEY, L. (1954): Proc. Roy. Entomol. Soc. (London) **A 29**, 119 (*100*).

BAKER, H. G., HURD, JR., P. D. (1968): Ann. Rev. Entomol. **13**, 385 (*109, 111*).

BAL, A. K., DE, D. N. (1961): Develop. Biol. **3**, 241 (*10, 43*).

BANKS, W., GREENWOOD, C. T., MUIR, D. D., WALKER, J. T. (1971): Stärke **23**, 97 (*133*).

BARBIER, M. (1966): Ann. Abeille **9**, 243 (*151, 153*).

BARBIER, M. (1971): Prog. Phytochem. **2**, 1 (*153*).

BARENDSE, G. W. M., RODRIGUES PEREIRA, A. S., BERKERS, P. A., DRIESSEN, F. M. VAN EYDEN-EMONS, A., LINSKENS, H. F. (1970): Acta Botan. Neerl. **19**, 175 (*253*).

BARKER, R. J. (1972): Ann. Entomol. Soc. Am. **65**, 270 (*105*).

BARKER, R. J., LEHNER, V. (1972): Am. Bee J. **112**, 336 (*100, 102*).

BARKSDALE, A. W., McMORRIS, T. C. (1967): Nature **215**, 320 (*256*).

BARNER, H., CHRISTIANSEN, H. (1958): Silvae Genet. **7**, 19 (*54*).

BARNES, D. K., CLEVELAND, R. W. (1963): Crop. Sci. **3**, 295 (*79*).

BARR, C. R., BALL, C. B., SELL, H. M. (1959): J. Am. Oil Chem. Soc. **36**, 303 (*147*).

BARSKAYA, E. I., BALINA, N. V. (1971): Fiziol. Rast. (Eng. Transl.) **18**, 605 (*15*).

BARTELS, G. (1960): Silvae Genet. **9**, 137 (*81, 196, 197*).

BATHURST, N. O. (1954): J. Exp. Botany **5**, 253 (*154*).

BATTAGLINI, M., BOSI, G. (1968): Apicolt. Ital. **35**, 37 (*148*).

BATTAGLINI, M., BOSI, G., D'ALBORE, G. R. (1970): Nuova Chim. **46**, 3 (*151*).

BATTAGLINI, M., BOSI, G., D'ALBORE, G. R. (1970): Nuova Chim. **46**, 84 (*93*).

BAUMANE, G., MARAUSKA, D., BAUMANIS, E. (1968): Latv. Psr. Zinat. Vestis. **3**, 133. (Chem. Abst. **68**, 112180; 1968.) (*154*).

BEARD, G. (1876): Hayfever, or summer catarrh, its nature and treatment. New York: Harper (*165*).

BECK, W. A., JOLY, R. A. (1940): Nature **145**, 226 (*224*).

BECK, W. A., JOLY, R. A. (1941): Stud. Inst. Divi. Thomae **3**, 81 (*72*).

BECQUEREL, P. (1929): Compt. Rend. **188**, 1308 (*59*).

BECQUEREL, P. (1932): Compt. Rend. **195**, 165 (*24*).

DE BEER, J. F. (1963): Influences of temperature on *Arachis hypogaea*. Wageningen: Centr. Landbouwpublik. en Landbouwdoc. (*47, 48*).

BEER, R. (1906): Beih. Botan. Cbl. A **19**, 286 (*13*).

BELIN, L. (1972): Int. Arch. Allergy **42**, 300 (*169*).

BELIN, L. (1972a): Int. Arch. Allergy **42**, 329 (*170*).

BELIN, L., ROWLEY, J. R. (1971): Int. Arch. Allergy **40**, 754 (*169, 171*).

BELING, I. (1931): Arch. Bienenk. **12**, 352 (*88*).

BELL, C. R. (1959): Amer. J. Bot. **46**, 621 (*28, 29*).

BELLARTZ, S. (1956): Planta **47**, 588 (*155, 200, 201*).

BENNETT, R. D., KO, S. T., HEFTMANN, E. (1966): Phytochemistry **5**, 231 (*151*).

BENT, A. C. (1940): Life histories of North American cuckoos, goatsuckers, hummingbirds and their allies. U.S. Natl. Mus., Smithsonian Inst. Bull. 176, Wash. D.C. (*111*).

BERGER, F. (1933): Beitr. Biol. Pflanz. **20**, 221 (*151*).

BERGQUIST, G., NILSEN, A. (1968): Acta Allergol. **23**, 363 (*178*).

BERI, S. M., ANAND, S. C. (1971): Euphytica **20**, 327 (*31*).

BERLEPSCH, VON, K. (1864): Nördlinger Bienenztg. **20**, 15 (*95, 98*).

BERRENS, L. (1967): Acta Allergol. **22**, 331 (*168*).

BERRENS, L. (1970): Conference on allergen standardization. VIII. Symposium Coll. Int. Allergol. Montreux (*175, 179*).

BERRENS, L. (1971): Mongr. Allergy **7**, 261 (*168*).

BERTRAND, G. (1940): Compt. Rend. **210**, 685 (*121*).

BERTRAND, G., POIRAULT, G. (1892): Compt. Rend. **115**, 828 (*223, 228*).

BERTRAND, G., SILBERSTEIN, L. (1938): C. R. Acad. Sci. Paris **205**, 796 (*122*).

BERTUGLIA, P. (1970): Minerva Med. **61**, Suppl. 30, 87 (*114*).

BESLIN, R., MILOSEVIC, M., KOSTIC, J. (1972): Arkiv Poljopriv. Nauk **25**, 40 (*113*).

BETTS, A. D. (1935): Bee World **16**, 111 (*93*, *96*).

BHADURI, P. N., BHANJA, P. K. (1962): Stain Technol. **37**, 351 (*81*).

BHOJWANI, S. S., COCKING, C. C. (1972): Nature New Biol. **239**, 29 (*42*).

BIEBERDORF, F. W., GROSS, A. L., EICHLEIN, R. W. (1961): Ann. Allergy **19**, 867 (*154*, *155*, *156*).

BIENZ, D. R. (1959): Am. Potato J. **36**, 292 (*80*).

BINDING, G. P. (1971): About pollen. London: Thorsons (*114*).

BINGHAM, W. E., KRUGMAN, S. L., ESTERMANN, E. F. (1964): Nature **202**, 923 (*160*).

BIOURGE, P. (1892): Cellule **8**, 47 (*138*).

BISHOP, C. J. (1949): Stain Technol. **24**, 9 (*72*).

BJÖRK, S. (1967): Folia Limnol. Scand. **14**, 1 (*27*).

BLACKLEY, C. H. (1873): Experimental researches on the causes and nature of *Catarrhus aesti-vus*—hay-fever or hay-asthma. London: Bailliere (*165*).

BODEN, R. W. (1958): Australian For. **22**, 73 (*68*).

BODMER, H. (1927): Flora **122**, 306 (*29*, *33*).

BONNER, J. (1965): In: BONNER, J., VARNER, J. (Eds.): Plant biochemistry, p. 665. New York: Academic Press (*226*).

BOPP-HASSENKAMP, G. (1959): Exp. Cell Res. **18**, 183 (*10*, *16*).

BORMOTOV, V. E. (1966): Acta Agron. Acad. Sci. Hung. **15**, 311 (*183*).

BOSIO, M. G. (1940): Nuovo Gior. Botan. Ital. **47**, 591 (*78*).

BOSTOCK, J. (1819): Med. Chir. Transact. (London) **10**, 161 (*165*).

BOUILLENNE, M., BOUILLENNE, R. (1931): Bull. Roy. Belg. Sci. Acad. **17**, 318 (*168*).

BOUVENG, H. O. (1965): Acta Chem. Scand. **19**, 953 (*136*).

BOUVENG, H. O., LUNDSTRÖM, H. (1965): Acta Chem. Scand. **19**, 1004 (*136*).

BOYER, W. D. (1970): Influences of temperature on date and duration of pollen shed by longleaf pine. Ph. D. Thesis, Duke University. Durham, N.C. (*24*).

BOYER, W. D. (1973): Ecology **54**, 420 (*24*).

BRACONNOT, H. (1829): Ann. Chim. Phys. Ser. 2. **42**, 91 (*119*, *138*).

BRAGG, L. H. (1969): Ecology **50**, 124 (*28*).

BRAMLETT, D. L. (1973): USDA For. Ser. Res. Paper SE-104, 7 pp. (*32*).

BRANSCHEIDT, P. (1939): Ber. Deut. Botan. Ges. **57**, 495 (*73*).

BRAUN, A. (1885): Monatsber. K. Preuss. Akad. Wiss. Berlin 378 (*143*).

BREDEMANN, G., GERBER, K., HARTECK, P., SUHR, KL. A. (1948): Naturwissenschaften **34**, 279 (*59*).

BREDEMEIJER, G. M. M. (1970): Acta Botan. Neerl. **19**, 481 (*208*).

BREDEMEIJER, G. M. M. (1971): Acta Botan. Neerl. **20**, 119 (*197*, *208*).

BREWBAKER, J. (1957): J. Heredity **48**, 271 (*10*, *11*).

BREWBAKER, J. L. (1959): Indian J. Genet. and Pl. Breed. **19**, 121 (*50*, *56*, *58*, *131*, *198*).

BREWBAKER, J. L. (1971): In: HESLOP-HARRISON, J. (Ed.): Pollen development and physiology, p. 156. London: Butterworths (*195*, *199*).

BREWBAKER, J., EMERY, G. C. (1962): Radiation Botany **1**, 101 (*50*).

BRIEGER, H., LaBELLE, C. B., GODDARD, J. W., ISRAEL, H. L. (1962): Arch. Environ. Med. **5**, 470 (*148*).

BRINK, R. A., MacGILLIVRAY, J. J., DEMEREC, M. (1926): Genetics **11**, 38 (*131*).

BRITIKOV, E. A., MUSATOVA, N. A. (1964): Fiziol. Rast. **11**, 464 (*154*).

BRITIKOV, E. A., MUSATOVA, N. A., VLADIMIRTSEVA, S. V., PROTSENKO, M. A. (1964): In: LINSKENS, H. F. (Ed.): Pollen physiology and fertilization, p. 77. Amsterdam: North Holland (*155*).

BRODIE, H. J. (1957): Proc. Indiana Acad. Sci. **66**, 65 (*36*).

BRONCKERS, F. (1968): Bull. Soc. Roy. Botan. Belg. **101**, 23 (*16*).

BROOKS, J. (1971): In: BROOKS, J., GRANT, P. R., MUIR, M., VAN GIJZEL, P., SHAW, G. (Eds.): Sporopollenin, p. 351. London: Academic Press (*139*).

BROOKS, J., GRANT, P. R., MUIR, M., VAN GIJZEL, P, SHAW, G. (1971): Sporopollenin. London: Academic Press (*138*).

BROOKS, J., SHAW, G. (1968): Nature **219**, 532 (*140*, *141*).

BROOKS, J., SHAW, G. (1971): In: HESLOP-HARRISON, J.: Pollen development and physiology, p. 99. London: Butterworths (*140*).

BROTHERTON, J. (1969): Cytobios **1 B**, 95 (*27*).

BROWN, A. G. (1958): Australian Forest. **22**, 10 (*71*).

BROWN, W. H., FELAUER, E. E., SMITH, M. V. (1962): Nature **195**, 75 (*103*).

BROWN, W. V. (1954): Bull. Torrey Botan. Club **81**, 127 (*81*).

BRUCKER, H. (1948): Physiol. Plantarum **1**, 343 (*82*).

BRUES, C. T. (1946): Insect dietary. Harvard Univ. Press (*111*).

BRYAN, H. D. (1951): Chromosoma **4**, 369 (*6, 9, 10*).

BUCHANAN, D. W., BIGGS, R. H. (1969): J. Am. Soc. Hort. Sci. **94**, 327 (*254*).

BÜNNING, E., HERDTLE, H. (1946): Z. Naturforsch. **1**, 93 (*58*).

BUIJTENEN, VAN J. P. (1952): Identification of pine pollens by chromatography of their free amino acids. M.S. Thesis. Berkeley: Univ. Calif. (*154, 155, 157*).

BULLOCK, R. M., OVERLEY, F. L. (1949): Proc. Am. Soc. Hort. Sci. **54**, 125 (*42*).

BULLOCK, R. M., SNYDER, J. C. (1946): Proc. Wash. State Hort. Ass. **1946**, 215 (*58*).

BURG, S. P., DIJKMAN, M. J. (1967): Plant Physiol. **42**, 1648 (*254*).

BURRELL, A. B., KING, G. E. (1931): Proc. Am. Soc. Hort. Sci. **28**, 85 (*42*).

BURRI, R. (1947): Schweiz. Bienen-Ztg. **70**, 273 (*99*).

BUR'YANOV, YA. I., EROSHINA, N. V., VAGABOVA, L. M., II'IN, A. V. (1972): Dokl. Akad. Nauk. SSSR **206**, 992 (*190*).

BUSSE, J. (1926): Thar. Forstl. Jb. **77**, 225 (*26, 34*).

BUTENANDT, A. (1933): Naturwissenschaften **21**, 49 (*257*).

BUTLER, C. G. (1954): The world of the honeybee. London: Collings (*87, 88, 89*).

BUTZ, L. W., FRAPS, R. M. (1945): Proc. Soc. Exp. Biol. Med. **60**, 213 (*150, 257*).

CAILLAS, A. (1959): Le Pollen. Orléans: (Loiret) L'Auteur (*42*).

CAILLON, M. (1958): Rev. Cytol. Biol. Végét. **19**, 63 (*13*).

CALLAHAM, R. Z., DUFFIELD, J. W. (1961): J. Forest. **59**, 204 (*42, 86*).

CALVINO, E. G. M. (1952): Nuovo Gior. Botan. Ital. **59**, 1 (*133, 146*).

CAMARGO, I. J. B., KITAJIMA, E. W., COSTA, A. S. (1969): Phytopath. Z. **64**, 282 (*100*).

CAMERONI, R. (1958): Riv. ital. essenze, profumi pianti offic, oli vegetal, saponi **40**, 208 (*225*).

CAMPBELL, J. H. (1946): Gladiolus Mag. **10**, 2 (*77*).

CAPOOR, S. P. (1937): Beih. Botan. Cbl. A. **57**, 233 (*10*).

CAPPELLETTI, C., TAPPI, G. (1948): Nuovo Gior. Botan. Ital. **55**, 150 (*246*).

CARLILE, B. (1971): Am. Bee J. **111**, 68 (*103*).

CARMICHAEL, J. W. (1970): Proc. Soil Crop Sci. Soc. Fla. **30**, 255 (*253*).

CARNIEL, C. (1963): Österr. Botan. Z. **110**, 145 (*15*).

CARON, B. (1972): Phopholipides et glycolipides du pollen et du style chez *Oenothera missouriensis* Sims espèce auto-incompatible. Ph. D. Thesis: Lille: Univ. Faculté Pharmacie (*149*).

CARR, D. J. (1972): Plant growth substances, 1970. Berlin: Springer. (*257*).

CASTEEL, D. B. (1912a): U.S. Entomol. Bur. Cir. **161**, 1 (*88*).

CASTEEL, D. B. (1912b): U.S. Entomol. Bur. Bull. **121**, 1 (*88*).

CEFULU, M., SMIRAGLIA, C. B. (1964): I pollini aerodiffusi e le pollinosi. Rome: G. Denaro (*27, 50*).

CHANDLER, C. (1958): Contr. Boyce Thompson Inst. **19**, 215 (*253*).

CHANG, C. W., KIVILAAN, A. (1964): Phytochemistry **3**, 693 (*182, 183*).

CHAUBAL, P. D., DEODIKAR, G. B. (1963): Grana Palynol. **4**, 393 (*105*).

CHAUVIN, R. (1957): Compt. Rend. **244**, 120 (*113*).

CHAUVIN, R., LAVIE, P. (1956): Ann. Inst. Pasteur **90**, 523 (*99*).

CHERVONENKO, T. A. (1959): Trudy Ukr. Nauk Issled. Selekt. Genet. **4**, 135 (*249*).

CHING, T. M., CHING, K. K. (1962): Science **138**, 890 (*147, 148, 149*).

CHING, T. M., CHING, K. K. (1966): Plant Physiol. **39**, 705 (*58, 62*).

CHING, T. M., SLABAUGH, W. H. (1966): Cryobiology **2**, 321 (*62*).

CHIRA, E. (1963): Lesnictoi Cas. Praha **9**, 821 (*48, 55*).

CHIRA, E. (1966): Sb. Vysoke Skoly Zemedel. Brne. **35**, 339 (*30*).

CHIRA, E. (1967): Biologia (Bratislava) **22**, 260 (*68*).

CHIRA, E., BERTA, F. (1965): Biologia (Bratislavia) **20**, 600 (*130, 132*).

CHRISTENSEN, B. B. (1946): Dan. Geol. Unders. Raekke 4, **3(2)**, 1 (*26*).

CHRISTENSEN, G. M. (1962): Nature **195**, 74 (*103*).

CLAUSEN, K. E. (1960): Pollen et Spores **2**, 299 (*29*).

CLAUSEN, K. E. (1962): Pollen et Spores **4**, 168 (*28*).

CLEMENTS, F. F., LONG, F. L. (1923): Experimental pollination. Washington: Carnegie Inst. of Wash. (*88*).

COCA, A. F. (1922): J. Immunol. **7**, 163 (*174*).

COCA, A. F., WALTZER, M., THOMMEN, A. (1931): Asthma and hayfever in theory and practice, p. 851. Springfield, Ill.: C. C. Thomas (*173*).

COLLDAHL, H. (1959): In: JAMAR, J. M. (Ed.): International textbook of allergy, p. 142. Springfield, Ill.: C. C. Thomas (*169*, *177*).

COLLDAHL, H., CARLSSON, G. (1968): Acta Allergol. **23**, 387 (*170*, *171*).

COLWELL, R. N. (1951): Am. J. Botany **38**, 511 (*127*).

COOK, S. A., STANLEY, R. G. (1960): Silvae Genet. **9**, 133 (*64*, *84*).

COOMBE, B. G. (1960): Plant Physiol. **35**, 241 (*253*, *257*).

COOPER, D. C. (1935): Am. J. Botany **22**, 453 (*10*).

COOPER, D. C. (1952): Am. Naturalist **86**, 219 (*188*).

COOPER, W. C. (1972): In: CARR, D. J. (Ed.): Plant growth substances, 1970, p. 543. Berlin-Heidelberg-New York: Springer (*254*).

COSTANTINI, F., D'ALBORE, G. R. (1971): Intl. Beekeeping Congr. **23**, 539 (*113*).

COUPIN, H. (1936): Compt. Rend. **203**, 408 (*73*).

COX, L. S. (1943): Rept. North. Nutgrowers Ass. **34**, 58 (*54*).

CRANE, J. C. (1969): Hort. Sci. **4**, 8 (*258*).

CRANG, R. E., HEIN, N. B. (1971): Cytologia **36**, 41 (*135*, *213*).

CRANG, R. E., MILES, G. B. (1969): Am. J. Botany **56**, 398 (*216*, *217*, *222*).

CRANG, R. E., MILLAY, M. A. (1971): Proc. 29th Elect. Micro. Soc. Amer. Meeting, 348 (*22*).

CRAWFORD, C. L. (1937): Proc. Am. Soc. Hort. Sci. **35**, 91 (*64*).

CRIFO, S., IANNETTI, G. (1969): Acta Allergol. **24**, 294 (*169*).

CRIFO, S., LOMBARDI, D., AMENDOLEA, L., IANNETTI, G. (1969): Boll. Soc. Ital. Biol. Sper. **45**, 1037 (*169*).

CRUDEN, R. W. (1972): Science **176**, 1439 (*111*).

CUMMINGS, M. M. (1959): J. Forestry **57**, 377 (*148*).

CUMMINGS, M. M. (1964): Acta Med. Scand. **176**, 284 (*148*).

CUMMINGS, M. M., DUNNER, E., SCHMIDT, JR., R. H., BARNWELL, J. B. (1956): Postgrad. Med. **19**, 437 (*148*).

CUMMINGS, M. M., HUDGINS, P. C. (1958): Am. J. Med. Sci. **236**, 311 (*148*).

CUMMINGS, W. C., RIGHTER, F. I. (1948): Methods used to control pollinate pines in sierra nevada of california. USDA Circ. 792 (*41*, *52*).

CURRIER, H. B. (1957): Am. J. Botany **44**, 478 (*134*).

CZAPEK, F. (1920): Biochemie der Pflanzen I. Jena: Gustav Fischer (*138*).

DAHL, A. O., ROWLEY, J. R., STEIN, O. L., WESTEDT, L. (1957): Exp. Cell Res. **13**, 31 (*10*).

DAHLGREN, K. V. O. (1915): Svensk. Botan. Tidskr. **9**, 1 (*9*).

D'ALCONTRES, G. S., TROZZI, M. (1957): Giorn. Biochem. **6**, 102 (*124*).

DALE, H. H. (1913): J. Pharmacol. Exp. Ther. **4**, 167 (*172*).

D'AMATO, F. (1947): G. Botan. Ital. **53**, 405 (*10*).

D'AMATO, F. M., DEVREUX, G. T., SCARASCIA MUGNOZZA, G. T. (1965): Caryologia **18**, 377 (*184*).

DANCKELMANN, B. (1898): Z. Forst. Jagdw. **30**, (*24*).

DANIEL, L. (1955): Novenytermeles **4**, 315 (*57*).

DARWIN, C. (1889): British and foreign orchards as fertilized by insects. (2nd Ed.). London: J. Murray (*247*).

DARWIN, C. (1904): The various contrivances by which orchards are fertilized by insects, 2nd Ed. London: John Murray (*247*).

DARWIN, C., DARWIN, F. (1880): The Power of Movement in Plants. London: J. Murray (*247*).

DASHEK, W. V., HARWOOD, H. I., ROSEN, W. G. (1970): In: HESLOP-HARRISON, J. (Ed.): Pollen Development and Physiology, p. 914. London: Butterworths (*132*, *155*).

DASHEK, W. V., JOHNSON, R. H., HAYWARD, D. M., HARWOOD, H. I. (1972): Am. J. Botany **59**, 649 (*155*).

DASHEK, W. V., ROSEN, W. G. (1966): Protoplasma **61**, 192 (*213, 215, 216, 217*).
DAVIES, M. D., DICKINSON, D. B. (1972). Arch. Biochem. Biophys. **152**, 53 (*196, 219*).
DAVIS, M. B. (1963): Am. J. Sci. **216**, 897 (*45*).
DAVIS, M. B. (1966): Ecology **47**, 310 (*107*).
DAVYDOVA, N. S. (1954): Uchen. Zap. Kishinev. Univers. **13**, 167 (*92*).
DEAN, C. E. (1964): Tobacco Sci. **8**, 60 (*75*).
DEBARR, G. L. (1969): J. Forest. **67**, 326 (*110*).
DEDIC, G. A. and KOCH, O. G. (1957): Phyton (Argent.) **9**, 65 (*121, 124*).
DEMIANOWICZ, Z. (1961): Pszczel. Zesz. Nauk. **5**, 95 (*107*).
DEMIANOWICZ, Z. (1963): Ann. Abeille. **6**, 249 (*107*).
DEMIANOWICZ, Z. (1964): Ann. Abeille. **7**, 273 (*107*).
DEMIANOWICZ, Z., DEMIANOWICZ, A. (1957): Pszczel. Zesz. Nauk. **1**, 96 (*106*).
DEMIANOWICZ, Z., JABLONSKI, B. (1959): Pszczel. Zesz. Nauk. **3**, 25 (*107*).
DEMIANOWICZ, Z., LECEWICZ, W., WARAKOMSKA, Z. (1966): Z. Bienenforsch. **8**, 148 (*106*).
DENGLER, A. (1955): Z. Forstgenet. **4**, 107 (*36*).
DENGLER, A., SCAMONI, A. (1944): Z. Genet. Forstwiss. **76/70**, 136 (*34*).
DENGLER, G., SCAMONI, A. (1939): Z. Forst. Jagdw. **71**, 1 (*73*).
DENIS, L. J. (1966): Acta Urol. Belg. **34**, 49 (*114*).
DENNEY, F. E., MILLER, L. P. (1935): Contrib. Boyce Thompson Inst. **7**, 97 (*254*).
DESBOROUGH, S., PELOQUIN, S. J. (1968): Theor. Appl. Genet. **38**, 327 (*161, 196, 205*).
DEVYS, M., ALCAIDE, A., BARBIER, M. (1969): Phytochemistry **8**, 1441 (*153*).
DEVYS, M., ALCAIDE, A., PINTE, F., BARBIER, M. (1970): Tetrahedron Letters **53**, 4621 (*153*).
DEVYS, M., BARBIER, M. (1965): Compt. Rend. **261**, 4901 (*152*).
DEVYS, M., BARBIER, M. (1967): Compt. Rend. **264**, 504 (*151*).
DEXHEIMER, J. (1968): Compt. Rend. **267**, 2126 (*190*).
DEXHEIMER, J. (1972): Rev. Cytol. Biol. Vegetales **35**, 17 (*188, 190*).
DIACONU, P. (1961): Probl. Agric. (Bucuresti) **13**, 61 (*84*).
DIACONU, P. (1968): T'sitol. Genet. **2**, 476 (*82*).
DICKINSON, D. B. (1967): Physiol. Plant. **20**, 118 (*197, 201, 207*).
DICKINSON, D. B., DAVIES, M. D. (1971): In: HESLOP-HARRISON, J. (Ed.): Pollen development and physiology. p. 190. London: Butterworths (*196, 197, 199, 206, 207, 220*).
DICKINSON, D. B., DAVIES, M. D. (1971a): Plant Cell. Physiol. **12**, 157 (*196, 212*).
DICKINSON, D. B., HOPPER, J. E., DAVIES, M. D. (1973): In: LOEWUS, F. (Ed.): Biosynthesis of plant cell wall polysaccharides, p. 29. New York: Academic Press (*196*).
DICKINSON, H. G, LEWIS, D. (1973): Proc. Roy. Soc. (Lond.) B **183**, 21 (*22*).
DICKSON, G. H., SMITH, M. V. (1958): Ontario Dept. Agr. Circ. 172 (*42*).
DIETZ, A. (1969): Ann. Entomol. Soc. Am. **62**, 43 (*100*).
DIJKMAN, M. J. (1938): Arch. Rubbercultuur **22**, 239 (*60*).
DIMBLEBY, G. W. (1957): New Phytol. **56**, 12 (*143*).
DJURBABIĆ, B.; VIDAKOVIĆ, M., KOLBAH, D. (1967): Experientia **23**, 296 (*154, 158. 159*).
DÖPP, W. (1931): Ber. Deut. Botan. Ges. **49**, 173 (*49*).
DOLCHER, T. (1967): G. Botan. Ital. **101**, 41 (*258*).
DOLLFUS, H. (1936): Planta **25**, 1 (*248*).
DOMANSKI, R. (1959): Acta Soc. Botan. Pol. **28**, 277 (*48*).
DOULL, K. M., STANDIFER, L. N. (1970): J. Apicult. Res. **9**, 129 (*93*).
DOWTY, B., LASETER, J. L., GRIFFIN, G. W., POLITZER, I. R., WALKINSHAW, C. H. (1973): Science **181**, 669 (*148*).
DUBEY, P. S., MALL, L. P. (1972): Experientia **28**, 600 (*49*).
DUCHAINE, J. (1959): In: JAMAR, J. M. (Ed.): International textbook of allergy, p. 154. Springfield, Ill.: C. C. Thomas (*166, 173*).
DUCLOUX, E. H. (1925): Rev. Facult. Ciencias **3**(2), 23 (*225*).
DUFFIELD, J. W. (1953): USDA, Calif. For. + Range Exp. Stat. Res. Note 85 (*46*).
DUFFIELD, J. W. (1954): Z. Forstgenetik **3**, 39 (*71, 74*).
DUFFIELD, J. W., CALLAHAM, R. Z. (1959): Silvae Genet. **8**, 22 (*59*).
DUFFIELD, J. W., SNOW, A. G. (1941): Am. J. Botany **28**, 175 (*60*).
DUNBAR, W. P. (1903): Zur Ursache und spezifischen Heilung des Heufiebers, p. 60. München: R. Oldenbourg (*165*).

DUNGWORTH,G., McCORMICK,A., POWELL,T.G., DOUGLAS,A.G. (1971): In: BROOKS,J., GRANT,P.R., MUIR,M., VAN GIJZEL,P., SHAW,G. (Eds.): Sporopollenin, p.512. London: Academic Press (*151*).

DURHAM,O.C. (1951): Econ. Botany **5**, 211 (*165, 173*).

DURRIEU-VABRE,A. (1960): Pollen et Spores **2**, 119 (*72*).

DURZAN,D.J. (1964): The nitrogen metabolism of *Picea glauca* (Moench) Voss. and *Pinus banksiana* L. with special reference to nutrition and environment. Ph. D. Thesis. Cornell Univ., Ithaca, N.Y. (*155*).

DUTCHER,R.A. (1918): J. Biol. Chem. **36**, 551 (*210*).

DYAKOWSKA,J. (1937): Bull. Akad. Pol. Sci. Ser. **B 8/10**, 155 (*33*).

DYAKOWSKA,J., ZURZYCKI,J. (1959): Bull. Acad. Pol. Sci. Cl. II, **7**, 11 (*35, 36*).

DŽAMIĆ,M., PEJKIĆ,B. (1970): Arkiv. Polyopriv. Nauke **23**, 97 (*129, 130, 131, 145, 156*).

DZHAPARIDZE,L.I. (1965): Sex in plant. Pt.2. Tbilisi: Acad. Sci. Georgian SSR, Inst. of Botany. Transl. U.S. Dept. of Commerce TT 68-50453 (*229*).

EAST,E.M. (1929): Bibliogr. Genet. **5**, 331 (*161*).

EATON,G.W. (1961): Can. J. Plant Sci. **41**, 740 (*49*).

EBEL,B.H. (1965): Ann. Entomol. Soc. Am. **58**, 623 (*111*).

EBEL,B.H. (1966): Ann. Entomol. Soc. Am. **59**, 227 (*110*).

ECHLIN,P. (1969): Ber. Deut. Botan. Ges. **81**, 462 (*19*).

ECHLIN,P., GODWIN,H. (1968): J. Cell Sci. **3**, 161 (*15*).

ECHLIN,P., GODWIN,H. (1969): J. Cell Sci. **5**, 459 (*15*).

ECHOLS,R.M., MERGEN,F. (1956): Forest Sci. **2**, 321 (*68*).

ECKERT,J.E. (1942): J. Econ. Entomol. **35**, 309 (*96*).

EDGEWORTH,M.P. (1879): Pollen. London: Hardwicke and Bogue (*51*).

EDWARDS,C.A., HEATH,G.W. (1964): The principles of agricultural economics. Springfield, I 11.: Chas. C. Thomas (*116*).

EFRON,Y. (1969): J. Histochem. Cytochem. **17**, 734 (*194*).

EGOROV,I.A., EGOFAROVA,R.K.H. (1971): Dokl. Akad. Nauk. SSSR **199**, 1439 (*150, 151*).

EIGSTI,O.J. (1940): Proc. Oklahoma Acad. Sci. **20**, 45 (*69*).

EIGSTI,O.J. (1966): Turtox News **44**, 162 (*72*).

EISENHUT,G. (1961): Forstwiss. Forsch. **15**, 1 (*27, 33, 50*).

ELLIOTSON,J. (1831): London Med. Gaz. **8**, 411 (*165*).

ELSER,E., GANZMÜLLER,J. (1930): Z. Physiol. Chem. **194**, 21 (*121, 197*).

ELSIK,W.C. (1971): In: BROOKS,J., GRANT,P.R., MUIR,M., VAN GIJZEL,P., SHAW,G. (Eds.): Sporopollenin, p.480. London: Academic Press (*143*).

ENGELS,F.M. (1973): Acta Botan. Neerl. **22**, 6 (*43, 213, 258*).

ENGELS,F.M. (1974): Acta Botan. Neerl. **23**, 81 (*131, 213, 216*).

ENGELS,F.M. (1974a): Acta Botan. Neerl. **23**, 209 (*131*).

ENGELS,F.M., KREGER,D.R. (1974): Naturwissenschaften **61**, 273 (*217, 218*).

ERDTMAN,G. (1952): Pollen morphology and plant taxonomy. I. Angiosperms, 2nd Ed. Waltham, Mass.: Chronica Botanica (*7, 15*).

ERDTMAN,G. (1969): Handbook of palynology. New York: Hafner (*7, 16, 45*).

ERLENMEYER,E. (1874): Sitzber. Akad. München Math. Phys. Kl. **4**, 204 (*195*).

ERVANDYAN,S.G. (1964): Izv. Akad. Nauk. Armen. S.S.R. Biol. Nauk. **17**, 79 (*47*).

ESCHRICH,W. (1956): Protoplasma **47**, 487 (*13, 143*).

ESCHRICH,W. (1962): Protoplasma **55**, 419 (*15*).

ESCHRICH,W. (1963): Protoplasma **56**, 718 (*15*).

ESCHRICH,W. (1964): In: LINSKENS,H.F. (Ed.): Pollen physiology and fertilization, p.48. Amsterdam: North Holland (*14, 135*).

ESCHRICH,W. (1966): Z. Pflanz. Physiol. **54**, 463 (*13, 15*).

ESCHRICH,W., CURRIER,H.C., YAMAGUCHI,S., McNAIRN,R.B. (1965): Planta **65**, 49 (*15*).

ESSER,K. (1953): Z. Indukt. Abst. Vererbungsl. **85**, 28 (*244*).

ESSER,K. (1963): Z. Botan. **51**, 32 (*15*).

ESSER,K., STRAUB,J. (1954): Biol. Zbl. **73**, 449 (*244*).

EULER,H.v., AHLSTRÖM,L., HÖGBERG,B., PETTERSSON,I. (1945): Arkiv. Kemi. Mineral. Geol. **19A(4)**, 1 (*150, 183, 228*).

EULER,H.v., EULER,J.V. (1949): Arkiv. Kemi Mineral. Geol. **26A(30)**, 1 (*196, 198*).

EULER, H. v., HELLER, L., HÖGBERG, K. G. (1948): Arkiv. Kemi Mineral. Geol. **26** A (21, 1 (*183, 196*).

EULER, H. v., RYDBOM, M. (1931): Arkiv. Kemi. Mineral. Geol. **10** B (3), 1 (*228, 229*).

FAEGRI, K., IVERSEN, H. (1964): Textbook of pollen analysis. Copenhagen: (*16, 27*).

FAEGRI, K., VAN DER PIJL, L. (1966): The principle of pollination ecology. Oxford: Pergamon (*25, 69*).

FARBER, E., LOUVIERE, C. D. (1956): J. Histochem. Cytochem. a4, 347 (*82*).

FATHIPOUR, A., SCHLENDER, K. K., SELL, H. M. (1967): Biochim. Biophys. Acta **144**, 476 (*146, 256*).

FAVRE-DUCHATRE, M. (1963): Ann. Sci. Nat. Botan., Paris 12 th Series **5**, 233 (*11*).

FAWCETT, P., GREEN, D., SHAW, G. (1971): Radiochem. Radioanal. Letters **8**, 37 (*127*).

FEDER, W. A. (1968): Science **160**, 1122 (*49*).

FEDOROVA, R. V. (1955): Zemledelie **12**, 109 (*30*).

FERGUSON, M. C. (1934): Genetics **19**, 394 (*223*).

FERNHOLZ, D. L., HINES, L. (1942): Proc. Am. Soc. Hort. Sci. **40**, 251 (*65, 99*).

FIELDING, J. M. (1957): Australian Forest **21**, 17 (*46*).

FILIPPOV, V. V. (1959): Byul. Glav. Bot. Sada S.S.S.R. **33**, 94 (*209*).

FINN, W. W. (1937): Trav. Inst. Rech. Sci. Biol. Univers. Kiev, 71 (cited in STEFFEN, 1963) (*9*).

FIRBAS, F., REMPE, H. (1936): Bioklimat. Beibl. Meteorol. Z. **3**, 50 (*34, 36*).

FIRBAS, F., SAGROMSKY, H. (1947): Biol. Zbl. **66**, 129 (*31*).

FITTING, H. (1909): Z. Botan. **1**, 1 (*247*).

FITTING, H. (1910): Z. Botan. **2**, 265 (*247*).

FODA, H. A. (1966): Ain Shams. Sci. Bull. **10**, 235 (*256*).

FORD, R. H. (1968): The cytology of *Caryota mitis* and *Caryota urens*. M.S., U. of Miami, Coral Gables, Fla. (*68*).

FOWDEN, L. (1965): In: BONNER, J., VARNER, J. E. (Eds.): Plant biochemistry, p. 361. New York: Academic Press (*155*).

FRANKE, W. W., HERTH, W., VANDERWOUDE, W. J. (1972): Planta **105**, 317 (*12, 213*).

FRANKEL, S., JOHNSON, M. C., SCHIELE, A. W., BUKANTZ, S. C., ALEXANDER, H. L. (1955): J. Allergy **26**, 33 (*169*).

FREE, J. B. (1970): Insect pollination of crops. New York: Academic Press (*87*).

FREE, J. B. (1971): Am. Bee J. **111**, 460 (*89*).

FREE, J. B., WILLIAMS, I. H. (1972): J. Appl. Ecol. **9**, 609 (*88*).

FREY-WYSSLING, A., EPPRECHT, W., KESSLER, G. (1957): Experientia **13**, 22 (*15*).

FREY-WYSSLING, A., GRIESHABER, E., MÜHLETHALER, K. (1963): J. Ultastruct. Res. **8**, 506 (*222*).

FREY-WYSSLING, A., MÜHLETHALER, K. (1965): Ultrastructural plant cytology. Amsterdam: Elsevier (*139*).

FREYTAG, K. (1958): Grana Palynol. **1**, 10 (*49*).

FREYTAG, K. (1964): Grana Palynol. **5**, 277 (*19*).

FRITZ; H. G. (1960): Beitrag zur Kenntnis der Polleninhaltsstoffe von Petunien. Ph. D. Thesis. Univ. Köln (*246*).

FRITZSCHE, J. (1837): Über den Pollen. L'Acad. Sci., St. Petersburg. Mem. Sav. Etrang. **3**, 649 (*45*).

FROLIK, E. F., MORRIS, R. (1950): Science **111**, 153 (*49*).

FUCHS, A. M., STRAUSS, M. B. (1959): J. Allergy **30**, 66 (*173*).

FUJITA, M., HISAMICHI, S., ANDO, T., MURAKAMI, N. (1960): Chem. Pharm. Bull. (Tokyo) **8**, 1124 (*238, 246*).

FUKUI, H. H., TEUBNER, F. G., WITTWER, S. W., SELL, H. M. (1958): Plant Physiol. **33**, 144 (*146, 256*).

FUNKE, C. (1956): Naturwissenschaften **43**, 66 (*28*).

GÄRTNER, K. F. (1844): Versuche und Beobachtungen über die Befruchtungsorgane der vollkommenen Gewächse über die natürliche und künstliche Befruchtung durch eigenen Pollen. Stuttgart: E. Schweizerbart (*56*).

GALSTON, A. W. (1969): In: HARBORNE, J. B., SWAIN, T. (Eds.): Perspectives in phytochemistry, p. 193. London: Academic Press (*246*).

GATES, R. R. (1911): Ann. Botany **25**, 909 (*16*).

GEITLER, L. (1935): Planta **24**, 361 (*10*).

GEITLER, L. (1942): Planta **32**, 187 (*11*).

GENCHEV, S. (1970): Dokl. Akad. Sel'skokhoz. Nauk. Bulg. **3**, 25 (*227*).

GEORGIEV, G. P. (1969): Ann. Rev. Genet. **3**, 155 (*8*).

GEORGIEVA, E., VASILEV, V. (1971): Sympos. on use of bee products in human and veterinary medicine. Intl. Beekeeping Congress **23**, summar (*114*).

GHAI, G., MODI, V. V. (1970): Biochem. Biophys. Res. Comm. **41**, 1088 (*229*).

GHERARDINI, G. L., HEALY, P. L. (1969): Nature **224**, 718 (*144, 204*).

GIERTYCH, M. M. (1964): Botan. Rev. **30**, 292 (*257*).

GIJZEL, VAN P. (1971): In: BROOKS, J., GRANT, P. R., MUIR, M., VAN GIJZEL, P., SHAW, G. (Eds.): Sporopollenin, p. 659. London: Academic Press (*141*).

GILBERT, L. E. (1972): Proc. Natl. Acad. Sci. (USA) **69**, 1403 (*111*).

GILLIAM, M., PREST, D. (1972): J. Invert. Pathol. **29**, 101 (*100*).

GIORDANO, E., BONECHI, R. (1956): Ital. Forest. Mart. **11**, 175 (*84*).

GLADYSHEV, B. N. (1962): Biokimiya **27**, 240 (*132*).

GLENK, H. O.., BLASCHKE, G., BAROCKA, K. H. (1969): Theor. Appl. Genet. **39**, 197 (*68*).

GLENK, H., WAGNER, W. (1960): Ber. Deut. Botan. Ges. **73**, 463 (*122, 212*).

GLENNIE, C. W., HARBORNE, J. B. (1971): Phytochemistry **10**, 1325 (*242*).

GMELIN, R., VIRTANEN, A. I. (1961): Suomen Kemist. **B 33**, 15 (*256*).

GOAS, M. (1954): C. R. Acad. Sci. Soc. (Paris) **239**, 1662 (*197, 222*).

GOEBEL, K. (1905): Organography of plants. Oxford: Clarendon Press (*11*).

GOFF, E. S. (1901): Rept. Wisc. Agric. Exp. Stat. **18**, 289 (*56*).

GOLD, A. H., SUNESCON, C. A., HOUSTON, B. R., OSWALD, J. W. (1954): Phytopathology **44**, 115 (*53, 100*).

GOLDFARB, A. R. (1968): J. Immunol. **100**, 902 (*173*).

GOLDSTEIN, S. (1960): Ecology **41**, 543 (*143*).

GOLLMICK, F. (1942): Angew. Botan. **24**, 221 (*56, 59*).

GOLUBINSKII, I. N. (1959): Sad i Ogorod **97**, 26 (*99*).

GOODMAN, D. H., HARRIS, J., MILLER, S. (1968): Ann. Allergy **26**, 463 (*174*).

GOODWIN, T. W. (1965): Chemistry and biochemistry of plant pigments. New York: Academic Press (*224*).

GÓRSKA-BRYLASS, A. (1962): Acta Soc. Botan. Polon. **31**, 409 (*145, 221, 222*).

GÓRSKA-BRYLASS, A. (1965): Acta Soc. Botan. Polon. **34**, 589 (*197, 201*).

GÓRSKA-BRYLASS, A. (1967): Naturwissenschaften **54** (9), 230 (*134, 197*).

GÓRSKA-BRYLASS, A. (1968): Acta Soc. Botan. Polon. **37**, 119 (*134*).

GOSS, J. A. (1968): Botan. Rev. **34**, 333 (*150, 154, 183*).

GOSS, J. A. (1971): Am. J. Botany **58**, 476 (*48*).

GOSS, J. A., PANCHAL, Y. C. (1965): Bioscience **15**, 38 (*201, 206*).

GOTOH, K. (1931): Memor. Fac. Sci. Agr., Taihoku Imp. Univ. **3**, 61 (*62, 66*).

GOULD, F. W., (1957): Brittonia **9**, 72 (*28*).

GRANO, C. X. (1958): For. Sci. **4**, 94 (*51*).

GRANT, V., GRANT, K. A. (1965): Flower pollination on the phlox family. New York: Columbia Univ. Press (*110*).

GREBINSKIJ, S. O., ROLIK, R. P. (1949): Dokl. Akad. Nauk SSSR **68**, 1109 (*65*).

GREEN, I., KINMAN, J. (1970): J. Immunol. **104**, 1094 (*179*).

GREEN, J. R. (1894): Ann. Botany **8**, 225 (*195, 220*).

GREEN, J. R. (1894a): Phil. Trans. Roy. Soc. London B **185**, 385 (*204*).

GRIGGS, W. H., VANSELL, G. H., IWAKIRI, B. T. (1953): Proc. Am. Soc. Hort. Sci. **62**, 304 (*62*).

GRIGGS, W. H., VANSELL, G. H., REINHARDT, J. F. (1950): J. Econ Entomol. **43**, 549 (*59*).

GRIGGS, W. H., VANSELL, G. H., REINHARDT, J. F. (1951): Am. Bee J. **91**, 470 (*65*).

GRINKEVICH, N. G. (1968): Byull. Gl. Botan. Sada **71**, 89 (*119*).

GRONEMEYER, W., FUCHS, E. (1959): Int. Arch. Allergy **14**, 217 (*177*).

DE GROOT, A. P. (1953): Physiol. Comp. Oecolog. **3**, 1 (*102, 105, 106*).

GROVE, E. F., COCA, A. F. (1923): Proc. Soc. Exp. Biol. Med. **21**, 48 (*168*).

GRUNT, E. V., DEVJATKIN, V. A., ZACHAROVA, M. (1948): Pcelovodstvo **10**, 52 (*210*).

GULLVÅG, B. M. (1967): Phytomorphology **16**, 211 (*213*).

GUNASEKARAN, M., ANDERSEN, W. R. (1973): Personel communication (*146, 147*).

GUSSIN, A. E. S., McCORMACK, J. H. (1970): Phytochemistry **9**, 1915 (*196, 197, 221*).

GUSSIN, A. E. S., McCORMACK, J. H., WAUNG, L. Y.-L., GLUCKIN, D. S. (1969): Plant Physiol. **44**, 1163 (*197*).

GUSTAFSON, F. G. (1937): Amer. J. Botany **24**, 102 (*249*).

GUSTAFSON, F. G. (1939): Amer. J. Botany **26**, 135 (*78, 79*).

GUSTAFSON, F. G. (1961): In: RUHLAND, W. (Ed.): Encyclopedia plant physiology, Vol. 14, p. 951 (*258*).
HAECKEL, A. (1951): Planta **39**, 431 (*58, 197, 201, 222*).
HAGEMANN, P. (1937): Gartenbauwissenschaften **11**, 144 (*249*).
HAGERUP, O. (1950): Kgl. Vidensk. Selsk. Biol. Medd. **18(4)**, 1 (*24*).
HAGIAYA, K. (1949): Botan. Mag. (Tokyo) **62**, 9 (*64*).
HAGMAN, M. (1964): In: LINSKENS, H. F. (Ed.): Pollen physiology and fertilization, p. 244. Amsterdam: North Holland (*162, 163*).
HAGMAN, M. (1971): Commun. Inst. Forest Fenn. **73**, (6), 1 (*163*).
HALL, G. C., FARMER, JR., R. E. (1971): Can. J. Botany **49**, 260 (*68*).
HALL, I. V., FORSYTH, F. R. (1967): Can. J. Botany **45**, 1163 (*254*).
HALLGREN, B., LARSSON, S. (1963): Acta Chem. Scand. **17**, 1822 (*150, 151*).
HAMILL, D. E., BREWBAKER, J. L. (1969): Physiol. Plantarum **22**, 945 (*204*).
HAMMER, O., JÖRGENSEN, E. G., MIKKELSEN, V. M. (1948): Tids. Planteavl. **52**, 293 (*106*).
HANSON, C. H., CAMPBELL, T. A. (1972): Crop Sci. **12**, 874 (*59*).
HARA, A., YAMAMOTO, M., HORITA, Y., WATANABE, T. (1972): Mem. Fac. Ag. Kagoshima Univ. **8**, (2) 27 (*219*).
HARA, A., YOSHIHARA, K., WATANABE, T. (1970): J. Agr. Chem. Soc. Japan. **44**, 385 (*197, 207*)
HARBORNE, J. B. (1967): Comparative biochemistry of the flavonoids. New York: Academic Press (*230, 240, 242, 243*).
HARE, R. C. (1970): Physiology and biochemistry of pine resistance to the fusiform rust fungus, *Cronartium fusiforme*. Ph. D. Thesis, Gainesville: Univ. of Florida (*206*).
HARRIS, W. F. (1956): New Zealand J. Sci. Technol. **37**, 635 (*28*).
HARRIS, W. F. (1956a): New Zealand J. Sci. Technol. **37**, 731 (*28*).
HARSH, S. F. (1946): Calif. Western Med. **64**, 245 (*50*).
HASSAN, A., WAFA, A. (1947): Nature **159**, 409 (*151*).
HATANO, K. (1955): Bull. Tokyo Univ. For. **48**, 149 (*158*).
HAUSER, E. J. P., MORRISON, J. H. (1964): Am. J. Botany **51**, 748 (*82, 84*).
HAVINGA, A. J. (1963): Med. Landbouwhogesch, Wageningen **63**, 1 (*142*).
HAVINGA, A. J. (1964): Pollen et Spores **6**, 621 (*142, 143*).
HAVINGA, A. J. (1967): Rev. Palaeobot. Palynol. **2**, 81 (*143*).
HAVINGA, A. J. (1971): In: BROOKS, J., GRANT, P. R., MUIR, M., VAN GIJZEL, P., SHAW, G. (Eds.): Sporopollenin, p. 446. London: Academic Press (*142, 143*).
HAVIVI, E., LEIBOWITZ, J. (1958): Bull. Res. Coun. of Israel **6 D**, 259 (*72*).
HAYDAK, M. H. (1933): Arch. Bienenk. **14**, 220 (*105*).
HAYDAK, M. H. (1939): J. Econ. Entomol. **33**, 397 (*105*).
HAYDAK, M. H. (1960): Bee World **41**, 292 (*103*).
HAYDAK, M. H. (1961): Am. Bee J. **101**, 354 (*103*).
HAYDAK, M. H., VIVINO, E. (1950): Ann. Entomol. Soc. Am. **43**, 361 (*100*).
HAZSLINSZKY, B. (1955): Emelmiszervizsgalati Közlemenyek **1**, 3 (*107*).
HECKER, R. J. (1963): J. Am. Soc. Sugar Beet Technol. **12**, 521 (*84*).
HEINEN, W. (1963): Acta Botan. Neerl. **12**, 51 (*143*).
HEINEN, W., LINSKENS, H. F. (1960): Naturwissenschaften **27**, 18 (*144*).
HEINEN, W., LINSKENS, H. F. (1961): Nature **191**, 1416 (*200*).
HEINRICH, B., RAVEN, R. H. (1972): Science **176**, 597 (*111*).
HEJTMANEK, J. (1943): Prievidz. (*99*).
HEJTMANEK, J. (1961): Mitt. Tschechoslovak. Akad. Landwirtsch. (Prag) **169**, (*102*).
HELANDER, E. (1960): Grana Palynol. **2**, 2 (*165*).
HELANDER, E. (1960a): Grana Palynol. **2**, 119 (*114*).
HELLMERS, H., MACHLIS, L. (1956): Plant Physiol. **31**, 284 (*43, 130, 201, 221*).
HELLSTRÖM, N. A. (1956): Grana Palynol. **1**, 20 (*39*).
HENDERSON, L. G. (1926): Cold Spring Harb. Monogr. **10**, 3 (*25*).
HENRIQUES, R. (1897): Ber. Deut. Chem. Ges. **30**, 1415 (*148*).
HERAPATH, T. J. (1848): J. Chem. Soc. London **1**, 1 (*138*).
HERICH, R. (1961): Züchter **31**, 48 (*29*).
HESEMANN, C. U. (1971): Theor. Appl. Genet. **41**, 338 (*184*).
HESLOP-HARRISON, J. (1963): Symp. Soc. Exp. Biol. **17**, 315 (*16, 17, 21*).

HESLOP-HARRISON, J. (1964): In: LINSKENS, H. F. (Ed.): Pollen physiology and fertilization, p. 39. Amsterdam: North Holland (15).

HESLOP-HARRISON, J. (1966): Endeavour 25, 65 (15, 16, 135).

HESLOP-HARRISON, J. (1966a): Ann. Botany 30, 221 (15, 16, 21).

HESLOP-HARRISON, J. (1968): Science 161, 230 (142).

HESLOP-HARRISON, J. (1968a): Develop. Biol. Suppl. 2, 118 (13, 139).

HESLOP-HARRISON, J. (1968b): Can. J. Botany 46, 1185 (17, 19, 140).

HESLOP-HARRISON, J. (1968c): New Phytol. 67, 779 (17, 140, 224).

HESLOP-HARRISON, J. (1968d): Nature 220, 605 (17).

HESLOP-HARRISON, J. (1971): In: BROOKS, J., GRANT, P. R., MUIR, M., VAN GIJZEL, P., SHAW, G. (Eds.): Sporopollenin, p. 1. London: Academic Press (22).

HESLOP-HARRISON, J., DICKINSON, H. G. (1967): Phytomorphology 17, 195 (222).

HESLOP-HARRISON, J., DICKINSON, H. G. (1969): Planta 84, 199 (18, 19).

HESLOP-HARRISON, J., HESLOP-HARRISON, Y. (1970): Stain Technol. 45, 115 (82).

HESLOP-HARRISON, J., MACKENZIE, A. (1967): J. Cell Sci. 2, 387 (135, 188).

HESS, D. (1963): Planta 59, 567 (240, 241).

HESS, D. (1964): Planta 61, 73 (242).

HESS, D. (1968): Biochemische Genetik. Berlin: Springer-Verlag (240).

HESSELMANN, H. (1919): Medd. Stat. Skogford 16, 27 (34).

HESSELTINE, C. W., SNYDER, E. B. (1958): Bull Torrey Botan. Club 85, 134 (59, 62).

HEYL, F. W. (1919): J. Am. Chem. Soc. 41, 1285 (238).

HEYL, F. W. (1923): J. Am. Pharm. Ass. 12, 669 (145, 150).

HEYL, F. W., HOPKINS, H. H. (1920): J. Am. Chem. Soc. 42, 1738 (168).

HILL, H. M., ROGERS, L. J. (1969): Biochem. J. 113, 31 P (226).

HIROSE, T. (1957): Sci. Rept. Saikyo Univ. 9, 5 (26).

HIRSCHFELDER, H. (1950): Z. Bienenforsch. 1, 67 (95, 96, 103).

HIRSCHFELDER, H. (1952): Imkerfreund 6, 366 (95, 96).

HIRST, J. M. (1952): Ann. Appl. Biol. 39, 257 (51).

HISAMICHI, S. (1961): J. Pharm. Soc. Japan. 81, 446 (239).

HISAMICHI, S., ABE, Y. (1967): Tohoku-Yak. Dai. Kenkyu Nem. 14, 47 (150).

HITCHCOCK, J. D. (1956): Am. Bee J. 96, 487 (100, 201, 205).

HIURA, M., HIURA, M. (1964): Rakuno Gakuen Daigaku Kojo 2, 30 (132).

HO, R. H., ROUSE, G. E. (1970): Can. J. Botany 48, 213 (68).

HOCKING, B., RICHARDS, W. R., TWINN, C. R. (1950): Can. J. Res. D 28, 58 (110).

HODGES, D. (1952): The pollen loads of the honeybee. London: Bee Res. Assoc. (88, 96, 97).

HODGES, D. (1967): Bee World 48, 58 (89).

HOEBERICHTS, J. A., LINSKENS, H. F. (1968): Acta Botan. Neerl. 17, 433 (148, 149, 151, 256).

HOEFERT, L. L. (1969): Am. J. Botany 56, 363 (17, 213).

HOEFERT, L. L. (1969a): Protoplasma 68, 237 (213).

HOEKSTRA, F. A. (1972): Ademhaling et vitaliteit van bi- en trinukleaat pollen. Ing. Thesis, Wageningen: Lab. Plantenfysiologie, 35 p. (68, 80).

HOEKSTRA, P. E. (1965): USDA, For. Serv. Res. Note SE-40, 3 pp. (51).

HOFFMANN (1871): Botan. Z. 29, 80, 97 (56).

HOFMEISTER, L. (1956): Protoplasma 46, 367 (10).

HOLMAN, R. M., BRUBAKER, F. (1926): Univ. Calif. Publ. Botan. 13, 179 (56, 58, 66).

HOPKINS, J. S. (1950): Ecology 31, 633 (50).

HOPKINS, C. Y., JAVANS, A. W., BOCH, R. (1969): Can. J. Biochem. 47, 433 (93, 147).

HÔPLA, C. E. (1965): Bull. Brooklyn Entomol. Soc. 59/60, 88 (110).

HOPPER, J. E., DICKINSON, D. B. (1972): Arch. Biochem. Biophys. (208).

HOTTA, Y., HECHT, N. (1971): Biochim. Biophys. Acta 238, 50 (190).

HOTTA, Y., ITO, M., STERN, H. (1966): Proc. Nat. Acad. Sci. (USA) 56, 1184 (7).

HOTTA, Y., STERN, H. (1963): J. Cell Biol. 16, 259–279 (187).

HOTTA, Y., STERN, H. (1971): J. Mol. Biol. 55, 337 (185).

HOWARD, H. W. (1958): Am. Potato J. 35, 676 (59).

HSIANG, T. T. (1951): Plant Physiol. 26, 708 (248).

HUBBE, W. (1954): Arch. Geflügelzucht u. Kleintierk. 3, 441 (88).

HUBSCHER, T., EISEN, A. H. (1972): Int. Arch. Allergy 42, 466 (168).

HÜGEL, M.-F. (1962): Ann. Abeille **5**, 97 (*93*).

HÜGEL, M.-F. (1965): Ann. Abeille **8**, 299 (*132, 152*).

HÜGEL, M.-F., VETTER, W., AUDIER, H., BARBIER, M., LEDERER, E. (1964): Phytochemistry **3**, 7 (*151*).

HUREL-PY (1934): Compt. Rend. **198**, 195 (*73*).

HYDE, H. A. (1950): New Phytol. **49**, 405 (*50*).

HYDE, H. A. (1951): J. Forest **45**, 172 (*32*).

IBRAHIM, S. H., SELIM, H. A. (1962): Agric. Res. Rev. **40**, 116 (*92*).

ICHIKAWA, C. (1936): J. Agr. Chem. Soc. Japan **12**, 117 (*125*).

ILLY, G., SOPENA, J. (1963): Rev. Forest Franc. **15**, 7 (*25, 46*).

INGLETT, G. E. (1956): Nature **178**, 1346 (*238*).

ISHIKAWA, S., WYATT, J. P. (1970): Fed. Proc. **29**, A 421 (*148*).

ISHIZAKA, K., ISHIZAKA, T. (1966): J. Allergy **37**, 169 (*172*).

ISTATKOV, S., SECENSKA, S., EDREVA, E. (1964): C. R. Acad. Bulg. Sci. **17**, 73 (*196*).

IVONIS, I. W. (1969): Fiziol. Rast **16**, 937 (Engl. Transl.) (*252*).

IWANAMI, Y. (1959): J. Yokohoma Municipal Univ. (116) **C-34**, 1 (*220*).

IWANAMI, Y. (1962): J. Botanique **3**, 61 (*63*).

IWANAMI, Y. (1971): Jap. J. Palynol. **8**, 39 (*63*).

IWANAMI, Y. (1972): Plant Cell Physiol. **13**, 1139 (*63*).

JACOBSEN, J. G.; SMITH, JR., L. H. (1968): Physiol. Rev. **42**, 424 (*155*).

JACOPINI, P. (1955): Riv. Ortoflorofruttic Ital. **38**, 433 (*82*).

JAESCHKE, J. (1935): Beih. Botan. Cbl. B. **52**, 622 (*27*).

JAFFE, M. J. (1971): Plant Physiol. **47**, Suppl. 49 (*201*).

JALOUZOT, R. (1969): Exp. Cell Res. **55**, 1 (*10, 184*).

JEFFREE, E. P. (1956): Insectes Sociaux **3**, 417 (*102*).

JEFFREE, E. P., ALLEN, M. D. (1957): J. Econ. Entomol. **50**, 211 (*92, 94*).

JENSEN, C. J. (1964): Arsskr. Vet.-Landbohojsk (Yearbook Royal Vet. + Ag. College, Copenhagen), 133 (*62*).

JENSEN, H. (1943): Konig. Lantbruks-Akad. Handl. och Tidskr. **82**, 330 (*54*).

JOHANSEN, C. J. (1956): J. Econ. Entomol. **49**. 825 (*39, 99*).

JOHANSSON, S. G. O., BENNICH, H. (1967): Immunology **13**, 381 (*172*).

JOHNSON, C. A., RAPPAPORT, B. Z. (1932): J. Infect. Dis. **50**, 290 (*168*).

JOHNSON, L. C., CRITCHFIELD, W. B. (1974): J. Heredity **65**, 123 (*245*).

JOHNSON, L. P. V. (1943): Can. J. Res., Sect. C, **21**, 332 (*60*).

JOHNSON, L. P. V. (1945): Forest. Chron. **21**, 130 (*54*).

JOHNSON, P., MARSH, D. G. (1965): Nature **206**, 935 (*170*).

JOHNSON, P., THORNE, H. V. (1958): Int. Arch. Allergy Appl. Immunol. **13**, 291 (*168*).

JOHRI, B. M., VASIL, I. K. (1961): Botan. Rev. **27**, 325 (*56, 76, 84, 133*).

JOLEY, L. E., HESSE, C. O. (1950): Proc. Am. Soc. Hort. Sci. **56**, 231 (*49*).

JONES, M. D., NEWELL, L. C. (1948): J. Am. Soc. Agron. **40**, 136 (*29*).

JORDE, W., LINSKENS, H. F. (1972): XII Kongr. Dtsch. Ges. Allergie u. Immunitätsforsch. Wiesbaden. Z. Immunitätsforsch. Suppl. **1**, 214 (WERNER, M., GRONEMEYER, W., Eds.) (*169, 179*).

JORDE, W., LINSKENS, H. F. (1974): Acta Allergol. **29**, 165 (*181*).

JOVANCEVIĆ, M. (1962): Nar. Sumar. Sarajevo **16**, 493 (Translation-U.S. Dept. Comm. TT 67-58009, 1968) (*67, 81*).

JUNGALWALA, F. B., CAMA, H. R. (1962): Biochem. J. **85**, 1 (*224*).

KAIENBURG, A. L. (1950): Planta **38**, 377 (*12, 43, 56*).

KAKHIDZE, N. T., MEDVEDEVA, G. E. (1956): Fiziol. Rast. **3**, 435 (*210*).

KAMIEŃSKA, A., PHARIS, R. P. (1971): Am. J. Botany **58**, 476 (*253*).

KAMMANN, O. (1912): Biochem. Z. **46**, 151 (*168, 195, 196, 200*).

KAMRA, O. P. (1960): Hereditas (Lund) **46**, 592 (*16*).

KANNO, T., HINATA, K. (1969): Plant Cell Physiol. **10**, 213 (*200*).

KANTA, K., MAHESHWARI, P. (1963): Phytomorphology **13**, 215 (*78*).

KANTOR, J., CHIRA, E. (1971): Acta Univ. Agr. (Brno) Ser. C, **40**, 15 (*130*).

KAPADIA, Z. J., GOULD, R. W. (1964): Am. J. Botany **51**, 166 (*28*).

KARPER, R. E. (1933): J. Heredity **24**, 257 (*132*).

KARRER, P., EUGSTER, C. H., FAUST, M. (1950): Helv. Chim. Acta **33**, 300 (*224*).
KARRER, P., LEUMANN, E. (1951): Helv. Chim. Acta **34**, 1412 (*225*).
KASAI, C., MIYA, K., KATSUMATA, T., TOGASAWA, Y. (1966): J. Fac. Agr. Iwate Univ. **8**, 89 (*184*).
KATSUMATA, T., TOGASAWA, Y. (1968): J. Agric. Chem. Soc. Japan. **42**, 8 (*196*).
KATSUMATA, T., TOGASAWA, Y. (1968a): J. Agric. Chem. Soc. Japan. **42**, 13 (*207*).
KATSUMATA, T., TOGASAWA, Y., OBATA, Y. (1963): J. Agric. Chem. Soc. Jap. **37**, 439 (*154*).
KAUFMAN, W. (1920): Bull. Acad. Polon. Sci. Letters Sci. Nat. **Ser. B.**, 191 (*133*).
KAUROV, I. A. (1957): Botan. Zhur. **42**, 267 (*48*).
KAVETSKAYA, A. A., TOKAŘ, L. O. (1963): Botan. Zhur. **48**, 580 (*78*).
KAWECKA, B. (1926): Bull. Acad. Pol. Sci. Letters Math. Nat. Ser. B **5/6**, 329 (*19, 29*).
KELLERMAN, M. (1915): Science **42**, 375 (*59*).
KELLY, J. W. (1928): U.S.D.A. Wash. D. C. Circ. 46. 9 pp. (*43*).
KENADY, R. M. (1968): Forest. Sci. **14**, 105 (*51*).
KERNER, VON MARILAUN, A. (1891): Pflanzenleben, 2. Band. Leipzig/Wien: Bibliogr. Inst. (*146*).
KERNER, A. J. (1904): The Natural History of Plants. London: Gresham (*25*).
KESSELER, E. V. (1930): Angew. Botan. **12**, 368 (*56*).
KESSLER, G. (1958): Ber. Schweiz. Botan. Ges. **68**, 5 (*15*).
KESSLER, G., FEINGOLD, D. S., HASSID, W. Z. (1960): Plant Physiol. **35**, 505 (*196, 199*).
KESZTYÜS, L. (1950): Kiserletes Orvost. **2**, 256 (*161*).
KESZTYÜS, L. (1957): Acta Physiol. Acad. Sci. Hung. **11**, 399 (*161, 165*).
KEULARTS, J. L. W., LINSKENS, H. F. (1968): Acta Botan. Neerl. **17**, 267 (*99*).
KHOO, U., STINSON, H. T. (1957): Proc. Natl. Acad. Sci (USA) **43**, 603 (*154*).
KIBALENKO, A. P., SIDORSHINA, T. N. (1971): Dopov. Akad. Nauk Ukr. RSR **B33**, 558 (*80*).
KIESEL, A. (1922): Z. Physiol. Chem. **120**, 85 (*120*).
KIESEL, A., RUBIN, B. (1929): Z. Physiol. Chem. **182**, 241 (*191*).
KIHARA, H. (1959): Proc. X Intern. Congr. Genetics **1**, 142 (*81*).
KINDL, H. VON (1968): Z. Physiol. Chem. **349**, 519 (*256*).
KING, J. R. (1955): Proc. Am. Soc. Hort. Sci. **66**, 155 (*53*).
KING, J. R. (1959): Bull. Torrey Botan. Club **86**, 383 (*59, 62*).
KING, J. R. (1960): Stain Technol. **35**, 225 (*82*).
KING, J. R. (1961): Econ. Botan. **15**, 91 (*62*).
KING, J. R., HESSE, C. O. (1938): Proc. Am. Soc. Hort. Sci. **36**, 310 (*60, 61*).
KING, T. P., NORMAN, P. S., CONNELL, J. T. (1964): Biochemistry **3**, 458 (*170*).
KIRCHHEIMER, F. (1933a): Botan. Arch. **35**, 134 (*142*).
KIRCHHEIMER, F. (1933b): Ber. Schweiz. Botan. Ges. **42**, 246 (*142*).
KIRCHHEIMER, F. (1935): Beih. Botan. Cbl. **A 53**, 389 (*142*).
KISIMOVA, L. A. (1966): Sel'skokhoz. Biol. **1**, 621 (*84*).
KIYOSAWA, S. (1962): Proc. Crop Sci. Soc. Japan **31**, 37 (*48*).
KLAEHN, F. U., NEU, R. L. (1960): Silvae Genet. **9**, 44 (*72*).
KLEBER, E. (1935): Z. vergl. Physiol. **22**, 221 (*47*).
KLYUKVINA, Y. V. (1963): Agrobiologica **5**, 782 (*72*).
KNIGHT, A. H., CROOKE, W. M., SHEPHERD, H. (1972): J. Sci. Food Agr. **23**, 263 (*21, 122, 124, 136, 137*).
KNIGHTS, B. A. (1968): Phytochemistry **7**, 1707 (*151*).
KNOLL, F. (1932): Forsch. Fortschr. **8**, 301 (*33, 35, 36*).
KNOWLTON, H. E. (1922): Mem. Cornell Agric. Exper. Stat. **52**, 747 (*59*).
KNOX, R. B. (1973): J. Cell Sci. **12**, 421 (*162*).
KNOX, R. B., HESLOP-HARRISON, J. (1969): Nature **223**, 92 (*142, 217*).
KNOX, R. B., HESLOP-HARRISON, J. (1970): J. Cell. Sci. **6**, 1 (*23, 134, 142, 160, 168, 201, 217*).
KNOX, R. B., HESLOP-HARRISON, J. (1971): Cytobios **4**, 49 (*161, 171*).
KNOX, R. B., WILLING, R. R., ASHFORD, A. E. (1972): Nature **237**, 381 (*162*).
KNUTH, P. (1906): Handbook of flower pollination. 3 Vol. (Transl. by J. R. Ainsworth Davis). Oxford Press. Handbuch der Blütenbiologie, Bd. III, Tl. 1 u. Tl. 2 (Ed. E. Loew). Leipzig: W. Engelmann (*109, 110*).
KOBABE, G. (1965): Züchter **35**, 299 (*41*).
KOBEL, F. (1951): Schweiz. Bienen-Z. **74**, 94 (*92*).
KÖGL, F., ERXLEBEN, H., HAAGEN-SMIT, A. J. (1934): Z. Physiol. Chem. **225**, 215 (*248*).

KOESSLER, J. H. (1918): J. Biol. Chem. **35**, 415 (*150*).

KOLLER, P. C. (1943): Proc. Roy. Soc. (Edinburgh) B **61**, 398 (*6, 9*).

KONAR, R. N. (1962): Phytomorphology **12**, 190 (*55*).

KONAR, R. N.; STANLEY, R. G. (1969): Planta **84**, 304 (*218*).

KOPCEWICZ, J. (1969) Rocz. Nauk Rosl. Ser. A **95**, 105 (*253*).

KOSAN, J. H. D. (1959): Culture of excised anthers of *Lilium longifolium*. (Part I) Incorporation of tritiated cytidine in excised anthers of *Lilium longiflorum* (Part II). Ph. D. Thesis. New York: Columbia Univ. (*55, 188*).

KOSKI, V. (1967): Finn. For. Res. Inst. P.L. 480-E 8-FS, 117 pp. (*34*).

KOSKI, V. (1967): Pollen dispersal and its significance in genetics and silviculture. Helsinki: Forest Res. Inst. (*51*).

KOSTRIUKOWA, K. (1939): J. Inst. Botan. Acad. Sci. RSS Ukraine **21**, 157 (*12*).

KOTAKE, M., ARAKAWA, H. (1956): Naturwissenschaften **43**, 327 (*238*).

KOTS, Z. P. (1971): Lesovod. Agrolesomol. **26**, 53 (*249*).

KOUL, A. K., PALIWAL, R. L. (1961): Agra Univ. J. Res. a10, 85 (*85*).

KOZUBOV, G. M. (1965): Botan. Zhur. **50**, 811 (*84*).

KOZUBOV, G. M. (1967): Botan. Zhur. **52**, 1156 (*81*).

KRAAL, A. (1962): Euphytica **11**, 53 (*99*).

KRAL'OVIC, J., KAULOVA, J. (1971): Agrochem. (Bratislava) **11**, 367 (*49*).

KREMER, J. C. (1949): Proc. Am. Soc. Hort. Sci. **53**, 153 (*99*).

KREMP, G. O. W. (1965): Morphological encylopedia of palynology. Tuscon, Arizona: Univ. of Arizona Press (*7*).

KRESSLING, K. (1891): Arch. Pharm. **229**, 389 (*120, 138, 145, 148*).

KROH, M. (1964): In: LINSKENS, H. F. (Ed.): Pollen physiology and fertilization, p. 221. Amsterdam: North Holland (*76, 207*).

KROH, M. (1967): Rev. Paleobot. Palynol. **3**, 197 (*217*).

KROH, M., LOEWUS, F. (1968): Science **160**, 1352 (*132*).

KROPACOVA, S., HASLBACHOVA, H., NOVAK, V. L. (1968): Acta Univ. Agr. Fac. Agron. **16**, 537 (*102*).

KRUGMAN, S. (1959): Forest. Sci. **5**, 169 (*239*).

KRUMBHOLZ, G. (1926): Jena. Z. Naturwiss. **62**, 187 (*29*).

KRUPKO, S. (1959): Acta Soc. Botan. Polan. **28**, 75 (*246*).

KRUPNIKOVA, T. A. (1970): Trudy Botan. Inst. Akad. Nauk. SSSR Ser. 4, **20**, 128 (*245*).

KRYLOVA, M. I. (1967): Botan. Zhur. **52**, 1340 (*227*).

KUBO, A. (1955): Japan. J. Botan. **15**, 15 (*72*).

KUBO, A. (1960): Botan. Mag. (Tokyo) **73**, 453 (*72*).

KÜHLWEIN, H. (1937): Botan. Zbl. **57**, 37 (*64*).

KÜHLWEIN, H. (1948): Planta **35**, 528 (*250*).

KUGLER, H. (1970): Blütenökologie, 2nd Ed. Stuttgart: Fischer (*30*).

KUHN, R. (1943): Wiener Chem. Z. **46**, 1 (*245*).

KUHN, R., JERCHEL, D. (1941): Ber. Deut. chem. Ges. **74 B**, 941 (*82*).

KUHN, R., LÖW, I. (1944): Chem. Ber. **77 B**, 196 (*232, 243*).

KUHN, R., LÖW, I. (1949): Chem. Ber. **82**, 474 (*131, 243, 244*).

KUHN, R., LÖW, I., MOEWUS, F. (1942): Naturwissenschaften **30**, 373 (*243*).

KUPRIJANOV, S. I. (1940): Jarouizacija **1**, 95 (*28*).

KURTZ, JR., E. B., LIVERMAN, J. L. (1958): Bull. Torrey Bot. Club **85**, 136 (*28, 29*).

KVANTA, E. (1968): Acta Chem. Scand. **22**, 2161 (*151*).

KVANTA, E. (1970): Acta Chem. Scand. **24**, 3672 (*132*).

KVANTA, E. (1972): Sympos. on Effect of Nutritive Supplement on Athletes. Helsingborg: Cernelle (*114*).

KWACK, B. H. (1965): Physiol. Plantarum **18**, 297 (*73*).

KWACK, B. H. (1967): Physiol. Plantarum **20**, 825 (*127*).

KWIATKOWSKI, A. (1964): Acta Soc. Botan. Polon. **33**, 547 (*146, 149, 150*).

KWIATKOWSKI, A., LUBLINER-MIANOWSKA, K. (1957): Acta Soc. Botan. Polon. **26**, 501 (*138, 142, 143*).

LACOUR, L. F. (1949): Heredity **3**, 319 (*9, 10*).

LAFOUNTAIN, JR., J. R., LAFOUNTAIN, K. L. (1973): Exp. Cell Res. **78**, 472 (*184*).

LAFOUNTAIN, K. L., MASCARENHAS, J. P. (1972): Exp. Cell Res. a73, 233 (*43, 190, 258*).

LAIBACH, F. (1932): Ber. Deut. Botan. Ges. **50**, 383 (*248*).

LAIBACH, F. (1933): Ber. Deut. Botan. Ges. **51**, 336 (*248*).
LAIBACH, F. (1933a): Ber. Deut. Botan. Ges. **51**, 386 (*248*).
LAIBACH, F., MASCHMANN, E. (1933): Jb. Wiss. Botan. **78**, 399 (*248, 249*).
LAKON, G. (1942): Ber. Deut. Botan. Ges. **60**, 299 (*82*).
LANGER, J. (1915): Bienenwirtschaftl. Z. **51**, 259 (*99*).
LANGSTROTH, L. L. (1863): A practical treatise on the hive and honeybee. 3rd Ed. (*87*).
LANNER, R. (1962): Silvae Genet. **11**, 114 (*56*).
LANNER, R. M. (1966): Silvae Genet. **15**, 50 (*51*).
LAMPORT, D. T. A. (1963): J. Biol. Chem. **238**, 1438 (*155*).
LAMPORT, D. T. A. (1970): Ann. Rev. Plant Physiol. **21**, 235 (*202, 216, 258*).
LAPRIORE, G. (1928): Ber. Deut. Botan. Ges. **46**, 413 (*81*).
LARSON, D. A. (1963): Nature **200**, 911 (*10, 135*).
LARSON, D. A. (1965): Am. J. Botany **52**, 139 (*43, 213*).
LARSON, D. A. (1965a): Int. Arch. Allergy **26**, 127 (*10*).
LARSON, D. A., LEWIS, C. W. (1962): 5th Intern'l. Cong. Eletronmicr. (Ed. S. BREESE). **8** Jr. Acad. Press., W, 11 (*43*).
LARSON, P. R. (1958): USDA, Lake St. For. Exp. Stat. Tech. Note 538 (*54*).
LARSON, R. L. (1971): Phytochemistry **10**, 3073 (*199, 242*).
LA RUE, C. D. (1936): Proc. Natl. Acad. Sci. (USA) **22**, 255 (*248*).
LAVIE, P. (1958): Thesis, Paris. (Abstract XVIII Internatl. Bienenzüchter Kongreß, Rome (*99*).
LAVIE, P. (1960): Ann. Abeille **3**, 201 (*99*).
LAYNE, R. E. C. (1963): I. Effect of vacuum-drying, freeze-drying and storage environment on the viability of pea pollen. II. Effect of boron, sucrose and agar on the germination of pea pollen. Ph. D. Thesis. Madison: Univ. Wisconsin (*62*).
LAYNE, R. E. C., HAGEDORN, D. J. (1963): Crop. Sci. **3**, 433 (*62*).
LEA, D. J., SEHON, A. H. (1962): Int'l. Arch. Allergy Appl. Immunol. **20**, 203 (*132*).
LEAL, F. J. (1964): Influence of storage methods on the viability of *Citrus* and *Poncirus* pollen. M. S. Thesis. Gainesville: Univ. Florida (*62*).
LEBEDEV, S. I. (1947): Dokl. Acad. Nauk SSSR **58**, 85 (*229*).
LEBEDEV, S. I. (1948): Dokl. Acad. Nauk SSSR **59**, 987 (*228, 229*).
LEBEDEV, S. I. (1955): Prioroda **6**, 42 (*228*).
LEE, D. W., FAIRBROTHERS, D. E. (1969): Brittonia **21**, 227 (*162*).
LEITCH, T. A. T. (1971): Oleagineux **26**, 305 (*52*).
LENDZIAN, K., SCHÄFER, E. (1973): Phytochemistry **12**, 1227 (*201*).
LEON, L. O. (1963): Investigations of the post-pollination phenomenon in the column of orchids. M. S. Thesis. Gainesville: Univ. of Fla., 41 pp. (*248*).
LEPAGE, M., BOCH, R. (1968): Lipids **3**, 530 (*93, 228*).
LESKOVCEVA, I. I., KORCKOV, V. S. (1965): Lesn. Hoz. **18**, 57 (*52*).
LEVIN, N., KNOWLTON, H. E. (1951): Ext. Bull. Utah State Agricul. Coll., Logan, No. **237**, 4 (*105*).
LEWIS, D. (1944): Nature **153**, 575 (*53*).
LEWIS, D. (1952): Proc. Roy. Soc. (London) **B140**, 127 (*161*).
LEWIS, D. (1954): Adv. Genet. **6**, 235 (*244*).
LEWIS, D. (1955): Sci. Prog. **43**, 590 (*79*).
LEWIS, D., BURRAGE, S., WALLS, D. (1967): J. Exp. Botany **18**, 371 (*22, 201, 207*).
LEWIS, W. (1759): In: The chemical works of Casper Neumann. Abridged and methodized, p. 231. London (*230*).
LICHTE, H. F. (1957): Angew. Bot. **31**, 1 (*62, 66*).
LIEFSTINGH, G. (1953): Pract. Verslag Inst. Veredeling Landbouwg. Wageningen, Netherlands (*59*).
LINDER, R., COUSTAUT, D. (1966): Compt. Rend. **263**, 1447 (*157*).
LINSKENS, H. F. (1955): Z. Botan. **43**, 1 (*43, 79*).
LINSKENS, H. F. (1958): Acta Botan. Neerl. **7**, 61 (*182, 183, 186, 188*).
LINSKENS, H. F. (1960): Z. Botan. **48**, 126 (*162*).
LINSKENS, H. F. (1964): Ann. Rev. Plant Physiol. **15**, 255 (*56*).
LINSKENS, H. F. (1966): Planta **69**, 79 (*8, 160, 196, 204*).
LINSKENS, H. F. (1967): Planta **73**, 194 (*43, 190, 213*).

LINSKENS, H. F. (1971): In: HESLOP-HARRISON, J. (Ed.): Pollen development and physiology, p. 232. London: Butterworths (*188, 190*).

LINSKENS, H. F., VAN BRONSWIJK, J. E. M. H. (1974): Biol. Rdsch. **12**, 4 (*44, 163*).

LINSKENS, H. F., ESSER, K. (1957): Naturwissenschaften **44**, 16 (*77, 134*).

LINSKENS, H. F., HAVEZ, R., LINDER, R., SALDEN, M., RANDOUX, A., LANIEZ, D., COUSTAUT, D. (1969): Compt. Rend. **269**, 1855 (*197, 216*).

LINSKENS, H. F.; HEINEN, W. (1962): Z. Botan. **50**, 338 (*22, 197, 200, 207*).

LINSKENS, H. F., JORDE, W. (1974): Naturwissenschaften **61**, 275 (*181*).

LINSKENS, H. F., KOCHUYT, A. S. L., SO, A. (1968): Planta **82**, 111 (*156*).

LINSKENS, H. F., KROH, M. (1967): In: RUHLAND, W. (Ed.): Encyclopedia of Plant Physiology, Vol. 18, p. 506. Berlin-Heidelberg-New York: Springer (*43, 73*).

LINSKENS, H. F., MULLENEERS, J. M. L. (1967): Acta Botan. Neerl. **16**, 132 (*84*).

LINSKENS, H. F., PFAHLER, P. L. (1973): Theor. Appl. Genet. **43**, 49 (*157*).

LINSKENS, H. F., SCHRAUWEN, J. (1964): Biol. Plant. (Praha) **5**, 239 (*8*).

LINSKENS, H. F., SCHRAUWEN, J. (1968a): Naturwissenschaften **55**, 91 (*8*).

LINSKENS, H. F., SCHRAUWEN, J. (1968b): Proc. Kon. Nederl. Akad. Wet. Ser. C, **71**, 267 (*8, 185, 187*).

LINSKENS, H. F., SCHRAUWEN, J. (1969): Acta Botan. Neerl. **18**, 605 (*154, 156, 196*).

LINSKENS, H. F., SCHRAUWEN, J. A. M., KONINGS, R. N. H. (1970): Planta **90**, 153 (*43*).

LLOYD, F. E., ULEHLA, V. (1926): Trans. Roy. Soc. Canada **20**-V, 45 (*85*).

LOEB, L. F. (1928): Klin. Wschr. **7**, 1078 (*168*).

LOERE, VAN, O. (1971): Apidologie **2**, 197 (*103*).

LÖVE, A. (1952): Hereditas (Lund) **38**, 11 (*28*).

LOEWUS, M. W., LOEWUS, F. (1971): Plant Physiol. **48**, 255 (*197, 203, 209*).

LOHNIS, M. P. (1940): Mededeel. Landbouwhog. Wageningen **44**, 3 (*123*).

LØKEN, A. (1958): Norsk Hogetid. **74**, 216 (*104*).

LOO, T. L., HWANG, T. C. (1944): Am. J. Botany **31**, 356 (*257*).

LOPRIORE, G. (1928): Ber. Deut. Botan. Ges. **46**, 413 (*196*).

LOUVEAUX, J. (1955): Physiol. Comp. Oecol. **4**, 1 (*93, 96*).

LOUVEAUX, J. (1958a): Ann. Abeille **1**, 113 (*92, 93*).

LOUVEAUX, J. (1958b): Ann. Abeille **1**, 197 (*92, 93*).

LOUVEAUX, J. (1959): Ann. Abeille **2**, 13 (*92, 93*).

LOUVEAUX, J. (1963): Ann. Nutr. (Paris) **17**, 313 (*103*).

LUBLINER-MIANOWSKA, K. (1955): Acta Soc. Botan. Polon. **24**, 609 (*225, 226, 228, 239*).

LUKOSCHUS, F. (1957): Z. Bienenforsch. **4**, 3 (*41, 88*).

LUND, H. A. (1956): Plant Physiol. **31**, 334 (*249, 257*).

LUNDÉN, R. (1954): Svensk Kem. Tidskr **66**, 201 (*121, 129, 132, 210*).

MACDANIELS, L. H., HILDEBRAND, E. M. (1939): Proc. Am. Soc. Hort. Sci. **37**, 137 (*49*).

MACDONALD, T. (1969): Isozyme studies of esterase 1-aspartate, 2-oxoglutarate, aminotransferase and carbohydrases in ZEA MAYS. Ph. D., Thesis. Honolulu: Univ. Hawaii (*196, 200*).

MACIEJEWSKA-POTAPCZYK, W., URBANEK, H., KULEC, I., PACYK, H. (1968): Zesz. Nauk. Univ. Lodz. Ser. 2, **30**, 63 (*183*).

MACKINNEY, G. (1968): In: GREENBERG, D. M. (Ed.): Metabolic pathways, p. 221. New York: Academic Press (*226, 228*).

MÄKINEN, Y., BREWBAKER, J. L. (1967): Physiol. Plantarum **20**, 477 (*196, 197, 207*).

MÄKINEN, Y., MACDONALD, T. (1968): Physiol. Plantarum **21**, 477 (*195, 200, 201*).

MAHADEVAN, S. (1963): Arch. Biochem. Biophys. **100**, 557 (*256*).

MAHESWARI, P. (1949): Bot. Rev. **15**, 1 (*12*).

MAHESWARI, P. (1958): (Ed.), Modern developments in plant physiology. Delhi: Univ. (*55*).

MAHESWARI, P., NARAYANASWAMI, S. (1952): J. Linnean Soc. **53**, 474 (*6*).

MAHESWARI, S. C., PRAKASH, R. (1965): Physiol. Plantarum **18**, 841 (*183*).

MAI, G. (1934): Jb. wiss. Bot. **79**, 681 (*248*).

MAKHANETS, A. (1968): Ukr. Botan. J. **25**, 38 (*186*).

MALLEY, A., REED, C. E., LIETZE, A. (1962): J. Allergy **33**, 84 (*170*).

MALIK, C. P., TEWARI, H. B., SOOD, P. P. (1970): Portug. Acta Biol. **11**, Ser. A, 245 (*197, 216*).

MALYUTIN, N. I. (1969): Botan. Zhur. **54**, 1050 (*245*).

MAMELI, E., CARRETTA, U. (1941): Ann. Chim. Farm. **1941**, 19 (*226*).

MANARESI, A. (1921): Stat. Sper. Agric. Ital. **45**, 807 (*69*).
MANARESI, A. (1924): Stat. Sper. Agr. Ital. **48**, 33 (*61*).
MANDAVA, N., MITCHELL, J. W. (1972): Chem. Ind. 930 (*252*).
MANDAVA, N., SIDWELL, B. A., MITCHELL, J. W., WORLEY, J. F. (1973): Ind. Eng. Chem. 138 (*252*).
MANDELL, M. (1967): Ann. Allergy **25**, 35 (*177*).
MANGELSDORF, P. C. (1932): J. Heredity **23**, 288 (*132*).
MANGIN, L. (1886): Bull. Soc. Botan. France **33**, 512 (*56*).
MANGIN, L. (1889): Bull. Soc. Botan. France **36**, 283 (*13*).
MANGIN, L. (1890): Compt. Rend. **110**, 644 (*56*).
MANZOS, A. M. (1960): Lesn. Z. Arhangel'sk **3**, 34 (*82*).
MARQUARDT, P., VOGG, G. (1952): Arzneimittel-Forsch. **2**, 267 (*99*, *155*).
MARSH, D. G., HADDAD, Z. H., CAMPBELL, D. H. (1970): J. Allergy **46**, 107 (*177*).
MARSH, D. G., LICHTENSTEIN, L. M., CAMPBELL, D. H. (1970a): Immunology **18**, 705 (*172*).
MARSH, D. G., MILNER, F. H., JOHNSON, P. (1966): Int. Arch. Allergy **29**, 521 (*169*, *170*).
MARTENS, N., VAN LAERE, O., PELERENTS, C. (1964): Biol. Jaarb. (Gent) **32**, 292 (*106*).
MARTENS, P., WATERKEYN, L. (1962): Cellule **62**, 173 (*135*, *137*).
MARTIN, E. C., MCGREGOR, S. E. (1973): Ann. Rev. Entomol. **18**, 207 (*92*).
MARTIN, F. W. (1959): Stain Technol. **34**, 125 (77).
MARTIN, F. W. (1968): Phyton **25**, 97 (*201*, *207*).
MARTIN, P. G. (1960): Heredity **14**, 125 (9).
MARTIN, P. S., SHARROCK, F. W. (1964): Am. Antiquity **30**, 168 (*45*).
MASCARENHAS, J. P. (1970): Biochem. Biophys. Res. Comm. **41**, 142 (*216*).
MASCARENHAS, J. P. (1971): In: HESLOP-HARRISON, J. (Ed.): Pollen development and physiology, p. 201. London: Butterworths (*187*, *188*, *190*, *203*).
MASCARENHAS, J. P., BELL, E. (1969): Biochim. Biophys. Acta **179**, 199 (*188*).
MASCARENHAS, J. P., BELL, E. (1970): Develop. Biol. **21**, 475 (*124*).
MASCARENHAS, J. P., MACHLIS, L. (1962): Vitamins and Hormones **20**, 374 (*257*).
MASKIN, S. I. (1960): Bot. Zhur. **45**, 547 (*54*).
MASSART, J. (1902): Bull. Jard. Botan. Bruxelles **1** (3), 1 (*247*).
MATHUR, J. M. S. (1969): Curr. Sci. **38**, 570 (*246*).
MATSUBARA, S. (1969): Botan. Mag. (Tokyo) (*257*).
MATSUBARA, S. (1971): In: HESLOP-HARRISON, J. (Ed.): Pollen development and physiology, p. 186. London: Butterworths (*153*, *257*).
MAURER, M. L., STRAUSS, M. B. (1961): J. Allergy **32**, 347 (*174*).
MAURIN, A. M., KAUROV, I. A. (1956): Botan. Zhur. **41**, 81 (*82*).
MAURIZIO, A. (1942): Schweiz. Bienen-Ztg. **65**, 524 (*97*).
MAURIZIO, A. (1944): Verhandl. Schweiz. Naturforsch. Ges. **124**, 128 (*99*).
MAURIZIO, A. (1945a): Beih. Schweiz. Bienen-Z. **1**, 337 (*104*).
MAURIZIO, A. (1945b): Beih. Schweiz. Bienen-Z. **1**, 430 (*104*).
MAURIZIO, A. (1949): Beih. Schweiz. Bienen-Z. **2**, 320 (*106*).
MAURIZIO, A. (1950): Schweiz. Bienen-Z. **73**, 58 (*103*).
MAURIZIO, A. (1953): Beih. Schweiz. Bienen-Z. **20**, 486 (*95*, *96*).
MAURIZIO, A. (1954): Bee World **35**, 49 (*44*).
MAURIZIO, A. (1955): Z. Bienenforsch. **3**, 32 (*103*, *106*).
MAURIZIO, A. (1958a): Ann. Abeille **2**, 93 (*99*, *106*).
MAURIZIO, A. (1958b): Z. Bienenforsch. **4**, 1 (*103*, *105*).
MAURIZIO, A. (1959): XVIII Internatl. Bienenzüchterkongreß (Bologna-Rome, 1958) 1 (*99*).
MAURIZIO, A. (1962): Dtsch. Bienenweide. **13**, 235 (*107*).
MAURIZIO, A., LOUVEAUX, J. (1965): Rev. Palaeobotan. Palynol. **3**, 291 (*106*).
MCGUIRE, D. C. (1952): Proc. Am. Soc. Hort. Sci. **60**, 419 (*58*).
MCILWAIN, D. L., BALLOU, C. E. (1966): Biochemistry **5**, 4054 (*149*).
MCMORRIS, T. C., BARKSDALE, A. W. (1967): Nature **215**, 320 (*153*).
MCWILLIAM, H. R. (1959): Silvae Genet. **8**, 11 (*36*).
MEEUSE, J. D. (1961): The story of pollination. New York: Ronald Press (*25*, *88*, *109*, *111*).
MELAM, H., PRUZANSKY, J. J., PATTERSON, R. (1970): J. Allergy **45**, 43 (*177*).
MERCER, E. H., DAY, M. R. (1952): Biol. Bull. **103**, 384 (*100*).

MERRILL, T. A., JOHNSON, S. (1939): Proc. Am. Soc. Hort. Sci. **37**, 617 (*54*).

MICHAELIS, P. (1928): Biol. Zbl. **48**, 370 (*29*).

MICHALSKI, L. (1958): Acta Soc. Botan. Polon. **27**, 75 (*249, 250*).

MICHALSKI, L. (1967): Acta Soc. Botan. Polon. **36**, 475 (*251, 252*).

MICHALSKI, L., CHROMINSKI, A. (1960): Zes. Nauk Univ. Torum Biol. **4**, 65 (*250*).

MICHELBACHER, A. E., SMITH, R. F., HURD, JR., P. D. (1964): Calif. Agr. **18**, 2 (*110*).

MICHEL-DURAND, E. (1938): Compt. Rend. **206**, 1673 (*140*).

MIKI-HIROSIGE, H. (1961): Mem. Coll. Sci. U. Kyoto. Ser. B. **28**, 375 (Translation-U.S. Dept. of Comm. OTS 60-51185) (*69*).

MIKKELSEN, V. M. (1949): Physiol. Plantarum **2**, 323 (*29*).

MILBORROW, B. V., PRYCE, R. J. (1973): Nature **243**, 46 (*252*).

MILLETT, M. R. O. (1944): Australian Forest. Bur. Leaflet **59**, 1 (*48*).

MILNER, F. H., DYBAS, B., FRASER, C. A. (1972): Clin. Allergy **2**, 79 (*169*).

MINAEVA, V. G., GORBALEVA, G. N. (1967): In: SOBDEVSKAYA, K. A: (Ed.): Polez. Rast. Prir. Flory. Sib. Novosibirisk, USSR: Nauk. Sib. Otd., 231 (Cited from Chem. Abst. **69**, 1039050, 1968) (*243, 246*).

MIRAVALLE, R. J. (1965): Empire Cotton Growing Rev. **42**, 287 (*72, 76*).

MIROV, N. T. (1967): The genus *Pinus*. New York: Ronald Press (*47*).

MITCHELL, J. W., MANDAVA, N., WORLEY, J. F., DOWNE, M. E. (1971): J. Agr. Food Chem. **19**, 391 (*252*).

MITCHELL, J. W., MANDAVA, N., WORLEY, J. F., FLIMMER, J. R., SMITH, M. V. (1970): Nature **225**, 1065 (*252*).

MITCHELL, J. W., WHITEHEAD, M. R. (1941): Botan. Gaz. **102**, 770 (*248, 256*).

MIYAKE, S. (1922): J. Biochem. (Japan) **2**, 27 (*120, 150*).

MIZUNO, T. (1958): Nippon Kagaku Zasschi (Japan. J. Pure Chem.) **79**, 192 (*120, 130, 136*).

MODLIBOWSKA, I. (1942): J. Heredity **33**, 187 (*78*).

MÖBIUS, M. (1923): Ber. Deut. Botan. Ges. **41**, 12 (*225*).

MOEWUS, F. (1950): Biol. Zbl. **69**, 181 (*243, 244*).

MOEWUS, F., SCHADER, E. (1952): Beitr. Biol. Pfl. **29**, 171 (*248*).

MOFFET, A. A. (1934): J. Pomol. Hort. Sci. **12**, 321 (*72, 74*).

MOHL, VON, H. (1835): Ann. Sci. Nat. Botan. **3**, 148 (*45*).

MOHL, VON, H. (1835): Ann. Sci. Nat. Botan. **3**, 304 (*71*).

MOLISCH, H. (1893): Sitzber. Akad. Wiss. Wien. Math.-Naturwiss. Kl. **102**, 423 (*56*).

MOORE, M. B., CROMWELL, H. W., MOORE, E. E. (1931): J. Allergy **2**, 6 (*238*).

MOORE, M. B., MOORE, C. E. (1931): J. Insect Physiol. **9**, 391 (*168, 238*).

MORGANDO, A. (1949): Sci. Genet. (Torino) **3**, 183 (*58*).

MORITA, K. (1918): Botan. Mag. (Tokyo) **32**, 39 (*247*).

MORRIS, R. F. (1951): Forest. Chron. **27**, 40 (*111*).

MORRIS, S. J., THOMPSON, R. H. (1963): Tetrahedron Letters **2**, 101 (*230*).

MOSES, M. J., TAYLOR, J. H. (1955): Exp. Cell Res. **9**, 474 (*185, 186*).

MOSS, G. S., HESLOP-HARRISON, J. (1967): Ann. Botany **31**, 554 (*186, 187*).

MOTOMURA, Y., WATANABE, T., ASO, K. (1962): Tohuku J. Agr. Res. **13**, 237 (*131*).

MÜHLDORF, A. (1939): Ber. Deut. Botan. Ges. **57**, 299 (*15*).

MÜHLETHALER, K. (1953): Mikroskopie **8**, 103 (*19*).

MÜHLETHALER, K. (1955): Planta **46**, 1 (*19*).

MÜHLETHALER, K., LINSKENS, H. F. (1956): Experientia **12**, 253 (*213*).

MÜLLER, R. (1953): Beitr. Biol. Pfl. **30**, 1 (*248*).

MÜLLER-STOLL, W. R. (1956): Grana Palynol. **1**, 38 (*218*).

MÜNTZING, A. (1928): Hereditas (Lund) **10**, 241 (*28*).

MUIR, R. M. (1947): Proc. Nat. Acad. Sci. U.S.A. **33**, 303 (*257*).

MUMFORD, R. A., LIPKE, H., LAUFER, D. A., FEDER, W. A. (1972): Environ. Sci. Technol. **6**, 427 (*49, 156*).

MURAVALLE, R. J. (1964): J. Heredity **55**, 276 (*53*).

MURPHY, J. B. (1973): Private Communication. (*198*).

NAGORAJAN, C. R., KRISHNAMURTHI, S., RAO, V. N. M. (1965): South Indian Hort. **13**, 1 (*81*).

NAIR, M. K., NARASIMHAN, R. (1963): Stain Technol. **38**, 341 (*76*).

NAKAMURA, G. R., BECKER, E. R. (1951): Arch. Biochem. Biophys. **33**, 78 (*206*).

NANDA, K. (1972): Structure, development and function of the anther tapetum. Ph. D. Thesis, Univ. of Delhi (*186, 187*).

NARASIMHAN, R. (1963): Stain Technol. **38**, 340 (*72*).

NATERMAN, H. L. (1957): J. Allergy **28**, 76 (*44*).

NATERMAN, H. L. (1965): J. Allergy **36**, 226 (*174*).

NATION, J. L., ROBINSON, F. A. (1966): Science **152**, 1765 (*106*).

NATION, J. L., ROBINSON, F. A. (1968): Ann. Entomol. Soc. Am. **61**, 514 (*106*).

NAUDIN, CH. (1858): Ann. Sci. Nat. Botan. IV **9**, 257 (*123*).

NAVARA, A., POSPÍSILOVÁ, D. (1962): Pol'nohospodarotvo. **9**, 659. [From Biol. Abst. 45(2), 2616, 1964.] (*131*).

NEAMTU, G. (1971): Stud. Cerc. Biochem. **14**, 215 (*229*).

NEAMTU, G., BODEA, C. (1970): Stud. Cerc. Biochem. **13**, 307 (*229*).

NEAMTU, G., ILLYES, G., BODEA, C. (1969): Stud. Cerc. Biochem. **12**, 77 (*227*).

NEBEL, B. R. (1931): Stain Technol. **6**, 27 (*76*).

NEBEL, B. R. (1939): Proc. Am. Soc. Hort. Sci. **37**, 130 (*58*).

NEBEL, B. R., RUTTLE, M. L. (1937): J. Pomol. **14**, 347 (*58, 64*).

DE NETTENCOURT, D., ERIKSSON, G. (1968): Hereditas **57**, 168 (*221*).

NEWCOMBER, E. H. (1939): Bull. Torrey Botan. Club **66**, 121 (*59*).

NEWELL, J. M. (1942): J. Allergy **13**, 177 (*171*).

NEWMARK, F. M., ITKIN, I. H. (1967): Ann. Allergy **25**, 251 (*148, 173*).

NIELSEN, N. (1956): Acta Chem. Scand. **10**, 332 (*64, 211*).

NIELSEN, N., GRÖMMER, L., LUNDÉN, R. (1955): Acta Chem. Scand. **9**, 1100 (*120, 129, 130, 155, 210*).

NIELSEN, N., HOLMSTRÖM, B. (1957): Acta Chem. Scand. **11**, 101 (*197, 203, 210*).

NIKOLAEVA, Z. V. (1962): Tashkent Gos. Univ. Lenin. Nauk Trudy **204**, 115 (Transl. U.S. Dept. Comm. TT 68-50430) (*68*).

NILSSON, M. (1956): Acta Chem. Scand. **10**, 413 (*132, 150*).

NILSSON, M., RYHAGE, R., SYDOW, v. E. (1957): Acta Chem. Scand. **11**, 634 (*150*).

NISHIDA, K. (1935): Bull. Agr. Chem. Soc. Japan **11**, 143 (*156, 160*).

NISSEN, O. (1950): Agron. J. **42**, 136 (*28*).

NITSCH, J. P., NITSCH, C. (1969): Science **163**, 85 (*193*).

NOON, L. (1911): Lancet **1**, 1572 (*168*).

NORMAN, P. S., LICHTENSTEIN, L. M., WINKENWERDER, W. L. (1970): J. Allergy **45**, (Abst.), 114 (*174*).

NORMAN, P. S., WINKENWERDER, W. L., LICHTENSTEIN, L. M. (1972): J. Allergy Clin. Immunol. **50**, 31 (*174*).

NORTON, J. D. (1966): Proc. Am. Soc. Hort. Sci. **89**, 132 (*84*).

NYGAARD, P. (1969): Fed. Europ. Biochem. Soc. 6th Meeting, Abst. 357 (*24*).

NYGAARD, P. (1972): Physiol. Plantarum **26**, 29 (*191*).

NYGAARD, P. (1973): Physiol. Plantarum **28**, 361 (*191, 202, 211*).

OBERLE, G. D., GOERTZEN, K. L. (1952): Proc. Am. Soc. Hort. Sci. **59**, 263 (*31*).

OBERLE, G. D., WATSON, R. (1953): Proc. Am. Soc. Hort. Sci. **61**, 299 (*84*).

OGDEN, E. C., HAYES, J. V., RAYNOR, G. S. (1969): Am. J. Botany **56**, 16 (*47*).

OGUR, M., ERICKSON, R. O., ROSEN, G. U., SAX, K. B., HOLDEN, C. (1951): Exp. Cell Res. **2**. 73 (*186, 187*).

OHTANI, M. (1955): Salk. Gabuj. Hokuku Nogaku **7**, 21 (*136*).

OKADA, I., OKA, A., SUGIYAMA, A. (1968): Tamag. Dai. Nog. Ken. Hokoku **7/8**, 175 (*99*).

O'KONUKI, K. (1933): Science **12**, 221 (*58*).

OKUNUKI, K. (1937): Acta Phytochem. **9**, 267 (*202*).

OKUNUKI, K. (1939): Acta Phytochem. **11**, 27 (*194, 196, 198, 199, 201, 211*).

OKUNUKI, K. (1939a): Acta Phytochem. **11**, 65 (*196*).

OKUNUKI, K. (1940): Acta Phytochem. **11**, 249 (*196*).

OLMO, H. P. (1942): Proc. Am. Soc. Hort. Sci. **41**, 219 (*58, 61*).

OSLER, A. G., LICHTENSTEIN, L. M., LEVY, D. A. (1968): Advan. Immunol. **8**, 183 (*178*).

OSTAPENKO, V. I. (1956): Dokl. Vsesoyuz. Akad. Sel'skokhoz. Nauk im V. I. Lenina **21**, 15 (*82*).

OSTAPENKO, V. I. (1960): Fizol. Rast. (Engl. Transl.) **7**, 444 (*196, 205*).

OVCINNIKOV, N. N. (1951): Selekt. Semenovodstvo **18**, 79 (*28*).

OVCINNIKOV, N. N. (1953): Acad. Selsko Khozicust. Nauk. Dokl. Vashnil **13**, 20 (*28*).

OVERLEY, F. L.; BULLOCK, R. M. (1947): Proc. Am. Soc. Hort. Sci. **49**, 163 (*79*).

OWCZARZAK, A. (1952): Stain Technol. **27**, 249 (*81*).

PAIN, J., MAUGENET, J. (1966): Ann. Abeille **9**, 209 (*99*).

PALUMBO, R. F. (1953): A cytological investigation of enzyme activities in the developing pollen of *Tradescantia paludosa* and *Lilium longiflorum* var. craft. Ph. D. Thesis. Seattle: Univ. of Washington. (*197, 201*).

PANDEY, K. K. (1962): Am. J. Botany **49**, 874 (*79*).

PANDEY, K. K., HENRY, R. D. (1958): Stain Technol. **34**, 19 (*77*).

PANKOW, H. (1957): Flora **146**, 240 (*18, 22*).

PANZANI, R. (1962): Rev. Franc. Allerg. **3**, 164 (*148*).

PARAG, Y., FEINBRUN, N., TAS, J. (1957): Bull. Res. Council Israel **60**, 5 (*50*).

PARCHMAN, L. G., LIN, K. C. (1972): Nature New Biol. **239**, 237 (*184*).

PARK, O. W. (1928): Bull. Agricul. Exp. Stat. Iowa **108**, 184 (*92*).

PARKER, R. L. (1926): Mem. Cornell Univ. Agric. Exp. Stat. No. **98**, 7 (*89, 95, 100, 102, 108*).

PARNELL, F. R. (1921): J. Genet. **11**, 209 (*131*).

PARTHASARATHY, M. V. (1970): Principes **14**, 55 (*22*).

PASSEGER, K. (1930): Gartenbauwissenschaften **3**, 201 (*65*).

PATEL, N. G., HAYDAK, M. H., LOVELL, R. (1961): Nature **191**, 362 (*103*).

PATON, J. B. (1921): Am. J. Botany **8**, 471 (*132, 197, 204*).

PAVLOVSKY, E. N., ZARIN, E. J. (1922): Quart. J. Microscop. Sci. **66**, 509 (*102*).

PAYNE, R. C., FAIRBROTHERS, D. E. (1973): Am. J. Botany **60**, 182 (*160, 204*).

PERCIVAL, M. S. (1947): New Phytol. **46**, 142 (*96, 97*).

PERCIVAL, M. S. (1950): New Phytol. **49**, 40 (*47, 96*).

PERCIVAL, M. S. (1955): New Phytol. **54**, 353 (*96, 97*).

PERRSON, A. (1955): Z. Forstgenet. **4**, 129 (*36*).

PETERKA, V. (1939): Schweiz. Bienen-Z. **62**, 143 (*99*).

PETRÙ, E., HRABĚTOVÁ; E., TUPÝ, J. (1964): Biol. Plant. (Praha) **6**, 68 (*53*).

PFAHLER, P. L. (1965): Crop. Sci. **5**, 597 (*68*).

PFAHLER, P. L. (1966): Can. J. Botany **45**, 839 (*68*).

PFAHLER, P. L. (1973): Radiation Botan. **13**, 13 (*65*).

PFAHLER, P. F., LINSKENS, H. F. (1970): Theor. Appl. Genet. **40**, 6 (*155, 157*).

PFAHLER, P. F., LINSKENS, H. F. (1972): Theor. Appl. Genet. **42**, 136 (*66*).

PFAHLER, P. F., LINSKENS, H. F. (1973): Planta **111**, 253 (*57*).

PFEIFFER, N. E. (1936): Contrib. Boyce Thompson Inst. **8**, 141 (*58, 59*).

PFEIFFER, N. E. (1938): Contrib. Boyce Thompson Inst. **9**, 199 (*59*).

PFEIFFER, N. E. (1939): Contrib. Boyce Thompson Inst. **10**, 429 (*62*).

PFEIFFER, N. E. (1944): Contrib. Boyce Thompson Inst. **13**, 281 (*59*).

PFEIFFER, N. E. (1955): Contrib. Boyce Thomp. Inst. **18**, 153 (*59, 64, 222*).

PFEIFFER, N. E. (1956): Ber. Deut. Botan. Ges. **69**, 223 (*219*).

PFIRSCH, R. (1953): Bull. Assoc. Philomathem. Alsace Lorraine **9**, 121 (*62*).

PFUNDT, M. (1910): Jahrb. wiss. Botan. **47**, 1 (*56, 58*).

PHILLIPS, G. L. (1967): In: SHELDON, J. M., LOVELL, R. G., MATHEWS, K. P. (Eds.): Clinical allergy, 2nd Ed., p. 507. Philadelphia: W. B. Saunders (*174*).

PHOEBUS, P. (1862): Der typische Frühsommerkatarrh oder das sogenannte Heufieber, Heuasthma. Gießen: Rieker (*165*).

PICADO, C. (1921): Ann. Inst. Pasteur **35**, 893 (*161*).

PIECH, K. (1922): Kosmos (Lwow) **47**, 412 (*29*).

PIJL, VAN DER, L. (1966): In: HAWKES, J. G. (Ed.): Reproductive biology and taxonomy of vascular plants. London: Pergamon Press (*92*).

PINTO-LOPES, J. (1948): Portugal. Acta Biol. A **2**, 237 (*10*).

PIPKIN, JR., J. L., LARSON, D. A. (1973): Exp. Cell Res. **79**, 28 (*184*).

PIRQUET, C. VON (1906): Münch. Med. Wschr. **53**, 1457 (*165*).

PLANTA, A. VON (1886): Landw. Versvabk. Stat. **32**, 215 (*119, 138*).

PLAUT, M. (1957): Proc. Int. Seed Test. Ass. **22**, 1 (*83*).

PODDUBNAYA-ARNOLDI, V. A. (1936): Planta **25**, 502 (*11*).

PODDUBNAYA-ARNOLDI, V.A. (1960): Phytomorphology **10**, 185 (*218*).

PODDUBNAYA-ARNOLDI, V.A., ZINGER, N.V., PETROVSKAYA, T.P., POLUNINA, N.N. (1959): Rec. Advan. Botan. **1**, 682 (*198*, *199*).

PODDUBNAYA-ARNOLDI, V.A., ZINGER, N.V., PETROVSKAYA, T.P., POLUNINA, N.N. (1961): Tr. Glav. Botan. Sada Akad. Nauk S.S.S.R. **8**, 162 (*198*, *217*).

POHL, F. (1937): Beih. Botan. Cbl. A. **57**, 112 (*27*, *31*).

POHL, R. (1951): Biol. Zbl. **70**, 119 (*258*).

POLUNIN, N. (1951): Canad. J. Botany **29**, 206 (*51*).

PONCOVA, J. (1959): Czech. Akad. Zemedel, Ved. Sborn. Rostl. Vyroba **32**, 1637 (*59*).

POPA, O., SĂLĂJAN, GH., MOSOLOVA, L., IOZON, D. (1970): Inst. Agron. "Dr. Petru Groza" Luc. Stiit. Ser. Zootech. **26**, 143 (*113*).

PORLINGH. I.X. (1956): Ph. D. Thesis. Thessaloniki: Aristoteleison Univ. (*84*).

PORTYANKO, V.F., KUDRYA, L.M. (1966): Fisiol. Rast. **13**, 1086 (*121*).

PORTYANKO, V.F., LYZHENKO, I.I., PORTYANKO, A.V. (1971): Fisiol. Rast. (Transl.) **18**, 177 (*121*, *126*).

POTONIÉ, R., REHNELT, K. (1971): In: BROOKS, J., GRANT, P.R., MUIR, M., VAN GIJZEL, P., SHAW, G. (Eds.): Sporopollenin, p. 295. London: Academic Press (*140*, *141*).

PRATVIEL-SOSA, F., PERCHERON, F. (1972): Phytochemistry **11**, 1809 (*238*).

PRAUSNITZ, C., KUESTNER, H. (1921): Zbl. Bakt. Abt. **86**, 160 (*172*).

PRINTZ-ERDTMAN, G. (1963): Grana Palynolog. **4**, 339 (*45*).

PRITSCH, G. (1958): Arch. Geflügelzucht u. Kleintierk. **7**, 282 (*107*).

PRUZSINSZKY, S. (1960): Sitzber. Österr. Akad. Wiss., Mathem.-Naturwiss. Kl., Abt. I **169**, 43 (*56*).

PRYCE-JONES, J. (1944): Proc. Linnean Soc. (London) **155**, 129 (*104*).

PUBOLS, M.H., AXELROD, B. (1959): Biochim. Biophys. Acta **36**, 582 (*196*, *197*, *202*).

PYLINEV, V.M., DIACONU, P. (1961): Dokl. Mosk. Sel'skokhoz. Akad. **62**, 163 (*84*).

QUADRIO, M. (1928): Riv. Biol. **10**, 708 (*133*).

RAGHAVEN, V., BARUAH, H.K. (1956): Phyton **7**, 77 (*68*).

RASMUSON, H. (1918): Botan. Not. **287**, (*223*).

RAYNOR, G.S., OGDEN, E.C., HAYES, J.V. (1972): Agric. Meteorol. **9**, 347 (*36*).

RAZMOLOGOV, V.P. (1963): Byul. Glav. Botan. Sada Akad. Nauk S.S.S.R. **49**, 70 (*188*, *199*).

READ, R.W. (1964): Stain. Technol. **39**, 99 (*68*).

REDDY, P.R., GOSS, J.A. (1971): Am. J. Botany **58**, 721 (*80*).

REDEMANN, C.T. (1949): Biochemical studies of pollen of *Zea mays*. Ph. D. Thesis. East Lansing: Michigan State College (*256*).

REDEMANN, C.T., WITTWER, S.H., BALL, C.D., SELL, H.M. (1950): Arch. Biochem. **25**, 277 (*234*).

REEVES, R.G. (1928): Am. J. Botany **15**, 114 (*15*).

REITER, R. (1947): Ohio J. Sci. **47**, 137 (*97*).

REITSMA, T. (1969): Rev. Palaeobot. Palynol. **9**, 175 (*27*).

REITSMA, T. (1970): Rev. Palaeobot. Palynol. **10**, 39 (*16*).

REMPE, H. (1937): Planta 27, 93 (*34*).

RENNER, O. (1919): Z. Botan. **11**, 305 (*28*, *29*).

RENNER, O. (1922): Z. Indukt. Abstamm. Vererbungsl. **27**, 235 (*121*).

RENNER, O. (1934): Ber. Sächs. Akad. Wiss., Math.-Physik. Kl. **86**, 241 (*12*).

RENNER, O. (1958): Z. Naturforsch. **13b**, 399 (*244*).

RESNIK, M.E. (1958): Rev. Invest. Agric. Buenos Aires **12**, 311 (*59*).

REZNIK, H. (1957): Biol. Zbl. **76**, 352 (*244*).

REZNIKOVA, S.A. (1971): Ann. Univers. et A.R.E.R.S. **9**, 145 (*186*).

RICHERT, M.-T. (1971): Oléagineux **26**, 261 (*152*).

RICHTER, M., HARTER, J.G., SCHON, A.H., ROSE, B. (1957): J. Immunol. **79**, 23 (*169*).

RICHTER, S. (1929): Planta **8**, 154 (*24*).

RICHTER-LANDMANN, W. (1959): Planta **53**, 162 (*10*).

RIDI, EL, M.S., WAFA, M.H.A. (1947): Egypt Med. Ass. **30**, 124 (*151*).

RIDI, M.S., ABOUL WAFA, M.H. (1950): J. Roy. Egypt. Med. Ass. **33**, 168 (*209*).

RIGHTER, F.I. (1939): J. Forest. **37**, 574 (*71*, *72*).

RILEY, C.V. (1892): Mo. Botan. Gard. Ann. Rept. **3**, 99 (*111*).

RILEY, R., BENNETT, M. D. (1971): Nature **230**, 182 (*185*).
RISUENO, M. C., GIMENEZ-MARTIN, G., LOPEZ-SAEZ, J. F., GARCIA, M. I. R. (1969): Protoplasma **67**, 361 (*19*).
RITTINGHAUS, P. (1886): Verhandl. Naturwiss. Ver. Rheinland **43**, 123 (*56*).
ROBBINS, K. C., WU, H., HSIEH, B. (1966): Immunochemistry **3**, 71 (*170*).
RODKIEWICZ, B. (1969): Acta Soc. Botan. Pol. **29**, 211 (*184*).
ROEMER, T. (1915): Z. Pflanzenzücht. **2**, 83 (*56*).
ROGGEN, H. P. J. R. (1967): Acta Botan. Neerl. **16**, 1 (*196, 197, 203*).
ROGGEN, H. P. J. R. (1974): Proc. Symp. „Fertilization in Higher Plants" (H. F. LINSKENS Ed.), ASP. Amsterdam (*22, 63, 64*)
ROGGEN, H. P. J. R., VAN DIJK, A. J., DORSMAN, C. (1972): Euphytica **21**, 181 (*43*).
ROGGEN, H. P. J. R., STANLEY, R. G. (1969): Planta **84**, 295 (*134, 218*).
ROGGEN, H. P. J. R., STANLEY, R. G. (1971): Physiol. Plantarum **24**, 80 (*199, 202, 218*).
ROLAND, F. (1971): Grana **11**, 101 (*135, 136*).
ROMASHOV, N. V. (1957): Botan. Zhur. **42**, 41 (*48*).
ROSEN, W. G. (1964): J. Roy. Microscop. Soc. Ser. 3: **83**, 337 (*217*).
ROSEN, W. G. (1968): Ann. Rev. Plant Physiol. **19**, 435 (*213, 217*).
ROSEN, W. G. (1971): In: HESLOP-HARRISON, J. (Ed.): Pollen development and physiology, p. 177. London: Butterworths (*12*).
ROSENTHAL, C. (1967): Apicultura **20**, 11 (*207*).
ROTH, R. R., NELSON, T. (1942): J. Allergy **13**, 283 (*168*).
ROTHSCHILD, LORD (1956): Fertilization, 51. London: Methuen (*244*).
ROWE, A. H. (1939): J. Allergy **10**, 377 (*148, 173*).
ROWLEY, J. R. (1959): Grana Palynol. **2**, 3 (*142*).
ROWLEY, J. R. (1963): Grana Palynol. **3**, 3 (*19*).
ROWLEY, J. R. (1964): In: LINSKENS, H. F. (Ed.): Pollen physiology and fertilization, p. 59. Amsterdam: North Holland (*19*).
ROWLEY, J. R. (1967): Rev. Palaeobotan. Palynol. **3**, 213 (*12*).
ROWLEY, J. R., DUNBAR, A. (1967): Svensk. Botan. Tidsk. **61**, 49 (*19*).
ROWLEY, J. R., FLYNN, J. R. (1971): Cytobiology **3**, 1 (*137*).
ROWLEY, J. R., MÜHLETHALER, K., FREY-WYSSLING, A. (1959): J. Biochem. Biophy. Cytol. **6**, 537 (*21, 142*).
ROWLEY, J. R., SOUTHWORTH, D. (1967): Nature **213**, 703 (*19*).
RYBAKOV, M. N. (1961): Pchelovdstovo **38**, 15 (*103*).
SAAD, S. I. (1963): Pollen et Spores **5**, 17 (*16*).
SADASIVAN, T. S. (1962): In: MAHESHWARI, P., JOHRI, B. M., VASIL, I. K. (Eds.): Proc. Summer School of Botany, p. 371. Dajeeling 1960 (*42*).
SADEN-KREHULA, M., TAJIĆ, M., KOLBAH, D. (1971): Experientia **27**, 108 (*153*).
SAGROMSKY, H. (1947): Biol. Zbl. **66**, 140 (*211*).
SAITO, Y. (1967): Clin. Exp. Med. **44**(6), 1 (*114*).
SAKAMOTO, Y. (1969): Agric. Biol. Chem. (Tokyo) **33**, 818 (*239, 244*).
SAKATA, K., MORIZYA, K., HIURA, M. (1961): J. Coll. Diary Agricult. (Hokkaido) **1**, 95 (*96*).
SĂLĂJAN, GH. (1970): Inst. Agron. "Dr. Petru Groza" Luc. Stiit. Ser. Zootech. **26**, 165 (*113*).
SĂLĂJAN, GH. (1972): Cercetari cu privire la eficienta utilizarii polenului de porumb ca supliment proteic si biostimulator in hrana tineretului porcin si aviar. Ph. D. Thesis. Iasi: Inst. Agron. "Ian Ionescu de la Brad" (*113*).
SAMORODOVA-BIANKI, G. B. (1959): Fiziol. Rast. **6**, 99 (*228*).
SAMPATH, S., RAMANATHAN, K. (1951): J. Indian Botan. Soc. **30**, 40 (*29*).
SAMSONOVA, I. A., BÖTTCHER, F. (1966): Wiss. Z. Ernst-Moritz-Arndt-Univ. Greifswald **15**(2/3), 137 (*55*).
SANDERS, H. J. (1970): Chem. Eng. News **48**, 84 (*165*).
SANDSTEN, E. P. (1909): Res. Bull. Univ. Wisc. Agric. Res. Stat. **4**, 149 (*56*).
SANGER, J. M., JACKSON, W. T. (1971): J. Cell Sci. **8**, 303 (*219*).
SANNA, A. (1931): Ann. Chim. Appl. **30**, 397 (*104*).
SANTAMOUR, JR., F. S., NIENSTAEDT, H. (1956): J. Forest. **54**, 269 (*54*).
SARTORIS, G. B. (1942): Am. J. Botany **29**, 395 (*59*).
SARVAS, R. (1955): Z. Forstgenet. **4**, 137 (*36*).

SARVELLA, P. (1964): J. Heredity. **55**, 154 (*82, 84*).
SASSEN, M. M. A. (1964): In: LINSKENS, H. F. (Ed.): Pollen physiology and fertilization, p. 167. Amsterdam: North Holland (*10, 12, 145*).
SASSEN, M. M. A. (1964a): Acta Botan. Neerl. **13**, 175 (*213*).
SATO, J., MUTO, K. (1955): Res. Bull. Coll. Exp. For., Hokkaido Univ. **17**, 967 (*69*).
SAULS, J. W., BIGGS; R. H. (1970): Am. Soc. Hort. Sci. 67th Ann. Mtg. Abst. 341 (*254*).
SAUTER, J. J. (1969): Z. Pflanzenphysiol. **61**, 1 (*187, 189*).
SAUTER, J. J. (1971): In: HESLOP-HARRISON, J. (Ed.): Pollen development and physiology, p. 3. London: Butterworths (*8, 187, 189*).
SAUTER, J. J., MARQUARDT, H. (1967): Z. Pflanzenphysiol. **58**, 126 (*8*).
SAUTER, J. J., MARQUARDT, H. (1970): Z. Pflanzenphysiol. **63**, 15 (*194*).
SAVAGE, J. R. K. (1957): Stain Technol. **283**, 283 (*72*).
SAVELLI, R. (1940): Compt. Rend. **210**, 705 (*73*).
SAX, K. (1942): Proc. Nat. Acad. Sci. (USA) **28**, 303 (*9*).
SAX, K., EDMONDS, H. W. (1933): Botan. Gaz. **95**, 156 (*8, 9*).
SAYLOR, L. C., SMITH, B. W. (1966): Am. J. Botany **53**, 453 (*79*).
SCAMONI, A. (1949): Forstwiss. Cbl. **68**, 735 (*34*).
SCAMONI, A. (1955): Z. Forstgenet. **4**, 113 (*34, 46*).
SCAMONI, A. (1955): Z. Forstgenet. **4**, 145 (*24, 51*).
SCANDALIOS, J. G. (1964): J. Heredity **55**, 281 (*204*).
SCEPOTJEV, F. L., PABEGAILO, A. I. (1954): Dokl. Akad. Nauk. SSSR **98**, 289 (*74*).
SCHADE, C. (1957): Phytopathol. Z. **30**, 225 (*65*).
SCHADEWALDT, H. (1967): In: HANSEN, K., WERNER, M. (Eds.): Lehrbuch der klinischen Allergie. Herausg. Stuttgart: Thieme (*165*).
SCHAEFER, C. W., FARRAR, C. L. (1941): U.S. Bur. Entomol. Plant Quar E 531, 7 pp. (*103*).
SCHEIBE, F., LOEWE, G. (1969): Allergie u. Asthma **15**, 141 (*169*).
SCHILDKNECHT, H., BENONI, H. (1963): Z. Naturforsch. **18 B**, 45 (*69*).
SCHMID, R. (1970): Principes **14**, 39 (*110*).
SCHMIDT, W. (1918): Österr. Botan. Z. **67**, 313 (*34*).
SCHMITT, F. O.; JOHNSON, G. T. (1938): Ann. Missouri Botan. Gard. **25**, 455 (*9*).
SCHMUCKER, TH. (1932): Naturwissenschaften **20**, 839 (*68, 69*).
SCHOCH-BODMER, H. (1927): Flora **122**, 307 (*98*).
SCHOCH-BODMER, H. (1936): Ber. Schweiz. Botan. Ges. **45**, 62 (*27*).
SCHOCH-BODMER, H. (1937): Verhandl. Schweiz. Naturforsch. Ges. **117**, 149 (*29*).
SCHOCH-BODMER, H. (1938): Flora **133**, 69 (*29*).
SCHOCH-BODMER, H. (1939): Vierteljahrsschr. Naturforsch. Ges. Zürich **84**, 225 (*232, 240*).
SCHOCH-BODMER, H. (1940): J. Genet. **40**, 393 (*28, 98*).
SCHOELLER, W., GOEBEL, H. (1932): Biochem. Z. **251**, 223 (*256*).
SCHRAUWEN, J., LINSKENS, H. F. (1967): Acta Botan. Neerl. **16**, 177 (*73, 74*).
SCHREIBER, M. (1956): Zool. Beitr. NF. **2**, 1 (*102*).
SCHREINER, TH. (1952): Z. Vergl. Physiol. **34**, 278 (*102*).
SCHWAN, B., MARTINOVS, A. (1954): Medd. Lantbrukshogsköl. Stat. Husjursförsök **57**, 5 (*96*).
SCHWANITZ, F. (1950): Züchter **20**, 53 (*29*).
SCHWANITZ, F. (1952): Züchter **22**, 273 (*28, 29*).
SCHWARZENBACH, F. H. (1951): Helv. Chim. Acta **34**, 1064 (*228, 229, 230*).
SCHWARZENBACH, F. H. (1952): Experienta **8**, 28 (*257*).
SCHWARZENBACH, F. H. (1953): Vierteljahrsschr. Naturforsch. Ges. Zürich **98**, 1 (*228, 229, 230*).
SCOTT, H. G., STOJANOVICH, C. J. (1963): Fla. Entomol. **46**, 189 (*110, 111, 203*).
SCOTT, R. A. (1960): Cotton Defoliat. Physiol. Conf. **14**, 4 (*49*).
SCOTT, R. W., STROHL, M. J. (1962): Phytochemistry **1**, 189 (*150*).
SEARCH, R. W., STANLEY, R. G. (1970): Phyton (Argent.) **27**, 35 (*254*).
SEARS, P. B., METCALF, E. (1926): J. Genet. **17**, 33 (*133*).
SEDOV, E. N. (1955): Agrobiology **3**, 134 (*49*).
SEKI, H. (1954): Res. Repts. Fac. Textiles and Sericult., Shinshu Univ. **4**, 5 (*133*).
SEKINE, H., LI, T. (1942): Rept. Jap. Ass. Advan. Sci. **16**, 563 From: Chem. Abst. **44**, 2606 d, 1950 (*210, 226*).
SEN, B., SAINI, J. P. (1969): Curr. Sci. **38**, 19 (*68*).
SEN, B., VERMA, G. (1958): In: MAHESHWARI, P. (Ed.): Modern methods in plant physiology, p. 118. New Delhi: Minst. Sci. Res. (*68*).

SEREISKII, S. (1936): Bull. Soc. Natur. (Moscow) Biol. **45**, 456 (*248*).

SERIAN-BACK, E. (1961): Z. Bienenforsch. **5**, 234 (*103, 105*).

SESHAGIRIAH, K. H. (1941): Curr. Sci. **10**, 30 (*258*).

SEXSMITH, J. J., FRYER, J. R. (1943): Sci. Agric. (Ottawa) **24**, 95 (*68*).

SFAKIOTAKIS, E. M., SIMONS, D. H., DILLEY, D. R. (1972): Plant Physiol. **49**, 963 (*254*).

SHAGINYAN, E. G. (1956): Pchelovodstvo **33** (**11**), 45 (*104*).

SHARADAKOV, V. S. (1940): Dokl. Akad. Nauk. Sci. USSR **26**, 267 (*82*).

SHAW, G. (1971): In: BROOKS, J., GRANT, P. R., MUIR, M., VAN GIJZEL, P., SHAW, G. (Eds.): Sporopollenin, p. 305. London: Academic Press (*19*).

SHAW, G., YEADON, A. (1964): Grana Palynol. **5**, 247 (*140, 141*).

SHAW, G., YEADON, A. (1966): J. Chem. Soc. (C), 16 (*19, 139*).

SHEESLEY, B., PODUSKA, B. (1969): Calif. Agricult. **23** (**10**), 14 (*102*).

SHELDON, J. M., LOVELL, R. G., MATHEWS, K. P. (1967): A manual of clinical allergy, 2nd Ed. Philadelphia: W. B. Saunders (*175*).

SHELLARD, E. J., JOLLIFFE, G. H. (1968): J. Chromatog. **38**, 257 (*156*).

SHELLARD, E. J., JOLLIFFE, G. H. (1969): J. Chromatog. **40**, 458 (*129*).

SHELLHORN, S. J., HULL, H. M., MARTIN, P. S. (1964): Nature **202**, 315 (*77*).

SHERIDAN, W. F. (1972): Z. Pflanzenphysiol. **68**, 450 (*43, 258*).

SHERIDAN, W. F., STERN, H. (1967): Exp. Cell Res. **45**, 322 (*8*).

SHERMAN, W. B. (1968): Hypersensitivity, mechanism and management: London: W. B. Saunders (*176*).

SILEN, R. R., COPES, D. L. (1972): J. Forest. **70**, 145 (*36*).

SIMPSON, J. (1955): Quart. J. Microscop. Sci. **96**, 117 (*102*).

SINGH, S. (1950): Mem. Cornell Univ. Exp. Stat. Ithaca No. **288**, (*92*).

SINGH, S., BOYNTON, D. (1949): Proc. Am. Soc. Hort. Sci. **53**, 148 (*99*).

SINKE, N. M., SIGENAPA, HIRAOKA, T. (1954): Mem. Coll. Sci. U. Kyoto Ser. B **21**, 63 (*82, 198*).

SISA, M. (1930): J. Sci. Agric. Soc. Jap. **323**, 88 (*73*).

SITTE, P. (1953): Mikroskopie **8**, 290 (*19*).

SITTE, P. (1959): Z. Naturforsch. **14b**, 575 (*18*).

SKARZINSKY, B. (1933): Nature **131**, 766 (*150*).

SKREBTZOVA, N. D. (1957): Pielowodstwo **4**, 1 (*88*).

SKVARLA, J. J., LARSON, D. A. (1963): Science **140**, 173 (*19*).

SKVARLA, J. J., LARSON, D. A. (1966): Am. J. Botany **53**, 1112 (*17, 142*).

SKVARLA, J. J., TURNER, B. L. (1966): Ann. Missouri Botan. Gar. **53**, 220 (*213*).

SLADEN, F. W. L. (1912): Nature **88**, 586 (*88*).

SLADEN, F. W. L. (1913): Queen-rearing in England. London: Madgwick, Houlston, + Coltd. (*88*).

SNODGRASS, R. E. (1925): Anatomy and physiology of the honey bee. New York-London: McGraw Hill (*100*).

SNOPE, A. J., ELLISON, J. H. (1963): New Jersey Agric. **45**, 8 (*62*).

SNYDER, E. B., CLAUSEN, K. E. (1973): In: Woody-plant seed manual, Misc. Publ. 654 Revised. Washington: U.S. Dept. Agric. (*30, 52, 54*).

SOLIMAN, F. A., SOLIMAN, L. (1957): Experientia **13**, 411 (*151*).

SONDHEIMER, E., LINSKENS, H. F. (1974): Proc. Kon. Nederl. Akad. Wet. (Amsterdam), Ser. C. **77**, 116 (*255*).

SOOD, P. P., MALIK, C. P., TEWARI, H. B. (1969): Z. Biol. **186**, 215 (*194*).

SOOD, P. P., TEWARI, H. B., MALIK, C. P. (1969a): Beitr. Biol. Pfl. **46**, 239 (*196*).

SOSA, A., SOSA-BOURDOUIL, C. (1952): Compt. Rend. **235**, 971 (*146, 147, 149*).

SOSA, F., PERCHERON, F. (1965): Compt. Rend. **261**, 4544 (*238*).

SOSA, F., PERCHERON, F. (1970): Phytochemistry **9**, 441 (*238*).

SOSA-BOURDOUIL, C. (1943): Compt. Rend. **217**, 617 (*121*).

SOSA-BOURDOUIL, C., SOSA, A. (1954): Bull. Soc. Chim. Biol. **36**, 393 (*149, 150, 182, 183*).

SOUTHWORTH, D. (1969): Grana Palynol. **9**, 1 (*139, 141*).

SOUTHWORTH, D. (1971): In: HESLOP-HARRISON, J. (Ed.): Pollen development and physiology, p. 115. London: Butterworths (*140*).

SOUTHWORTH, D. (1973): J. Histochem. Cytochem. **21**, 73 (*23, 135, 136*)

SPADA, A., CAMERONI, R. (1955): Gazz. Chim. Ital. **85**, 1034 (*234*).

SPADA, A., CAMERONI, R. (1956): Gazz. Chim. Ital. **86**, 965 (*234*).

SPADA, A., COPPINI, D., MONZANI, A. (1958): Ann. Chim. (Rome) **48**, 181 (*150*).

SPARROW, JR., F. K. (1960): Aquatic Phycomycetes. 2nd Ed. Ann. Arbor: Univ. Michigan (*14, 111*).

SPASOJEVIĆ, V. (1942): Züchter **14**, 215 (*28*).

SPRENGEL, C. K. (1793): Das entdeckte Geheimnis der Natur in Bau und in der Befruchtung der Blumen. Berlin: Vieweg (*92*).

SPRENGEL, K. (1817): Anleitung zur Kenntnis der Gewächse. Halle: Kümmel (*85*).

STAIRS, G. R. (1964): Forest. Sci. **10**, 397 (*49*).

STANDIFER, L. N. (1966): Ann. Entomol. Soc. Amer. **59**, 1005 (*103, 147, 148*).

STANDIFER, L. N. (1966a): J. Apic. Res. **5**, 93 (*146, 147*).

STANDIFER, L. N. (1967): Insectes Soc. **14**, 415 (*160*).

STANDIFER, L. N., DEVYS, M., BARBIER, M. (1968): Phytochemistry **7**, 1361 (*151*).

STANDIFER, L. N., HAYDAK, M. H., MILLS, J. P., LEVIN, M. D. (1973): Am. Bee J. **113**, 94 (*93, 105*).

STANLEY, R. G. (1958): In: THIMANN, K. V. (Ed.) Physiology of forest trees, p. 583. New York: Roland Press (*201, 202, 203, 213*).

STANLEY, R. G. (1962): Silvae Genet. **11**, 164 (*43, 64, 76*).

STANLEY, R. G. (1971): In: HESLOP-HARRISON, J. (Ed.): Pollen development and physiology, p. 131. London: Butterworths (*128, 257*).

STANLEY, R. G. (1971a): In: HESLOP-HARRISON, J. (Ed.): Pollen development and physiology, p. 186. London: Butterworths (*153*).

STANLEY, R. G., BRUNE, A., PETER, J. (1974): (in prep.) (*126*).

STANLEY, R., KIRBY, E. G. (1973): In: KOZLOWSKI, T. (Ed.): Shedding of plant parts, p. 295. New York: Academic Press (*33, 46*).

STANLEY, R. G., LINSKENS, H. F. (1964): Nature **203**, 542 (*22, 142*).

STANLEY, R. G., LINSKENS, H. F. (1965): Physiol. Plantarum **18**, 47 (*22, 111, 142, 160*).

STANLEY, R., LOEWUS, F. A. (1963): In: LINSKENS, H. F. (Ed.): Pollen physiology and fertilization, p. 128. Amsterdam: North Holland (*132, 199, 257*).

STANLEY, R. G., PETERSON, J., MIROV, N. T. (1960): Pac. S. W. Forest and Range Exp. Sta. Tech. Note **173**, 5 pp. (*76*).

STANLEY, R. G., POOSTCHI, I. (1962): Silvae Genet. **11**, 1 (*64, 129, 145*).

STANLEY, R. G., SEARCH, R. W. (1970): Phyton **27**, 35 (*142*).

STANLEY, R. G., SEARCH, R. W. (1971): In: HESLOP-HARRISON, J. (Ed.): Pollen development and physiology, p. 174. London: Butterworths (*22, 126, 168*).

STANLEY, R. G., YEE, A. W. G. (1966): Nature **210**, 181 (*43, 183, 190, 212*).

STANLEY, R. G., YOUNG, L., GRAHAM, J. (1958): Nature **182**, 738 (*192, 195, 197, 201, 206*).

STARRATT, A. N., BOCH, R. (1971): Canad. J. Biochem. **49**, 251 (*93*).

STAUDT, G., KASSRAWI, M. (1972): Vitis **11**, 269 (*80*).

STEFFEN, K. (1963): Rec. Adv. Embryol. Angiosperms (Ed. P. MAHESHWARI), 15. Delhi: Univ. Delhi (*9, 10*).

STEFFEN, K., LANDMANN, W. (1958): Planta **51**, 30 (*10*).

STEFFENSEN, D. M. (1966): Exp. Cell Res. **44**, 1 (*8*).

STEFFENSEN, D. M. (1971): In: HESLOP-HARRISON, J. (Ed.): Pollen development and physiology, p. 223. London: Butterworths (*190, 203*).

STEFFENSEN, D., BERGERON, J. A. (1959): J. Biophys. Biochem. Cytol. **6**, 339 (*127*).

STEIN, H., GABELMAN, W. H. (1959): J. Am. Soc. Sugar Beet Technol. **10**, 612 (*242*).

STEINBERG, R. A. (1957): Tobacco **145**(**1**), 20 (*67*).

STEINBERG, R. A. (1959): Tobacco **149**(**10**), 20 (*78*).

STENLID, G., SAMORODOVA-BIANKI, G. B. (1969): Lantbruks-Hoeysk. Ann. **35**, 837 (*245*).

STERLING, C. (1963): Biol. Rev. **38**, 167 (*13*).

STERN, H. (1960): Ann. N. Y. Acad. Sci. **90**, 440 (*185, 186, 187*).

STERN, H. (1961): J. Biophys. Biochem. Cytol. **9**, 271 (*15*).

STERN, H., HOTTA, Y. (1968): Curr. Topics Develop. Biol. **3**, 37 (*13, 15*).

STERN, H., HOTTA, Y. (1969): Genet. Suppl. **61**, 27 (*185, 186*).

STEUCKARDT, R. (1965): Züchter **35**, 66 (*41*).

STEVENS, F. A., MORRE, D., BAER, H. (1951): J. Allergy **22**, 165 (*238*).

STIX, E. (1971): In: LANDES, E., ZIEGLER, H. (Eds.): Allergien durch Pollen und Sporen, S. 44. Bonn: Dtsch. Forschungsgemeinschaft (*167*).

STONE, C. L., JONES, L. E., WHITEHOUSE, W. E. (1943): Proc. Am. Soc. Hort. Sci. **42**, 305 (*59*).

STONE, G. C. H., HARKAVY, J., BROOKS, C. (1947): Ann. Allergy **5**, 546 (*168*).

STOW, I. (1930): Cytologia **1**, 417 (*29*).

STRAIN, H. H. (1935): J. Biol. Chem. **111**, 85 (*226*).

STRASBURGER, E. (1886): Jahrb. wiss. Botan. **17**, 50 (*195*).

STRASBURGER, E. (1892): Histol. Beitr. **4**, 1 (*11*).

STRAUSS, M. B. (1960): In: PIRGAL, S. J. (Ed.): Fundamentals of modern allergy, p.625. New York: McGraw Hill (*173*).

STROHL, M. J.; SEIKEL, M. K. (1965): Phytochemistry **4**, 383 (*145, 234, 237, 238, 241*).

STULL, A., COOKE, R. A. (1932): J. Allergy **4**, 87 (*174*).

STULL, A., COOKE, R. A., TENNANT, J. (1933): J. Allergy **4**, 455 (*174*).

SÜSS, J. (1970): Biol. Plant. (Praha) **12**, 332 (*182*).

SÜSS, J. (1971): Rev. Roum. Biochim. **8**, 149 (*182*).

SÜSS, J. (1971a): Biol. Plant. (Praha) **13**, 368 (*182*).

SÜSS, J. (1972): Biol. Plant. (Praha) **14**, 385 (*182*).

SUNDERLAND, N., WICKS, F. M. (1971): J. Exp. Botan. **22**, 213 (*193*).

SVOBODA, J. (1935): Cesky Vcelar **69**, 177 (*99*).

SVOLBA, F. (1942): Gartenbauwissenschaften **17**, 95 (*73*).

SWADA, Y. (1958): Bot. Mag. (Tokyo) **71**, 218 (*74*).

SWAIN, T. (1965): In: GOODWIN, T. W. (Ed.): Chemistry and biochemistry of plant pigments, p.211. New York: Academic Press (*232*).

SWEET, G., LEWIS, P. N. (1971): New Zealand J. Botan. **9**, 146 (*254, 255*).

SWIFT, H. H. (1950): Proc. Nat. Acad. Sci. (USA) **36**, 643 (*184*).

SYKUT, A. (1965): Bull. Acad. Pol. Sci. Biol. **13**, 257 (*224*).

SYLVÉN, N. (1909): Mitt. Forstl. Versuchs. Schwedens **6**, (*39*).

SYNGE, A. D. (1947): J. Animal Ecol. **16**, 122 (*95*).

SZAWBOWICZ, A. (1971): Acta Soc. Botan. Pol. **40**, 91 (*29*).

TAKAO, N., SHIMODA, C., YANAGISHIMA, N. (1970): Devel. Growth Different. **12**, 199 (*256*).

TAKATS, S. T. (1959): Chromosoma **10**, 430 (*16, 188*).

TAKATS, S. T. (1962): Am. J. Botany **49**, 748 (*188*).

TAKATS, S. T., WEVER, G. H. (1971): Exp. Cell Res. **69**, 25 (*185*).

TAKIGUCHI, K., HOTTA, K. (1960): Vitamins (Kyoto) **21**, 503 (*209, 210*).

TAMAS, V., SALAJAN, G. H., BODEA, C. (1970): Stud. Cerc. Biochem. **13**, 423 (*226*).

TANAKA, K. (1958): Sci. Repts. Tohoku Univ. Biol. **24**, 45 (*250, 254, 255, 256*).

TANAKA, K. (1960): Sci. Repts. Tohoku Univ. Biol. **26**, 259 (*254, 255*).

TANAKA, K. (1964): Sci. Repts. Tohoku Univ. Biol. **30**, 21 (*254, 255*).

TANAKA, K. (1964a): Sci. Repts. Tohoku Univ. Biol. **30**, 211 (*254, 255*).

TANO, S., TAKAHASHI, H. (1964): J. Biochem. **56**, 578 (*189*).

TAPPI, G. (1947/1948): Atti Accad. Sci. Torino **83**, 99 (*234*).

TAPPI, G. (1949/1950): Atti Accad. Sci. Torino **84**, 97 (*225*).

TAPPI, G., MENZIANA, E. (1955): Gazz. Chem. Ital. **85**, 694 (*238*).

TAPPI, G., MENZIANA, E. (1956): Atti Soc. Nat. Mat. Modena **32–33**, 28 (*225*).

TAPPI, G., MONZANI, A. (1955): Gazz. Chim. Ital. **85**, 725 (*226*).

TAPPI, G., MONZANI, A. (1955a): Gazz. Chim. Ital. **85**, 732 (*239*).

TAPPI, G., SPADA, A., CAMERONI, R. (1955): Gazz. Chim. Ital. **85**, 703 (*234*).

TARANOVA, E. A. (1965): Ioniz Izluch. Biol. Riga: Akad. Nauk Latv. SSR 103 (*49*).

TAUBÖCK, K. (1942): Naturwissenschaften **30**, 439 (*245*).

TAYLOR, J. H. (1950): Am. J. Botany **37**, 137 (*55*).

TAYLOR, J. H. (1959): Am. J. Botany **46**, 447 (*185, 186, 187*).

TAYLOR, J. J., MCMASTER, R. D. (1954): Chromosoma **6**, 489 (*7, 10, 185, 186*).

THIEN, L. B. (1969): Am. J. Botany **56**, 232 (*110*).

THIMANN, K. V. (1934): J. Gen. Physiol. **18**, 23 (*249*).

THIMANN, K. V., MAHADEVAN, S. (1958): Nature **181**, 1466 (*256*).

THOMAS, W. H. (1952): Boron contents of floral parts and the effects of boron on pollen germination and tube growth of *Lilium* species. M.S. Thesis. Univ. Maryland (*122*).

THOMSON, T. (1838): Chemistry of organic bodies. Vegetables. London: J. B. Bailliere (*119, 122*).

THUNBURG, T. (1925): Skand. Arch. Physiol. **40**, 137 (*196*).

VAN TIEGHEM, P. H. (1869): Ann. Sci. Nat. Botan. **5**, Ser T. XII, 314 (*69, 195, 201*).

VAN TIEGHEM, P. E. (1884): Bull. Fr. Soc. Botan. **31**, 67 (*223*).

TISCHER, J. (1941): Z. Physiol. Chem. **267**, 14 (*228*).

TISCHER, J., ANTONI, W. (1938): Z. Physiol. Chem. **252**, 234 (*120, 124*).

TISCHLER, G. (1910): Jahrb. wiss. Botan. **47**, 219 (*133*).

TISSUT, M., EGGER, K. (1969): Compt. Rend. Ser. D. **269**, 642 (*238*).

TODD, F. E., BRETHERICK, O. (1942): J. Econ. Entomol. **35**, 312 (*96, 119, 120, 121, 129, 132, 146, 160*).

TODD, F. E., REED, C. B. (1969): J. Econ. Entomol. **62**, 865 (*104*).

TOGASAWA, Y., KATSUMATA, T., KAWAJIRI, H., ONODERA, N. (1966): J. Agric. Chem. Soc. Japan. **40**, 461 (*232*).

TOGASAWA, Y., KATSUMATA, T., MOTOI, M. F. (1967): J. Agr. Chem. Soc. Japan. **41**, 184 (*210, 226, 228*).

TOGASAWA, Y., KATSUMATA, T., OTA, T. (1967a): J. Agr. Chem. Soc. Japan. **41**, 178 (*210*).

TOGASAWA, Y., KATSUMATA, T., YAMAMURA, Y., YAMANOUCHI, Y. (1968): J. Fac. Agr. Iwate Univ. **8**, 155 (*183, 190*).

TOMOZEI, I., SCUMPU, N. (1964): Lucrar. Str. Inst. Agron. Ion Ionescu de la Brad. Iasi 43 (*84*).

TOUSIMIS, A. J. (1964): Amer. Soc. Test. Mater. Tech. Publ. **349**, 193 (*128*).

TOWNSEND, S. F., RIDDELL, R. T., SMITH, M. V. (1958): Can. J. Pl. Sci. **38**, 39 (*42*).

TROLL, W. (1928): Flora **23**, 321 (*225*).

TSELUIKO, N. A. (1968): Fiziol. Rast. (Engl. Transl.) **15**, 159 (*154*).

TSINGER (ZINGER), V. N. (1961): Tr. Glav. Bot. Sada Akad. Nauk. S.S.S.R. **8**, 149 (*198, 199*).

TSINGER, N. W., PETROVSKAYA-BARANOVA, T. P. (1961): Dokl. Akad. Nauk SSSR **138**, 466 (*16, 23, 141, 142, 162, 218*).

TSINGER, V. N., PETROVSKAYA-BARANOVA, T. P. (1965): Dokl. Akad. Nauk SSSR (Engl. Transl.) **165**, 158 (*222*).

TSINGER, V. N., PETROVSKAYA-BARANOVA, T. P. (1965a): Dokl. Akad. Nauk SSSR (Engl. Transl.) **165**, 417 (*43*).

TSINGER, V. N., PETROVSKAYA-BARANOVA, T. P. (1967): Fiziol. Rast. **14**, 477 (*145*).

TSINGER, V. N., PODDUBNAYA-ARNOL'DI, V. A. (1954): Dokl. Akad. Nauk SSSR **110**, 157 (*229*).

TSUKAMOTO, Y., MATSUBARA, S. (1968): Plant Cell Physiol . **9**, 227 (*68*).

TULECKE, W. R. (1954): Bull. Torrey Botan. Club **81**, 509 (*53, 65*).

TUPÝ, J. (1963): Biol. Plant. (Praha) **5**, 154 (*154, 157, 158*).

TUPÝ, J. (1964): In: LINSKENS, H. F. (Ed.): Pollen physiology and fertilization, p. 86. Amsterdam: North Holland (*155*).

TUPÝ, J. (1966): Biol. Plant. (Praha) **8**, 398 (*207*).

TURBIN, N. V., BORMOTOV, V. E., SAVCHENKO, V. K., MATOSHKO, I. V. (1965): Dokl. Akad. Nauk. SSSR **161**, 463 (*182, 183*).

UEDA, M., EHMANN, A., BANDURSKI, S. (1970): Plant Physiol. **46**, 715 (*257*).

UENO, S. (1954): Kagaku (Sci.) **24**, 90 (*130*).

UMEBAYASHI, M. (1968): Plant Cell Physiol. **9**, 583 (*197, 212*).

UNDERDOWN, B. J., GOODFRIEND, L. (1969): Biochemistry **8**, 980 (*170, 171*).

USHIROZAWA, K., SHIBUKAWA, J. (1952): Aomori Apple Expt. Stat., 4 pp. (*61*).

VANDERWOUDE, W. J., MORRÉ, D. J. (1968): Proc. Indiana Acad. Sci. **77**, 164 (*216, 218*).

VANDERWOUDE, W. J., MORRÉ, D. J., BRACKER, C. E. (1969): Abst. XI Internatl. Bot. Congr. 226 (*131*).

VANDER WOUDE, W. J., MORRÉ, D. J., BRACKER, C. E. (1971): J. Cell Sci. **8**, 331 (*43, 131, 213, 216, 217, 258*).

VANSELL, G. H., WATKINS, W. G. (1933): J. Econ. Entomol. **26**, 168 (*104*).

VANSELL, G. H., WATKINS, W. G. (1934): J. Econ. Entomol. **27**, 635 (*104*).

VANYUSHIN, B. F., BELOZERSKII, A. N. (1959): Dokl. Akad. Nauk. SSSR, Biochem. Sect. **127**, 196 (*190*).

VANYUSHIN, B. F., FAIS, K. (1961): Biokimiya **26**, 1034 (*183, 190*).

VASIL, I. K. (1958): In: MAHESHWARI, P. (Ed.): Modern developments in plant physiology, p. 123. Delhi: Univ. (*47, 55*).

VASIL, I. K. (1959): Science **129**, 1487 (*188*).

VASIL, I. K. (1962): J. Indian Botan. Soc. **41**, 178 (*64, 65*).

VASIL, I. K. (1967): Biol. Rev. **42**, 327 (*3, 14, 55*).

VASIL, I. K., ALDRICH, H. C. (1971): In: HESLOP-HARRISON, J. (Ed.): Pollen development and physiology, p. 70. London: Butterworths (*16*).

VAZART, B. (1958): Protoplasmatologia **7**, 1 (*16*).

VAZHNITSKAYA, E. F. (1960): Trudy Aspirantio i Molodykl Nak. sotrundnikov Leningrad. **249** (*81*).

VEIDENBERG, A. E., SAFONOV, V. I. (1968): Dokl. Nauk. SSSR **180**, 1242 (*204*).

VENGRENOVSKII, S. I., DZHELALI, N. I. (1962): Vest. Sel'skokhoz. Nauki **10**, 103 (*59*).

VENKATASUBRAMANIAN, M. K. (1953): Madras Agr. J. **40**, 395 (*197*).

VENKATESH, C. S. (1956): Phytomorphology **6**, 168 (*24*).

VIEITEZ, E. (1952): Anales Edafol y Fisiol. Vegetal **11**, 297 (*84*).

VINSON, C. G. (1927): J. Agric. Res. **35**, 261 (*197*).

VIRTANEN, A. E., KARI, S. (1955): Acta Chem. Scand. **9**, 1548 (*154, 155*).

VISSER, T. (1951): Mededel. Directeur. Tuinbouw **14**, 707 (*53*).

VISSER, T. (1955): Med. Landbouwhogeschool Wageningen **55**(1), 1 (*56, 59, 64, 78, 79, 80*).

VISSER, T. (1955a): Med. Landbouwhogeschool Plantenfysiol. Dir. Tuinbouw **18**, 856 (*57, 78*).

VISSER, T. (1956): Proc. Kon. Nederl. Akad. Wetensch. Ser. C **59**, 685 (*123, 244*).

VIVINO, A. E., PALMER, L. S. (1944): Arch. Biochem. **4**, 129 (*100, 119, 159, 210, 226*).

VOGEL, S. (1968): Flora **157 B**, 562 (*111*).

VOGEL, S. (1971): Naturwissenschaften **58**, 58 (*93*).

VOISEY, P. W., BASSETT, I. J. (1961): Can. J. Pl. Sci. **41**, 849. (*51*).

VOLKHEIMER, G. (1972): Persorption (Gastroenterologie und Stoffwechsel, Vol. 2). Stuttgart: G. Thieme (*179*).

VOLKHEIMER, G., SCHULZ, F. J. (1968): Digestion **1**, 213 (*179*).

VORWOHL, G. (1968): Z. Bienenforsch. **9**, 224 (*106*).

VREDENBURCH, C. L. H., VAN LAAR, G. (1967): Nederl. Bosb. Tijdschr. **39**, 492 (*41*).

VRIES, DE, A. P. L. (1971): Euphytica **20**, 152 (*25*).

WAFA, M. H. A. (1951): Some biologically active substances in the pollen grains of date palm, *Dactyliferae palmae*. L. Ph. D. Thesis. Cairo: Faroud I. Univ. (*232*).

WAGENITZ, G. (1955): Ber. Deut. Botan. Ges. **68**, 297 (*29, 30*).

WAHL, O. (1954): Insectes Sociaux **1**, 285 (*105*).

WAHL, O. (1956): Südwestdeutscher Imker **8**, 358 (*103*).

WAHL, O. (1963): Z. Bienenforsch. **6**, 209 (*105*).

WAHL, O., BACK, E. (1955): Naturwissenschaften **42**, 103 (*103*).

WALDEN, D. B. (1967): Crop. Sci. **7**, 441 (*59*).

WALDEN, D. B., EVERETT, H. L. (1961): Crop. Sci. **1**, 21 (*78*).

WALLGREN, S. (1958): Nord. Med. **60**, 1194 (*148*).

WATANABE, K. (1955): Botan. Mag. (Tokyo) **68**, 40 (*22*).

WATANABE, M., OBA, K. (1964): J. Japan. For. Soc. **46**, 42 (*49*).

WATANABE, R., CHORNEY, W., SKOK, J., WENDER, S. H. (1964): Phytochemistry **3**, 391 (*245*).

WATANABE, T., MOTOMURA, Y., ASO, K. (1961): Tohoku J. Agr. Res. **12**, 173 (*120, 131, 238*).

WATERKEYN, L., BIENFAIT, A. (1971): In: HESLOP-HARRISON, J. (Ed.): Pollen development and physiology, p. 125. London: Butterworths (*21*).

WEDMORE, E. B. (1932): Manual of beekeeping for English-speaking beekeepers. New York: Longmans, Green (*108*).

WEIPPL, T. (1928): Arch. Bienenk. **9**, 7 (*96*).

WENT, F. W. (1928): Rec. Trav. Botan. Néerl. **25**, 1 (*248*).

WERFFT, R. (1951): Umschau **51**, 622 (*49, 65, 245*).

WERFFT, R. (1951a): Biol. Zbl. **70**, 354 (*245*).

WERNER, M., GRONEMEYER, W., FUCHS, E. (1970): Deut. Med. Wschr. **95**, 877 (*176*).

WERNER, M., RUPPERT, V. (Hrsg.) (1968): Praktische Allergie-Diagnostik. Methoden des direkten Allergennachweises. Stuttgart: G. Thieme (*44, 176*).

WETTSTEIN, VON, D. (1965): Encycl. Plant Physiol. **XV/1**, 298 (*9*).

WEVER, G., TAKATS, S. T. (1970): Biochim. Biophys. Acta **199**, 8 (*185, 196, 213*).

WEYGAND, R., HOFMANN, H. (1950): Chem. Ber. **83**, 405 (*145, 210*).

WHITCOMB, JR., W., WILSON, H. F. (1929): Res. Bull. Agric. Exp. Stat. Univ. of Wisconsin, Madison, No. **92**, 1 (*100, 101*).

WHITEHEAD, R. A. (1962): Nature **196**, 190 (*62*).

WIDE, L., BENNICH, H., JOHANSSON, S. G. O. (1967): Lancet **2**, 1105 (*172*).

WIERMANN, R. (1968): Ber. Deut. Bot. Ges. **81**, 3 (*232, 233, 234—237, 238*).

WIERMANN, R. (1969). Planta **88**, 311 (*240*).

WIERMANN, R. (1970): Planta **95**, 133 (*232, 240, 241, 242*).

WIERMANN, R. (1972): Planta **102**, 55 (*197, 232, 241*).

WIERMANN, R. (1972 a): Z. Physiol. **353**, 129 (*203, 240*).

WIERMANN, R. (1972 b): Z. Pflanzenphysiol. **66**, 215 (*240*).

WIERMANN, R., WEINERT, H. (1969): Z. Pflanzenphysiol. **61**, 173 (*240*).

WILMS, H., CARMICHAEL, J. W., SHANKS, S. C. (1970): Crop Sci. **10**, 309 (*79*).

WILSON, A. F., NOVEY, H. S., BERKE, R. A., SURPRENANT, E. L. (1973): New Engl. J. Med. **288**, 1056 (*179*).

WITHERELL, P. C. (1972): Am. Bee J. **112**, 129 (*89*).

WITTGENSTEIN, E., SAWICKI, E. (1970): Microchem. Acta, 765 (*225*).

WITTIG, H. J., WELTON, W. A., BURRELL, R. (1970): A primer on immunologic disorders. Springfield, I11.: C. C. Thomas (*163*).

WITTMANN; G., WALKER, D. (1965): Pollen et Spores **7**, 443 (*16*).

WODEHOUSE, R. P. (1935): Pollen grains. New York. McGraw Hill (*7, 27, 39, 45, 50*).

WODEHOUSE, R. P. (1945): Hayfever plants. Waltham, Mass.: Chronica Botanica (*27*).

WODEHOUSE, R. P. (1954): Int. Arch. Allergy **5**, 337 (*175*).

WODEHOUSE, R. P. (1955): Int. Arch. Allergy **6**, 65 (*170*).

WODEHOUSE, R. P. (1955 a): Ann. Allergy **13**, 39 (*175*).

WOLFF-EISNER, A. (1907): Derm. Zbl. **10**, 164 (*165*).

WOOD, G. W., BARKER, W. G. (1964): Can. J. Plant Sci. **44**, 387 (*62*).

WOODARD, J. W. (1958): J. Biophys. Biochem. Cytol. **4**, 383 (*184*).

WORLEY, J. F., MITCHELL, J. W. (1971): J. Am. Soc. Hort. Sci. **96**, 270 (*252*).

WORSLEY, R. G. F. (1959): Silvae Genet. **8**, 173 (*52, 72*).

WORSLEY, R. G. F. (1960): Silvae Genet. **9**, 143 (*74*).

WRIGHT, J. W. (1952): N.E. For. Expt. Stat., USDA Paper **46** (*51*).

WRIGHT, J. W. (1962): Genetics of forest tree improvement. F.A.O. Series No. **16**, 399 pp. Rome (*39, 47*).

WULFF, H. D. (1939): Jahrb. wiss. Botan. **88**, 141 (*6*).

WYMAN, M. (1872): Autumnal catarrh (Hay-Fever). Cambridge, Mass.: Hurd + Houghton, (*50, 165*).

YAKUSCHKINA, N. I. (1947): Dokl. Akad. Nauk SSSR **56**, 549 (*249, 250*).

YAMADA, Y. (1958): Kagaka (Science-Tokyo) **28**, 257 (*212*).

YAMADA, Y. (1965): Jap. J. Botan. **19**, 69 (*258*).

YAMADA, I., HASEGAWA, H. (1959): Proc. Crop Sci. Soc. Japan **28**, 157 (*48*).

YASUDA, S. (1934): Jap. J. Gent. Genet. **9**, 118 (*249*).

YOUNGER, V. B. (1961): Science **133**, 577 (*48*).

YUNGINGER, J. W., GLEICH, G. J. (1972): J. Allergy Clin. Immunol. **50**, 326 (*176*).

ZANDER, E. (1935): Die Bienenweide, Hb. Bienenkunde **7** (*108*).

ZETZSCHE, F., HUGGLER, K. (1928): Ann. Chemie **461**, 89 (*138*).

ZETZSCHE, F., KÄLIN, O. (1931): Helv. Chim. Acta **14**, 517 (*142, 143*).

ZETZSCHE, F., VICARI, H. (1931): Helvet. Chim. Acta **14**, 58 (*19*).

ZETZSCHE, F., VICARI, H. (1931 a): Helv. Chim. Acta **14**, 63 (*139*).

ZHEREBKIN, M. V. (1967): Proc. Internatl. Beekeeping Cong. **21**, 476 (*102*).

ZHUKOVSKII, P. M., MEDVEDEV, ZH. (1949): Dokl. Akad. Nauk. SSSR **66**, 965 (*229*).

ZIELINSKI, Q. B., THOMPSON, N. M. (1966): Euphytica **15**, 195 (*9*).

ZOLOTOVITCH, G., SEČENSKÁ, M. (1962): C. R. Acad. Bulg. Sci. **15**, 639 (*155, 223*).

ZOLOTOVITCH, G., SEČENSKÁ, M. (1963): C. R. Acad. Bulg. Sci. **16**, 105 (*130, 226*).

ZOLOTOVITCH, G., SEČENSKÁ, M., DEČEVA, R. (1964): C. R. Acad. Bulg. Sci. **17**, 295 (*130*).

ZUBER, M. S., DEATHERAGE, W. L., MACMASTERS, M. M., FERGASON, V. L. (1960): Agron. J. **52**, 411 (*133*).

Subject Index

L. van der Pijl

Principles of Dispersal in Higher Plants

Second Edition

By Dr. **Leendert van der Pijl,** Emeritus Professor
of Botany, University of Indonesia,
Extraord. Professor at University of Nijmegen

26 figs.
VIII, 162 pages. 1972
Cloth DM 39,50

As this book represents a functional and
ecological investigation of seeds and fruits,
it provides insights into general and
evolutionary botany. The second edition
contains numerous revisions and new material.

Distribution rights
for India:
Allied Publishers,
New Delhi

From the reviews of first edition:
". . . are so fascinating and excellent in scope
and treatment that they should appeal to a
broad spectrum of amateur, teaching and
professional botanists and biologists of all
kinds everywhere . . . The illustrations are
especially valuable since they cover so many
more examples than the limited few that get
repeated from one botany text to another . . .
Actually, the book is really a gem . . ."

Phytologia (U.S.A)

Springer-Verlag
Berlin Heidelberg New York

Lectures on
Photo
morphogenesis

By Professor Dr. Hans Mohr,
Biological Institute II
University of Freiburg

With 219 figures. XII, 237 pages. 1972
Soft cover DM 46,60
Distribution rights for India:
Universal Book Stall (UBS), New Delhi

One of outstandig achievements of modern biology
has been the physiological discovery of the reversible
light red-dark red photoreaction system of morpho-
genesis in plants and the subsequent identification
and isolation of the photoreceptor, which was named
phytochrome.

■ Prospectus on request

The early work was done mainly by a research team
led by H. A. Borthwick and S. B. Hendricks at the Plant
Industry Station in Beltsville, Md. The "Beltsville
group" went on to develop the theoretical concept of
the phytochrome system which forms the basis of the
modern interpretation of photomorphogenesis. The
classic phase of phytochrome research (approx.
1952-1960) is described in a series of excellent articles
but an adequate compilation of more recent advances
in this field has been lacking. The author of the present work undertook to sum up the current state of the
art in a series of 24 lectures, omitting the historical
aspects unless they were vital to an understanding of
the present situation.

Springer-Verlag
Berlin
Heidelberg
New York

The emphasis is on developmental physiology and the
significance of phytochrome research within molecular biology in general. The lectures on which the text
is based were given at the University of Massachusetts
in the fall and winter of 1971.